—NETWORKS—
—ON—
—CHIPS

Theory
and Practice

NETWORKS-ON-CHIPS

Theory and Practice

Edited by
FAYEZ GEBALI
HAYTHAM ELMILIGI
MOHAMED WATHEQ EL-KHARASHI

CRC Press
Taylor & Francis Group
Boca Raton London New York

CRC Press is an imprint of the
Taylor & Francis Group, an **informa** business

CRC Press
Taylor & Francis Group
6000 Broken Sound Parkway NW, Suite 300
Boca Raton, FL 33487-2742

First issued in paperback 2017

© 2009 by Taylor & Francis Group, LLC
CRC Press is an imprint of Taylor & Francis Group, an Informa business

No claim to original U.S. Government works

ISBN-13: 978-1-4200-7978-4 (hbk)
ISBN-13: 978-1-138-11272-8 (pbk)

Library of Congress Cataloging-in-Publication Data

Networks-on-chips : theory and practice / editors, Fayez Gebali, Haytham Elmiligi, Mohamed Watheq El-Kharashi.
    p. cm.
"A CRC title."
Includes bibliographical references and index.
ISBN 978-1-4200-7978-4 (hardcover : alk. paper)
  1. Networks on a chip. I. Gebali, Fayez. II. Elmiligi, Haytham. III. El-Kharashi, Mohamed Watheq. IV. Title.

TK5105.546.N48 2009
621.3815'31--dc22   2009000684

Visit the Taylor & Francis Web site at
http://www.taylorandfrancis.com

and the CRC Press Web site at
http://www.crcpress.com

Contents

Preface

Networks-on-chip (NoC) is the latest development in VLSI integration. Increasing levels of integration resulted in systems with different types of applications, each having its own I/O traffic characteristics. Since the early days of VLSI, communication within the chip dominated the die area and dictated clock speed and power consumption. Using buses is becoming less desirable, especially with the ever growing complexity of single-die multiprocessor systems. As a consequence, the main feature of NoC is the use of networking technology to establish data exchange within the chip.

Using this NoC paradigm has several advantages, the main being the separation of IP design and functionality from chip communication requirements and interfacing. This has a side benefit of allowing the designer to use different IPs without worrying about IP interfacing because wrapper modules can be used to interface IPs to the communication network. Needless to say, the design of complex systems, such as NoC-based applications, involves many disciplines and specializations spanning the range of system design methodologies, CAD tool development, system testing, communication protocol design, and physical design such as using photonics.

This book addresses many challenging topics related to the NoC research area. The book starts by studying 3D NoC architectures and progresses to a discussion on NoC resource allocation, processor traffic modeling, and formal verification. NoC protocols are examined at different layers of abstraction. Several emerging research issues in NoC are highlighted such as NoC quality of service (QoS), testing and verification methodologies, NoC security requirements, and real-time monitoring. The book also tackles power and energy issues in NoC-based designs, as power constraints are currently considered among the bottlenecks that limit embedding more processing elements on a single chip. Following that, the CHAIN®works, an industrial design flow from Silistix, is introduced to address the complexity issues of combining various design techniques using NoC technology. A case study of Multiprocessor SoC (MPSoC) for video coding applications is presented using Arteris NoC. The proposed MPSoC is a flexible platform, which allows designers to easily implement other multimedia applications and evaluate the future video encoding standards.

This book is organized as follows. Chapter 1 discusses the design of 3D NoCs, which are multi-layer-architecture networks with each layer designed as a 2D NoC grid. The chapter explores the design space of 3D NoCs, taking into account consumed energy, packet latency, and area overhead as cost factors. Aiming at the best performance for incoming traffic, the authors present a methodology for designing heterogeneous 3D NoC topologies with a combination of 2D and 3D routers and vertical links.

Chapter 2 studies resource allocation schemes that provide shared NoC communication resources, where well-defined QoS characteristics are analyzed. The chapter considers delay, throughput, and jitter as the performance measures. The authors consider three main categories for resource allocation techniques: circuit switching, time division multiplexing (TDM), and aggregate resource allocation. The first technique, circuit switching, allocates all necessary resources during the lifetime of a connection. The second technique, TDM, allocates resources to a specific user during well-defined time periods, whereas the third one, aggregate resource allocation, provides a flexible allocation scheme. The chapter also elaborates on some aspects of priority schemes and fairness of resource allocation. As a case study, an example of a complex telecom system is presented at the end of the chapter.

Chapter 3 deals with NoC protocol issues such as switching, routing, and flow control. These issues are vital for any on-chip interconnection network because they affect transfer latency, silicon area, power consumption, and overall performance. Switch-to-switch and end-to-end flow control techniques are discussed with emphasis on switching and channel buffer management. Different algorithms are also explained with a focus on performance metrics. The chapter concludes with a detailed list of practical issues including a discussion on research trends in relevant areas. Following are the trends discussed: reliability and fault tolerance, power consumption and its relation to routing algorithms, and advanced flow control mechanisms.

Chapter 4 investigates on-chip processor traffic modeling to evaluate NoC performance. Predictable communication schemes are required for traffic modeling and generation of dedicated IPs (e.g., for multimedia and signal processing applications). Precise traffic modeling is essential to build an efficient tool for predicting communication performance. Although it is possible to generate traffic that is similar to that produced by an application IP, it is much more difficult to model processor traffic because of the difficulty in predicting cache behavior and operating system interrupts. A common way to model communication performance is using traffic generators instead of real IPs. This chapter discusses the details of traffic generators. It first details various steps involved in the design of traffic generation environment. Then, as an example, an MPEG environment is presented.

Chapter 5 discusses NoC security issues. NoC advantages in terms of scalability, efficiency, and reliability could be undermined by a security weakness. However, NoCs could contribute to the overall security of any system by providing additional means to monitor system behavior and detect specific attacks. The chapter presents and analyzes security solutions to counteract various security threats. It overviews typical attacks that could be carried out against the communication subsystem of an embedded system. The authors focus on three main aspects: data protection for NoC-based systems, security in NoC-based reconfigurable architectures, and protection from side-channel attacks.

Chapter 6 addresses the validation of communications in on-chip networks with an emphasis on the application of formal methods. The authors formalize two dimensions of the NoC design space: the communication infrastructure and the communication paradigm as a functional model in the ACL2 logic. For each essential design decision—topology, routing algorithm, and scheduling policy—a meta-model is given. Meta-model properties and constraints are identified to guarantee the overall correctness of the message delivery over the NoC. Results presented are general and thus application-independent. To ensure correct message delivery on a particular NoC design, one has to instantiate the meta-model with the specific topology, routing, and scheduling, and demonstrate that each one of these main instantiated functions satisfies the expected properties and constraints.

Chapter 7 studies test and fault tolerance of NoC infrastructures. Due to their particular nature, NoCs are exposed to a range of faults that can escape the classic test procedures. Among such faults: crosstalk, faults in the buffers of the NoC routers, and higher-level faults such as packet misrouting and data scrambling. These fault types add to the classic faults that must be tested postfabrication for all ICs. Moreover, an issue of concern in the case of communication-intensive platforms, such as NoCs, is the integrity of the communication infrastructure. By incorporating novel error correcting codes (ECC), it is possible to protect the NoC communication fabric against transient errors and at the same time lower the energy dissipation.

Chapter 8 adapts the concepts of network monitoring to NoC structures. Network monitoring is the process of extracting information regarding the operation of a network for purposes that range from management functions to debugging and diagnostics. NoC monitoring faces a number of challenges, including the volume of information to be monitored and the distributed operation of the system. The chapter details the objectives and opportunities of network monitoring and the required interfaces to extract information from the distributed monitor points. It then describes the overall NoC monitoring architecture and the implementation issues of monitoring in NoCs, such as cost, the effects on the design process, etc. A case study is presented, where several approaches to provide complete NoC monitoring services are discussed.

Chapter 9 covers energy and power issues in NoC. Power sources, including dynamic and static power consumptions, and the energy model for NoC are studied. The techniques for managing power and energy consumption on NoC are discussed, starting with micro-architectural-level techniques, followed by system-level power and energy optimizations. Micro-architectural-level power-reduction methodologies are highlighted based on the power model for CMOS technology. Parameters such as low-swing signaling, link encoding, RTL optimization, multi-threshold voltage, buffer allocation, and performance enhancement of a switch are investigated to reduce the power consumption of the network. On the other hand, system-level approaches, such as dynamic voltage scaling (DVS), on–off links, topology selection, and application mapping, are addressed. For each technique, recent efforts to solve the power problem in NoC are presented. To evaluate the dissipation of communication energy in NoC, energy models for each NoC component are used.

Power modeling methodologies, which are capable of providing a cycle accurate power profile and enable power exploration at the system level, are also introduced in this chapter.

Chapter 10 presents CHAIN®works—a suite of software tools and clockless NoC IP blocks that fit into the existing ASIC flows and are used for the design and synthesis of CHAIN® networks that meet the critical challenges in complex devices. This chapter takes the reader on a guided tour through the steps involved in the design of an NoC-based system using the CHAIN®works tool suite. As part of this process, aspects of the vast range of trade-offs possible in building an NoC-based design are investigated. Also, some of the additional challenges and benefits of using a self-timed NoC to achieve true top-level asynchrony between endpoint blocks are highlighted in this chapter.

Chapter 11 presents an MPSoC platform, developed at the Interuniversity Microelectronics Center (IMEC), Leuven, Belgium in partnership with Samsung Electronics and Freescale, using Arteris NoC as communication infrastructure. This MPSoC platform is dedicated to high-performance HDTV image resolution, low-power, real-time video coding applications using state-of-the-art video encoding algorithms such as MPEG-4, AVC/H.264, and Scalable Video Coding (SVC). The presented MPSoC platform is built using six Coarse Grain Array ADRES processors, also developed at IMEC, four on-chip memory nodes, one external memory interface, one control processor, one node that handles input and output of the video stream, and Arteris NoC as communication infrastructure. The proposed MPSoC platform is designed to be flexible, allowing easy implementation of different multimedia applications, and scalable to the future evolutions of video encoding standards and other mobile applications in general.

The editors would like to give special thanks to all authors who contributed to this book. Also, special thanks to Nora Konopka and Jill Jurgensen from Taylor & Francis Group for their ongoing help and support.

Fayez Gebali
Haytham El-Miligi
M. Watheq El-Kharashi
Victoria, BC, Canada

About the Editors

Fayez Gebali received a B.Sc. degree in electrical engineering (first class honors) from Cairo University, Cairo, Egypt, a B.Sc. degree in applied mathematics from Ain Shams University, Cairo, Egypt, and a Ph.D. degree in electrical engineering from the University of British Columbia, Vancouver, BC, Canada, in 1972, 1974, and 1979, respectively. For the Ph.D. degree he was a holder of an NSERC postgraduate scholarship. He is currently a professor in the Department of Electrical and Computer Engineering, University of Victoria, Victoria, BC, Canada. He joined the department at its inception in 1984, where he was an assistant professor from 1984 to 1986, associate professor from 1986 to 1991, and professor from 1991 to the present. Gebali is a registered professional engineer in the Province of British Columbia, Canada, since 1985 and a senior member of the IEEE since 1983. His research interests include networks-on-chips, computer communications, computer arithmetic, computer security, parallel algorithms, processor array design for DSP, and optical holographic systems.

Haytham Elmiligi is a Ph.D. candidate at the Electrical and Computer Engineering Department, University of Victoria, Victoria, BC, Canada, since January 2006. His research interests include Networks-on-Chip (NoC) modeling, optimization, and performance analysis and reconfigurable Systems-on-Chip (SoC) design. Elmiligi worked in the industry for four years as a hardware design engineer. He also acted as an advisory committee member for the Wighton Engineering Product Development Fund (Spring 2008) at the University of Victoria, a publication chair for the 2007 IEEE Pacific Rim Conference on Communications, Computers and Signal Processing (PACRIM'07), Victoria, BC, Canada, and a reviewer for the *International Journal of Communication Networks and Distributed Systems* (IJCNDS), *Journal of Circuits, Systems and Computers* (JCSC), and *Transactions on HiPEAC*.

M. Watheq El-Kharashi received a Ph.D. degree in computer engineering from the University of Victoria, Victoria, BC, Canada, in 2002, and B.Sc. (first class honors) and M.Sc. degrees in computer engineering from Ain Shams University, Cairo, Egypt, in 1992 and 1996, respectively. He is currently an associate professor in the Department of Computer and Systems Engineering, Ain Shams University, Cairo, Egypt and an adjunct assistant professor in the Department of Electrical and Computer Engineering, University of Victoria, Victoria, BC, Canada. His research interests include advanced microprocessor design, simulation, performance evaluation, and testability, Systems-on-Chip (SoC), Networks-on-Chip (NoC), and computer architecture and computer networks education. El-Kharashi has published about 70 papers in refereed international journals and conferences.

Contributors

Nader Bagherzadeh
The Henry Samueli School
    of Engineering
University of California
Irvine, California
nadir@uci.edu

John Bainbridge
Silistix, Inc.
Armstrong House
Manchester Technology Centre
Manchester, United Kingdom
John.bainbridge@silistix.com

Alexandros Bartzas
VLSI Design and Testing Center
Department of Electrical
    and Computer Engineering
Democritus University of Thrace
Thrace, Greece
ampartza@ee.duth.gr

Dominique Borrione
TIMA Laboratory, VDS Group
Grenoble Cedex, France
Dominique.Borrione@imag.fr

Leandro Fiorin
ALaRI, Faculty of Informatics
University of Lugano
Lugano, Switzerland
fiorin@alari.ch

Antoine Fraboulet
Université de Lyon
INRIA
INSA-Lyon, France
Antoine.Fraboulet@insa-lyon.fr

Amlan Ganguly
Washington State University
Pullman, Washington
ganguly@eecs.wsu.edu

Cristian Grecu
University of British Columbia
British Columbia, Canada
grecuc@ece.ubc.ca

Amr Helmy
TIMA Laboratory, VDS Group
Grenoble Cedex, France
Amr.helmy@imag.fr

Andre Ivanov
University of British Columbia
British Columbia, Canada
ivanov@ece.ubc.ca

Axel Jantsch
Royal Institute of Technology,
Stockholm, Sweden
axel@kth.se

Michihiro Koibuchi
Information Systems Architecture
    Research Division
National Institute of Informatics
Chiyoda-ka, Tokyo, Japan
koibuchi@nii.ac.jp

George Kornaros
Technical University of Crete
Kounoupidiana, Crete, Greece
Technological Educational Institute
Heraklion, Crete, Greece
kornaros@epp.teiher.gr

Seung Eun Lee
The Henry Samueli School
    of Engineering
University of California
Irvine, California
seunglee@uci.edu

Anthony Leroy
Université Libre de Bruxelles—ULB
Brussells, Belgium
anleroy@ulb.ac.be

Zhonghai Lu
Royal Institute of Technology
Stockholm, Sweden
zhonghai@kth.se

Philippe Martin
Arteris S.A.
Parc Ariane Immeuble
Mercure, France
Philippe.martin@arteris.com

Hiroki Matsutani
Department of Information
    and Computer Science
Keio University
Minato, Tokyo, Japan
matutani@am.ics.keio.ac.jp

Dragomir Milojevic
Université Libre de Bruxelles—ULB
Brussells, Belgium
dragomir.milojevic@ulb.ac.be

Gianluca Palermo
Dipartimento di Electronica
    e Inforazione
Politecnico di Milano
Milano, Italy
gpalermo@elet.polimi.it

Partha Pratim Pande
Washington State University
Pullman, Washington
pande@eecs.wsu.edu

Ioannis Papaeystathiou
Technical University of Crete
Kounoupidiana, Chania, Greece
ygp@mhl.tuc.gr

Laurence Pierre
TIMA Laboratory, VDS Group
Grenoble Cedex, France
Laurence.Pierre@imag.fr

Dionysios Pnevmatikatos
Technical University of Crete
Kounoupidiana, Chania, Greece
pnevmati@ece.tuc.gr

Tanguy Risset
Université de Lyon
INRIA
INSA–Lyon, France
Tanguy.Risset@insa-lyon.fr

Frederic Robert
Université Libre de Bruxelles—ULB
Brussells, Belgium
frederic.robert@ulb.ac.be

Resve Saleh
University of British Columbia
British Columbia, Canada
res@ece.ubc.ca

Mariagiovanna Sami
Dipartimento di Electronica
    e Inforazione
Politecnico di Milano
Milano, Italy
sami@elet.polimi.it

Antoine Scherrer
Laboratoire de Physique
Université de Lyon
ENS-Lyon, France
antoine.scherrer@ens-lyon.fr

Julien Schmaltz
Radboud University Nijmegen
Institute for Computing and
    Information Sciences
Heijendaalseweg, The Netherlands
julien@cs.ru.nl

Cristina Silvano
Dipartimento di Electronica
    e Inforazione
Politecnico di Milano
Milano, Italy
silvano@elet.polimi.it

Kostas Siozios
VLSI Design and Testing Center
Department of Electrical
    and Computer Engineering
Democritus University of Thrace
Thrace, Greece
ksiop@ee.duth.gr

Dimitrios Soudris
VLSI Design and Testing Center
Department of Electrical
    and Computer Engineering
Democritus University of Thrace
Thrace, Greece
dsoudris@ee.duth.gr

Diederik Verkest
Interuniversity Microelectronics
    Centre - IMEC
Leuven, Belgium
Diederik.Verkest@imec.be

1

Three-Dimensional Networks-on-Chip Architectures

Alexandros Bartzas, Kostas Siozios, and Dimitrios Soudris

CONTENTS

1.1 Introduction

Future integrated systems will contain billions of transistors [1], composing tens to hundreds of IP cores. These IP cores, implementing emerging complex multimedia and network applications, should be able to deliver rich multimedia and networking services. An efficient cooperation among these IP cores (e.g., efficient data transfers) can be achieved through innovations of on-chip communication strategies.

The design of such complex systems includes several challenges. One challenge is designing on-chip interconnection networks that efficiently connect the IP cores. Another challenge is application mapping that makes efficient use of available hardware resources [2,3]. An architecture that is able to accommodate such a high number of cores, satisfying the need for communication and data transfers, is the networks-on-chip (NoC) architecture [4,5]. For these reasons NoC became a popular choice for designing the on-chip interconnect. The industry has initiated different NoC-based designs such as the Æthereal NoC [6] from Philips, the STNoC [7] from STMicroelectronics, and an 80-core NoC from Intel [8]. The key design challenges of emerging NoC designs, as presented by Ogras and Marculescu [9], are (a) the communication infrastructure, (b) the communication paradigm selection, and (c) the application mapping optimization.

The type of IP cores, as well as the topology and interconnection scheme, plays an important role in determining how efficiently an NoC will perform for a certain application or a set of applications. Furthermore, the application features (e.g., data transfers, communication, and computation needs) play an equally important role in the overall performance of the NoC system. An overview of the cost considerations for the design of NoCs is given by Bolotin et al. [10].

Up to now NoC designs were limited to two dimensions. But emerging 3D integration technology exhibits two major advantages, namely, higher performance and smaller energy consumption [11]. A survey of the existing 3D fabrication technologies is presented by Beyne [12]. The survey shows the available 3D interconnection architectures and illustrates the main research issues in current and future 3D technologies. Through process/integration technology advances, it is feasible to design and manufacture NoCs that will expand in the third dimension (3D NoCs). Thus, it is expected that 3D integration will satisfy the demands of the emerging systems for scaling, performance, and functionality. A considerable reduction in the number and length of global interconnect using 3D integration is expected [13].

In this chapter, we present a methodology for designing alternative 3D NoC architectures. We define 3D NoCs as architectures that use several active silicon planes. Each plane is divided into a grid where 2D or 3D router modules are placed. The main objective of the methodology is to derive 3D NoC topologies with a mix of 2D and 3D routers and vertical link interconnection patterns that offer best performance for the given chip traffic. The cost factors we consider are (i) energy consumption, (ii) average packet latency, and (iii) total switch block area. We make comparisons with an NoC in which all the routers are 3D ones. We have employed and extended the Worm_Sim NoC simulator [14], which is able to model these heterogeneous architectures and simulate them, gathering information on their performance. The heterogeneous NoC architecture can be achieved using a combined implementation of 2D and 3D routers in each layer.

The rest of the chapter is organized as follows: In Section 1.2 the related work is described. In Section 1.3 we present the 3D NoC topologies under consideration, whereas in Section 1.4 the proposed methodology is introduced. In Section 1.5 the simulation process and the achieved results are presented. Finally, in Section 1.6 the conclusions are drawn and future work is outlined.

1.2 Related Work

On-chip interconnection is a widely studied research field and good overviews are presented [15,16], which illustrate the various interconnection schemes available for present ICs and emerging Multiprocessor Systems-on-Chip (MPSoC) architectures. An NoC-based interconnection is able to provide an efficient and scalable infrastructure, which is able to handle the increased communication needs. Lee et al. [17] present a quantitative evaluation of 2D point-to-point, bus, and NoC interconnection approaches. In this work, an MPEG-2 implementation is studied and it proved that the NoC-based solution scales very well in terms of area, performance, and power consumption.

To evaluate NoC designs, a number of simulators has been developed, such as the Nostrum [18], Polaris [19], XPipes [20], and Worm_Sim [14], using C++ and/or SystemC [21]. To provide adequate input/stimuli to an NoC design, synthetic traffic is usually used. Several synthetic traffic generators have been proposed in several texts [22–25] to provide adequate inputs to NoC simulators for evaluation and exploration of proposed designs.

A methodology that synthesizes NoC architectures is proposed by Ogras, Hu, and Marculescu [26] where long-range links are inserted on top of a mesh network. In this methodology, the NoC design is addressed using an application specific approach, but it is limited to two dimensions. Li et al. [27] presented a mesh-based 3D network-in-memory architecture, using a hybrid NoC/bus interconnection fabric, to accommodate efficiently processors and L2 cache memories in 3D NoCs. It is demonstrated that by using a 3D L2 memory architecture, better results are achieved compared to 2D designs.

Koyanagi et al. [28] presented a 3D integration technique of vertical stacking and gluing of several wafers. By utilizing this technology, the authors were able to increase the connectivity while reducing the number of long interconnections. A fabricated 3D shared memory is presented by Lee et al. [29]. The memory module has three planes and can perform wafer stacking using the following technologies: (i) formation of buried interconnection, (ii) microbumps, (iii) wafer thinning, (iv) wafer alignment, and (v) wafer bonding. Another 3D integration scheme is proposed by Iwata et al. [30], where wireless interconnections are employed to offer connectivity.

An overview of the available interconnect solutions for Systems-on-Chip (SoC) are presented by Meindl [31]. This study includes interconnects for 3D ICs and shows that 3D integration reduces the length of the longest global interconnects [32] and reduces the total required wire length, and thus the dissipated energy [33].

Benkart et al. [34] presented an overview of the 3D chip stacking technology using throughchip interconnects. In their work, the trade-off between the high number of vertical interconnects versus the circuit density is highlighted. Furthermore, Davis et al. [35] show the implementation of an FFT in a 3D IC achieving 33% reduction in maximum wire length, thereby proving that the move to 3D ICs is beneficial. However, the heat dissipation is highlighted as one of the limiting factors.

The placement and routing in 3D integrated circuits are studied by Ababei et al. [36]. Also, a system on package solution for 3D network is presented by Lim [37]. However, the heat dissipation of 3D circuits remains a big challenge [38]. To tackle this challenge, several analysis techniques have been proposed [39–41]. One approach is to perform thermal-aware placement and mapping for 3D NoCs, such as the work presented by Quaye [42]. Furthermore, the insertion of thermal vias can lower the chip temperature as illustrated in several texts [43,44].

A generalized NoC router model is presented; based on that, Ogras and Marculescu performed NoC performance analysis. Using the aforementioned router model, it is feasible to perform NoC evaluation, which is significantly faster than performing simulation. Additionally, Pande et al. [46] presented an evaluation methodology to compare the performance and other metrics of a variety of NoC architectures. But, this comparison is made only among 2D NoC architectures. The work of Feero and Pande [47] extended the aforementioned work considering 3D NoCs, and illustrated that the 3D NoCs are advantageous when compared to 2D ones (with both having the same number of components in total). It is demonstrated that besides reducing the footprint in a fabricated design, 3D network structures provide a better performance compared to traditional 2D architectures. This work shows that despite the cost of a small area penalty, 3D NoCs achieve significant gains in terms of energy, latency, and throughput.

Pavlidis and Friedman [48] presented and evaluated various 3D NoC topologies. They also proposed an analytic model for 3D NoCs where a mesh topology is considered under a zero-load latency. Kim et al. [49] presented an exploration of communication architectures on 3D NoCs. A dimensionally decomposed router and its comparison with a hop-by-hop router connection and hybrid NoC-bus architecture is presented. The aforementioned works, both from the physical level as well as adding more communication architectures, such as full 3D crossbar and bus-based communication, are complementary to the one presented here and can be used for the extension of the methodology.

The main difference between the related work and the one presented here is that we do not assume full vertical interconnection (as shown in Figure 1.1), but rather a heterogeneous interconnection fabric, composed of a mix of 3D and 2D routers. An additional motivation for this heterogeneous design is not only for the reduction of total interconnection network length, but also to get the reduced size of the 2D routers when compared to the 3D ones [47]. Reducing the number of vertical interconnection links simplifies the fabrication of the design and frees up more active chip area for available logic/memory blocks. Two-dimensional routers are routers that have connections with neighboring ones of the same grid. By comparison, a 3D router has direct, hop-by-hop connections with neighboring routers belonging to the same grid and those belonging to the adjacent planes. This difference between

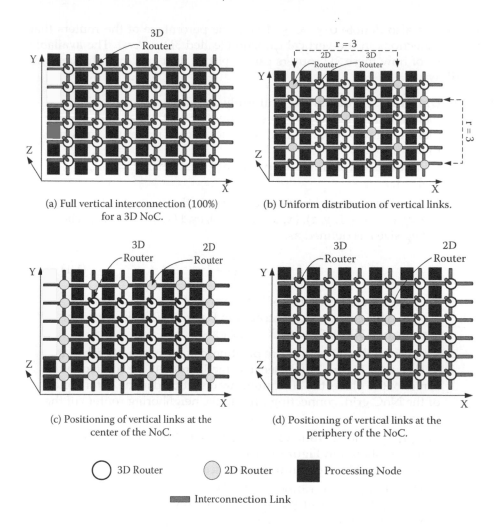

(a) Full vertical interconnection (100%) for a 3D NoC.

(b) Uniform distribution of vertical links.

(c) Positioning of vertical links at the center of the NoC.

(d) Positioning of vertical links at the periphery of the NoC.

FIGURE 1.1
Positioning of the vertical interconnection links, for each plane of the 3D NoC (each plane is a 6 × 6) grid.

2D and 3D routers for a 3D mesh NoC is illustrated in Figure 1.1. The figure shows a grid that belongs to a 3D NoC where several 2D and 3D routers exist.

1.3 Alternative Vertical Interconnection Topologies

We consider four different groups of interconnection patterns, as well as 10 vertical interconnection topologies in the context of this work. Consider a 3D NoC composed of Z 2D active silicon planes. Each 2D plane has dimensions

$X \times Y$. We also denote $0 \leq K \leq 100$ as the percentage of the routers that have connections in the vertical direction (called *3D routers*). The available scenarios of how these 3D routers can be placed on a grid in each plane are as follows:

1. **Uniform:** 3D routers are uniformly distributed over the different planes. Using this scheme, we "spread" the 3D routers along every plane of the 3D NoC. To find the place of each router we work like this:

   - Place the first 3D router at position $(0, 0, z)$ where $z = 0, 1, \cdots,$ $Z - 1$.
   - Place the four neighboring 2D routers in the positions $(x+r+1, y, z)$, $(x - r - 1, y, z)$, $(x, y + r + 1, z)$, and $(x, y - r - 1, z)$. The step size r is defined as:

   $$r = \frac{1}{K} - 1 \tag{1.1}$$

   r represents the number of 2D routers between consecutive 3D ones. This scheme is illustrated in Figure 1.1(b), showing one plane of a 3D NoC, with $K = 25\%$ and $r = 3$.

2. **Center:** All the 3D routers are positioned at the center of each plane, as shown in Figure 1.1(c). Because the 3D routers are located in the center of the plane, the 2D routers are distributed in the outer region of the NoC grid, connecting only to the neighboring routers of the same plane.

3. **Periphery:** The 3D routers are positioned at the periphery of each plane [as shown in Figure 1.1(d)]. In this case, the NoC is focused on serving best the communication needs of the outer cores.

4. **Full custom:** The position of the 3D routers is fully customized matching the needs of the application with the NoC architecture. This solution fits best the needs of the application, while it minimizes the number of 3D routers. However, derivation of a full custom solution requires high design time, because this exploration is going to be performed for every application. Furthermore, this will create a nonregular design that will not adjust well to the potential change of functionality, the number of applications that are going to be executed, etc.

The aforementioned patterns are based on the 3D FPGAs work presented by Siozios et al. [50]. To perform an exploration toward full customized interconnection schemes, real applications and/or application traces are needed. In this chapter, we adopt various types of synthetic traffic, so the exploration for full customized interconnections schemes is out of the scope. More specifically, we focus on pattern-based vertical interconnection topologies (categories 1–3). We consider 10 different vertical link interconnection topologies. For each of these topologies, the number of 3D routers is given and the value of K given in parentheses. For a $4 \times 4 \times 4$ NoC architecture we use the notation 64 (K).

- **Full:** Where all the routers of the NoC are 3D ones [number of 3D routers: 64 (100%)].
- **Uniform based:** Pattern-based topologies with r value equal to three [*by_three* pattern, as shown in Figure 1.1(b)], four (*by_four*), and five (*by_five*). Correspondingly, the number of 3D routers is 44 (68.75%), 48 (75%), and 52 (81.25%).
- **Odd:** In this pattern, all the routers belonging to the same row are of the same type. Two adjacent rows never have the same type of router [number of 3D routers: 32 (50%)].
- **Edges:** Where the center (dimensions $x \times y$) of the 3D NoC has only 2D routers [number of 3D routers: 48 (75%)].
- **Center:** Where only the center (dimensions $x \times y$) of the 3D NoC has 3D routers [number of 3D routers: 16 (25%)].
- **Side based:** Where a side (e.g., outer row) of each plane has 2D routers. Patterns evaluated have one (*one_side*), two (*two_side*), or three (*three_side*) sides as "2D routers only." The number of 3D routers for each pattern is 48 (75%), 36 (56.25%), and 24 (37.5%), respectively.

Each of the aforementioned vertical interconnection schemes has advantages and disadvantages. These schemes perform on the basis of the behavior of the applications that are implemented on the NoC. Experimental results in Section 1.5 show that a wrong choice may diminish the gains of using a 3D architecture.

1.4 Overview of the Exploration Methodology

An overview of the proposed methodology is shown in Figure 1.2. To perform the exploration of alternative topologies for 3D NoC architectures, the Worm_Sim NoC simulator [14], which utilizes wormhole switching, is used [51] (this is the center block in Figure 1.2).

To support 3D architectures/topologies, we have extended this simulator to adapt to the provided routing schemes, and be compatible with the Trident traffic format [23]. As shown in Figure 1.2, the simulator now supports 3D NoC architectures (3D mesh and 3D torus, as shown in Figure 1.3) and vertical link interconnection patterns. Each of these 3D architectures is composed of many grids, and each grid is composed of tiles that are connected to each other using mesh or torus interconnection networks. Each tile is composed of a processing core and a router. Because we are considering 3D architectures, the router is connected to the neighboring tiles and its

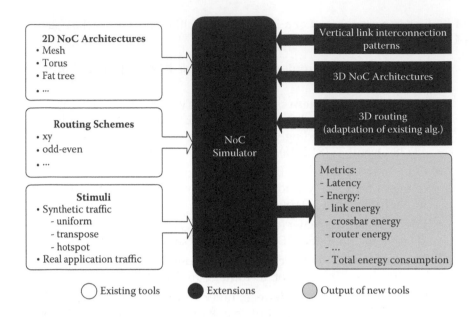

FIGURE 1.2
An overview of the exploration methodology of alternative topologies for 3D Networks-on-Chip.

local processing core via channels, consisting of bidirectional point-to-point links.

The NoC simulator can be configured using the following parameters (shown in Figure 1.2):

1. The NoC architecture (2D or 3D mesh or torus) as well as defining the specific x, y, and z parameters

2. The type of input traffic (uniform, transpose, or hotspot) as well as how heavy the traffic load will be

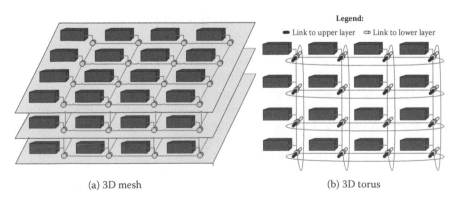

(a) 3D mesh   (b) 3D torus

FIGURE 1.3
3D NoC architectures.

3. The routing scheme

4. The vertical link configuration file, which defines the locations of the vertical links

5. The router model as well as the models used to calculate the energy and delay figures

The output of the simulation is a log file that contains the relevant evaluated cost factors, such as overall latency, average latency per packet, and the energy breakdown of the NoC, providing values for link energy consumption, crossbar and router energy consumption, etc. From these energy figures, we calculate the total energy consumption of the 3D NoCs.

The 3D architectures to be explored may have a mix of 2D and 3D routers, ranging from very few 3D routers to only 3D routers. To steer the exploration, we use different patterns (as presented in Section 1.3). The proposed 3D NoCs can be constructed by placing a number of identical 2D NoCs on individual planes, providing communication by interplane vias among vertically adjacent routers. This means that the position of silicon vias is exactly the same for each plane. Hence, the router configuration is extended to the third dimension, whereas the structure of the individual logic blocks (IP cores) remains unchanged.

1.5 Evaluation—Experimental Results

The main objective of the methodology and the exploration process is to find alternative irregular 3D NoC topologies with a mix of 2D and 3D routers. The new topologies exhibit vertical link interconnection patterns that acquire the best performance. Our primary cost function is the energy consumption, with the other cost factors being the average packet latency and total switch block area. We compare these patterns against the fully vertically interconnected 3D NoC as well as the 2D one (all having the same number of nodes).

1.5.1 Experimental Setup

The 3D router used here has a 7×7 crossbar switch, whereas the 2D router uses a 5×5 crossbar switch. Additionally, each router has a routing table and based on the source/destination address, the routing table decides which output link the outgoing packet should use. The routing table is built using the algorithm described in Figure 1.4.

The NoC simulator uses the Ebit energy model, proposed by Benini and de Micheli [52]. We make the assumption (based on the work presented by Reif et al. [53]) *that the vertical communication links between the planes are electrically equivalent to horizontal routing tracks with the same length.* Based on this assumption, the energy consumption of a vertical link between two routers

```
1:    function ROUTINGXYZ
2:        src : type Node; //this is the source node
3:        dst : type Node; //this is the destination node (final)
4:
5:        findCoordinates(); //returns src.x, src.y, src.z, dst.x, dst.y and dst.z
6:
7:        for all plane ∈ NoC do
8:            if packet passes from plane then
9:                findTmpDestination(); //find a temporary destination of the packet for each plane
            of the NoC that the packet passes from
10:           end if
11:       end for
12:       while tmpDestination NOT dst do //if we have not reached the final destination...
13:           packet.header = tmpDestination;
14:       end while
15:   end function

16:   function FINDTMPDESTINATION //for each plane that the packet is going to traverse
17:       tmpDestination.x = dst.x
18:       tmpDestination.y = dst.y
19:       tmpDestination.z = src.z //for xyz routing
20:
21:       for all validNodes ∈ plane do
22:           if link NOT valid //if vertical link does not exist. This information is obtained through
            the vertical interconnections patterns input file.
23:               newLink = computeManhattanDistance(); //returns the position of a verical link
            with the smallest Manhattan distance
24:               tmpDestination = newLink;
25:           else
26:               tmpDestination = link;
27:           end if
28:       end for
29:   end function
```

FIGURE 1.4
Routing algorithm modifications. (// denotes a comment in the algorithm)

equals the consumption of a link between two neighboring routers at the same plane *(if they have the same length)*.

More specifically because the 3D integration technology, which provides communication among layers using through-silicon vias (TSVs), has not been explored sufficiently yet, 3D-based systems design still needs to be addressed. Due to the large variation of the 3D TSV parameters, such as diameter, length, dielectric thickness, and fill material among alternative process technologies, a wide range of measured resistances, capacitances, and inductances have been reported in the literature. Typical values for the size (diameter) of TSVs is about 4×4 μm, with a minimum pitch around 8–10 μm, whereas their total length starting from plane T1 and terminating on plane T3 is 17.94 μm, implying wafer thinning of planes T2 and T3 to approximately 10–15 μm [54–56].

The different TSV fabrication processes lead to a high variation in the corresponding electrical characteristics. The resistance of a single 3D via varies from 20 mΩ to as high as 600 mΩ [55,56], with a feasible value (in terms of fabrication) around 30 mΩ. Regarding the capacitances of these vias, their values vary from 40 fF to over 1 pF [57], with feasible value for fabrication to be around 180 fF. In the context of this work, we assume a resistance of 350 mΩ and a capacitance of 2.5 fF.

Using our extended version of the NoC simulator, we have performed simulations involving a 64-node and a 144-node architecture with 3D mesh and torus topologies with synthetic traffic patterns. The configuration files describing the corresponding link patterns are supplied to the simulator as an input. The sizes of the 3D NoCs we simulated were $4 \times 4 \times 4$ and $6 \times 6 \times 4$, whereas the equivalent 2D ones were 8×8 and 12×12. We have used three types of input (synthetic traffic) and three traffic loads (heavy, normal, and low). The traffic schemes used are as follows:

- **Uniform:** Where we have uniform distribution of the traffic across the NoC with the nodes receiving approximately the same number of packets.

- **Transpose:** In this traffic scheme, packets originating from node (a, b, c) is destined to node $(X - a, Y - b, Z - c)$, where X, Y, and Z are the dimensions of the 3D NoC.

- **Hotspot:** Where some nodes (a minority) receive more packets than the majority of the nodes. The hotspot nodes in the 2D grids are positioned in the middle of every quadrant, where the size of the quadrant is specified by the dimensions of each plane in the 3D NoC architecture under simulation, whereas in the 3D NoC, a hotspot is located in the middle of each plane.

We have used the three routing schemes presented in Worm_Sim [14], and extended them in order to function in a 3D NoC as follows:

- **XYZ-old:** Which is an extended version of XY routing.
- **XYZ:** Which is based on XY routing but routes the packet along the direction with least delay.
- **Odd-even:** Which is the odd-even routing scheme presented by Chiu [58]. In this scheme, the packets take some turns in order to avoid deadlock situations.

From the simulations performed, we have extracted figures regarding the energy consumption (in joules) and the average packet latency (in clock cycles). Additionally, for each vertical interconnection pattern, as well as for the 2D NoC, we calculated the occupied area of the switching block, based on the gate equivalent of the switching fabric presented by Feero and Pande [47]. A good design is the one that exhibits lower values in the aforementioned metrics when compared to the 2D NoC as well as to the 3D NoC which has full vertical connectivity (all the routers are 3D ones). Furthermore, all the simulation measurements were taken for the same number of operational cycles (200,000 cycles).

1.5.2 Routing Procedure

To route packets over the 3D topologies, we modified the routing procedure, as shown in Figure 1.4. The modified routing procedure is valid for all routing schemes. This modification allows customization of the routing scheme to efficiently cope with the heterogeneous topologies, based on vertical link connectivity patterns.

The steps of the routing algorithm are as follows:

1. For each packet, we know the source and destination nodes and can find the positions of these nodes in the topology. The on-chip "coordinates" of the nodes for the destination one are `dst.x`, `dst.y`, `dst.z` and for the source one are `src.x`, `src.y`, `src.z`.

2. By doing so, we can formulate the *temporary destinations*, one for each plane. For the number of planes a packet has to traverse to arrive at its final destination, the algorithm initially sets the route to a temporary destination located at position `dst.x`, `dst.y`, `src.z`. The algorithm takes into consideration the "direction" that the packet is going to follow across the planes (i.e., if it is going to an upper or lower plane according to its "source" plane) and finds the nearest *valid link* at each plane. This outputs, as an outcome to update properly, the `z` coefficient of the temporary destination's position. *Valid link* is every vertical interconnection link available in the plane in which the packet traverses. This information is obtained from the vertical interconnection patterns file. A link is uniquely identified by the node that is connected and its direction. So, for all the specified valid links that are located at the same plane, the header flit of the packet checks if the desired route is matched to the destination up or down link.

3. If there is no match between them, compute the Manhattan distance (in case of 3D torus topology, we have modified it to produce the correct Manhattan distance between the two nodes).

4. Finally, the valid link with the smallest Manhattan distance is chosen, and its corresponding node is chosen to be the temporary destination at each plane the packet is going to traverse.

5. After finding a set of temporary destinations (each one located at a different plane), they are stored into the header flit of the packet. The aforementioned temporary destinations may or may not be used, as the packet is being routed during the simulation, so they are "candidate" temporary destinations. The decision of being just a candidate or the actual destination per plane is taken based on one of two scenarios: (1) if a set of vertical links, which exhibited relatively high utilization during a previous simulation with the same network parameters, achieved the desired minimum link communication volume or (2) according to a given vertical link pattern such as the one presented in Section 1.1.

The modification of the algorithm essentially checks if a vertical link exists in the temporary destination of the packet, otherwise the closest router with such a link is chosen. Thus the routing complexity is kept low.

1.5.3 Impact of Traffic Load

Three different traffic loads were used (heavy, medium/normal, low). In this way, by altering the packet generation rate, it is possible to test the performance of the NoC. The heavy load has 50% increased traffic, whereas the low one has 90% decreased traffic compared to the medium load, respectively. The behavior of the NoCs in terms of the average packet latency is shown in Figure 1.5. In this figure, the latency is normalized to the average packet latency of the *full_connectivity* 3D NoC *under medium load and for each traffic scheme*. The impact of the traffic load (latency increases as the load increases) can be observed, and also we can see that NoCs can cope with the increased traffic as well as the differences between different traffic schemes.

Mesh topologies exhibit similar behavior, though the latency figures are higher due to the decreased connectivity when compared to torus topologies. This is shown in Figure 1.6 where the latency of 64-node mesh and torus NoCs are compared (the basis for the latency normalization is the average packet latency of the *full_connectivity* 3D torus). From this comparison, it is shown that the mesh topologies have an increased packet latency of 34% compared to the torus ones (for the same traffic scheme, load, and routing algorithm).

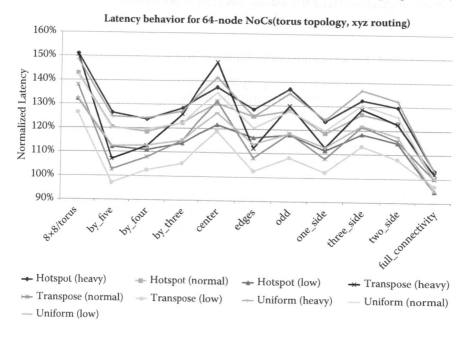

FIGURE 1.5
Impact of traffic load on 2D and 3D NoCs (for all different types of traffic used).

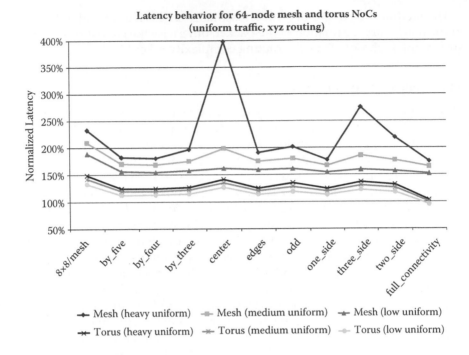

FIGURE 1.6
Impact of traffic load on 2D and 3D mesh and torus NoCs (for uniform traffic).

1.5.4   3D NoC Performance under Uniform Traffic

Figure 1.7 shows the results of employing a nonfully vertical link connectivity to 3D mesh networks by using uniform traffic, medium load, and *xyz*-old routing. We compared the total energy consumption, average packet latency, total area of the switching blocks (routers), and the percentage of 2D routers (having 5 I/O ports instead of 7) under $4 \times 4 \times 4$ [Figure 1.7(a)] and $6 \times 6 \times 4$ [Figure 1.7(b)] mesh architectures. In the *x*-axis all the interconnection patterns are presented. In the *y*-axis, and in a normalized manner (used as the basis for the figures of the full vertically interconnected 3D NoC), the cost factors for total energy consumption, average packet latency, total switching block area, and percentage of vertical links are presented.

The advantages of 3D NoCs when compared to 2D ones are shown in Figure 1.7(a). In this case, the 8×8 mesh dissipates 39% more energy and has 29% higher packet delivery latency. However, the switching area is 71% of the area of the fully interconnected 3D NoC because all its routers are 2D ones. Employing the *by_five* link pattern results in 3% reduction in energy and 5% increase in latency. In this pattern, only 81% of the routers are 3D ones so the area of the switching logic is reduced by 5% (when compared to the area of the fully interconnected 3D NoC). Figure 1.7(b) shows that more patterns exhibit

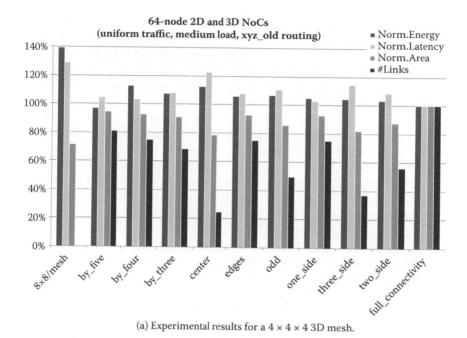

(a) Experimental results for a 4 × 4 × 4 3D mesh.

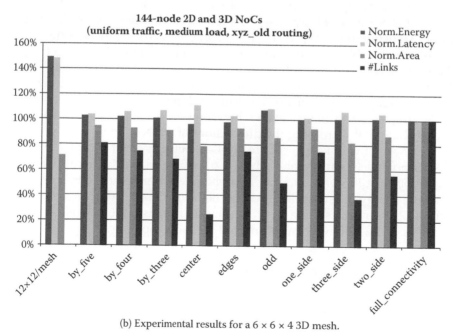

(b) Experimental results for a 6 × 6 × 4 3D mesh.

FIGURE 1.7
Uniform traffic (medium load) on a 3D NoC for alternative interconnection topologies.

better results. It is worth noticing that the overall performance of the 2D NoC significantly decreases, exhibiting around 50% increase in energy and latency.

When we increase the traffic load by increasing the packet generation rate by 50%, we see that all patterns have worse behavior than the *full_connectivity* 3D NoC. The reason is that by using a pattern-based 3D NoC, we decrease the number of 3D routers by decreasing the number of vertical links, thereby reducing the connectivity within the NoC. As expected, this reduced connectivity has a negative impact in cases where there is an increased traffic.

For low traffic load NoC, the patterns can become beneficial because there is not that high need for communication resources. This effect is illustrated in Figure 1.8. The figure shows the experimental results for 64- and 144-node 2D and 3D NoCs under low uniform traffic and *xyz* routing. The exception is the *edges* pattern in the 64-node 3D NoC [Figure 1.8(a)], where all the 3D routers reside on the edges of each plane of the 3D NoC. This results in a 7% increase in the packet latency. Again it is worth noticing that as the NoC dimensions increase, the performance of the 2D NoC decreases. This can be clearly seen in Figure 1.8(b), where the 2D NoC has 38% increased energy dissipation.

We have also compared the performance of the proposed approach against that achievable with a torus network, which provides wraparound links added in a systematic manner. Note that the vertical links connecting the bottom with the upper planes are not removed, as this is the additional feature of the torus topology when compared to the mesh. Our simulations show that using the transpose traffic scheme, the vertical link patterns exhibit notable results; this pattern continues as the dimensions of the NoC get bigger. The explanation is that the flow of packets between a source and a destination follows a diagonal course among the nodes at each plane. At the same time, the wraparound links of the torus topology play a significant role in preserving the performance even when some vertical links are removed. The results show that increasing the dimensions of the NoC increases the energy savings, when the link patterns are applied. But, this is not true for the case of mesh topology. In particular, in the $6 \times 6 \times 4$ 3D torus architecture, using the *by_five*, *by_four*, *by_three*, *one_side*, and *two_side* patterns show better results as far as the energy consumption is concerned. For instance, the *two_side* pattern exhibits 7.5% energy savings and 32.84 cycles increased latency relative to the 30 cycles of the fully vertical connected 3D torus topology.

1.5.5 3D NoC Performance under Hotspot Traffic

In the case of hotspot traffic (Figure 1.9), testing the $4 \times 4 \times 4$ 3D mesh architecture, seven out of the nine link patterns perform better relative to the fully vertically connected topology. For instance, the *two_side* pattern exhibits 2% decrease in network energy consumption, whereas the increase in latency is 2.5 cycles. Note that only 56.25% of the vertical links are present. The hotspot traffic in 3D mesh topologies favors cube topologies (e.g., $6 \times 6 \times 6$). Even so, in $6 \times 6 \times 4$ mesh architecture, the *center* and *two_side* patterns exhibit similar performance regarding average cycles per packet compared to that of fully

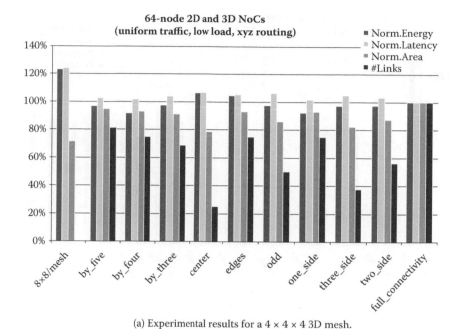

(a) Experimental results for a 4 × 4 × 4 3D mesh.

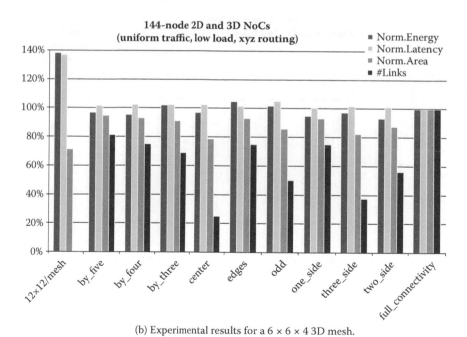

(b) Experimental results for a 6 × 6 × 4 3D mesh.

FIGURE 1.8
Uniform traffic (low load) on a 3D NoC for alternative interconnection topologies.

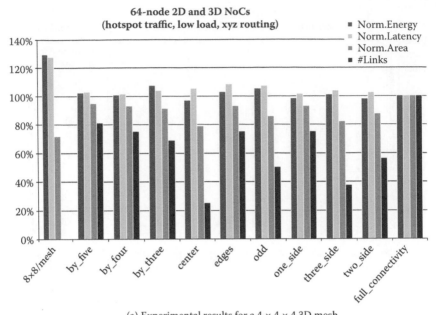

(a) Experimental results for a 4 × 4 × 4 3D mesh.

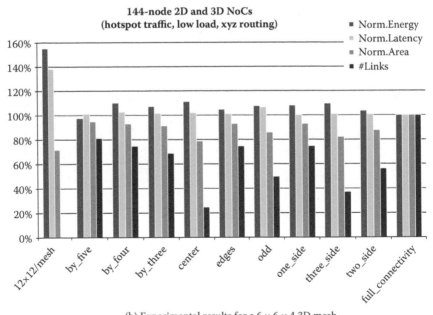

(b) Experimental results for a 6 × 6 × 4 3D mesh.

FIGURE 1.9
Hotspot traffic (low load) on a 3D NoC for alternative interconnection topologies.

vertical connected architecture (that was expected due to the location where the hotspot nodes were positioned).

In Figure 1.10, the simulation results for the two 3D NoC architectures when triggered by a hotspot-type traffic are presented. Figures 1.10(a) and 1.10(b) present the results for the mesh and torus architectures, respectively, showing gains in energy consumption and area, with a negligible penalty in latency. Again, the architectures where congestion is experienced are highlighted.

These results are also compared to their equivalent 2D architectures. For the 8×8 2D NoC (same number of cores as the $4 \times 4 \times 4$ architecture), it shows 25% increased latency and 40% increased energy consumption compared to the *one_side* link pattern, whereas the 12×12 mesh (same number of cores as the $6 \times 6 \times 4$ architecture) shows 46% increase in latency and 49% increase in energy consumption compared to the same pattern using uniform traffic. In addition, comparing the *by_four* pattern on the 64-node architecture under transpose traffic shows 31% and 18% reduced latency and total network consumption, respectively. However, in the case of hotspot traffic and employing the *two_side* link pattern, these numbers change to 24% reduced latency and 56% reduced energy consumption.

1.5.6 3D NoC Performance under Transpose Traffic

Under the transpose traffic scheme, the *by_four* link pattern adopted shows 6.5% decrease in total network energy consumption at the expense of 3 cycles increased latency. In Figure 1.11, the simulation results for the 3D $4 \times 4 \times 4$ mesh and $6 \times 6 \times 4$ torus NoCs are presented for transpose traffic. In Figure 1.11(a), we can see that we have a 4% gain in the energy consumption of the 3D NoCs with a 5% increase in the packet latency. Additionally, we gain 6% in the area occupied by the switching blocks of the NoC. Comparing these patterns to the 2D NoC (having the same number of nodes) we can have on average a 14% decrease in energy consumption, a 33% decrease in total packet latency. But on the area, the cost of the 3D NoC is higher by 23%. In Figure 1.11(b), we can see that the 2D NoC experiences traffic contention and not being able to cope with that amount of traffic (the actual value of the latency is close to 5000 cycles per packet). Additionally, 47% gains achieved in energy consumption. When this torus architecture is compared to the "full" 3D one, it shows 5% gains in energy consumption with 8% increased latency and 9% reduced switching block area.

1.5.7 Energy Dissipation Breakdown

The analytical results of the Ebit [52] energy model indicate that, when moving to 3D architectures, the energy consumption of the links, crossbars, arbiters, and buffer read energy decreases, whereas there is an increase in the energy consumed when writing to the buffer and taking the routing decisions.

On average, the link energy consumption accounts for 8% of the total energy, the crossbar 6%, the buffer's read energy 23%, and the buffer's write

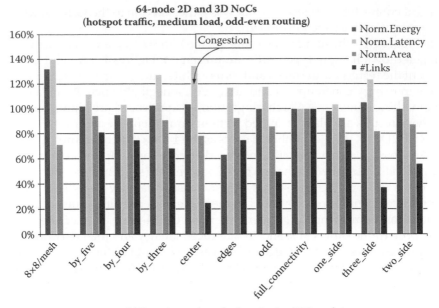

(a) Experimental results for a 4 × 4 × 4 3D mesh.

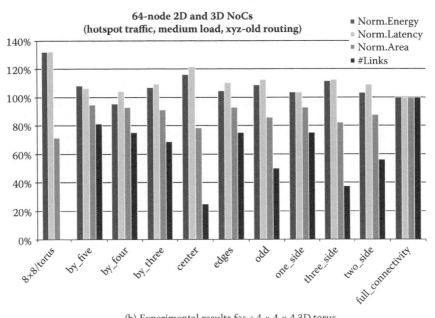

(b) Experimental results for a 4 × 4 × 4 3D torus.

FIGURE 1.10
Hotspot traffic (medium load) on a 3D NoC for alternative interconnection topologies.

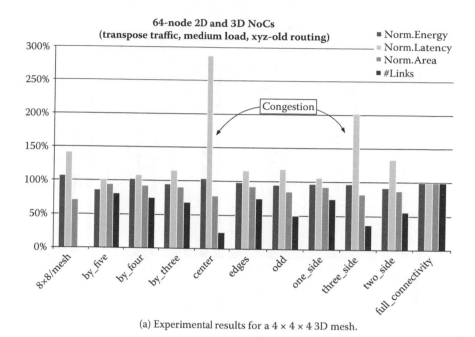

(a) Experimental results for a 4 × 4 × 4 3D mesh.

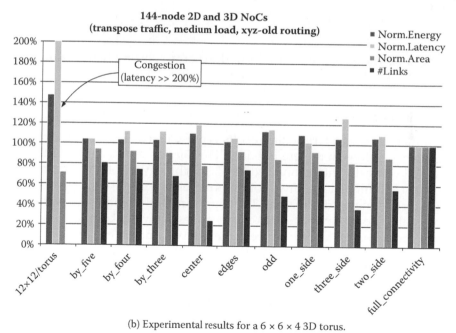

(b) Experimental results for a 6 × 6 × 4 3D torus.

FIGURE 1.11
Transpose traffic on a 3D NoC for alternative interconnection topologies.

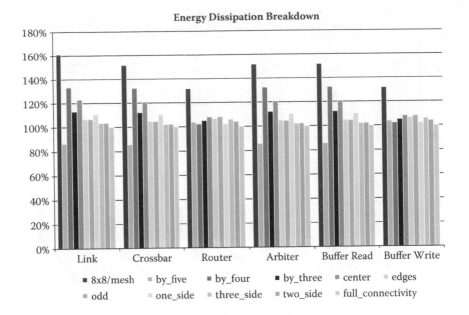

FIGURE 1.12
An overview of the energy breakdown in a 3D NoC (4 × 4 × 4 3D mesh, uniform traffic, xyz-old routing).

energy 62%. The normalized results of the energy consumption for a uniform traffic on a 4 × 4 × 4 NoC are presented in Figure 1.12.

1.5.8 Summary

A summary of the experimental results is presented in Table 1.1. The energy and latency values that were obtained are compared to the ones of the 3D mesh full vertically interconnected NoC. The three types of traffic are shown in the first column. The next two columns present the gains [min to max values (in%)] for the energy dissipation. The fourth and fifth columns show the min to max values for the average packet latency, respectively. It can been seen that energy reduction up to 29% can be achieved. But gains in energy dissipation

TABLE 1.1

Experimental Results: Min-Max Impact on Costs
(Energy and Latency) with Medium Traffic Load

	Normalized			
	Energy		Latency	
Traffic Patterns	**min**	**max**	**min**	**max**
Uniform	92%	108%	98%	113%
Transpose	88%	116%	100%	354%
Hotspot	71%	116%	100%	134%

cannot be reached without paying a penalty in average packet latency. It is the responsibility of the designer, utilizing this exploration methodology, to choose a 3D NoC topology and vertical interconnection patterns that best meet the requirements of the system.

1.6 Conclusions

Networks-on-Chips are becoming more and more popular as a solution able to accommodate large numbers of IP cores, offering an efficient and scalable interconnection network. Three-dimensional NoCs are taking advantage of the progress of integration and packaging technologies offering advantages when compared to 2D ones. Existing 3D NoCs assume that every router of a grid can communicate directly with the neighboring routers of the same grid and with the ones of the adjacent planes. This communication can be achieved by employing wire bonding, microbumb, or through-silicon vias [35].

All of these technologies have their advantages and disadvantages. Reducing the number of vertical connections makes the design and final fabrication of 3D systems easier. The goal of the proposed methodology is to find heterogeneous 3D NoC topologies with a mix of 2D and 3D routers and vertical link interconnection patterns that performs best to the incoming traffic. In this way, the exploration process evaluates the incoming traffic and the interconnection network, proposing an incoming traffic-specific alternative 3D NoC. Aiming in this direction, we have presented a methodology that shows by employing an alternative 3D NoC vertical link interconnection network, in essence proposing an NoC with less vertical links, we can achieve gains in energy consumption (up to 29%), in the average packet latency (up to 2%), and in the area occupied by the routers of the NoC (up to 18%).

Extensions of this work could include not only more heterogeneous 3D architectures but also different router architectures, providing better adaptive routing algorithms and performing further customizations targeting heterogeneous NoC architectures. In this way it would be able to create even more heterogeneous 3D NoCs. For providing stimuli to the NoCs, a move toward using real applications would be useful apart from using even more types of synthetic traffic. By doing so, it would become feasible to propose application-domain-specific 3D NoC architectures.

Acknowledgments

The authors would like to thank Dr. Antonis Papanikolaou (IMEC vzw., Belgium) for his helpful comments and suggestions. This research is supported by the 03ED593 research project, implemented within the framework of the "Reinforcement Program of Human Research Manpower" (PENED) and cofinanced by national and community funds (75% from European Union—European Social Fund and 25% from the Greek Ministry of Development—General Secretariat of Research and Technology).

References

1. Semiconductor Industry Association, "International technology roadmap for semiconductors," 2006. [Online]. Available: http://www.itrs.net/Links/2006Update/2006UpdateFinal.htm.

2. S. Murali and G. D. Micheli, "Bandwidth-constrained mapping of cores onto NoC architectures," In *Proc. of DATE*. Washington, DC: IEEE Computer Society, 2004, 896–901.

3. J. Hu and R. Marculescu, "Energy- and performance-aware mapping for regular NoC architectures," *IEEE Transactions on Computer-Aided Design of Integrated Circuits and Systems* 24 (2005) (4): 551–562.

4. L. Benini and G. de Micheli, "Networks on chips: a new SoC paradigm," *Computer* 35 (2002) (1): 70–78.

5. A. Jantsch and H. Tenhunen, eds., *Networks on Chip*. New York: Kluwer Academic Publishers, 2003.

6. K. Goossens, J. Dielissen, and A. Radulescu, "The Æthereal network on chip: Concepts, architectures, and implementations," *IEEE Des. Test*, 22 (2005) (5): 414–421.

7. STMicroelectronics, "STNoC: Building a new system-on-chip paradigm," White Paper, 2005.

8. S. Vangal, J. Howard, G. Ruhl, S. Dighe, H. Wilson, J. Tschanz, D. Finan, et al., "An 80-tile 1.28 TFLOPS network-on-chip in 65nm CMOS," In *Proc. of International Solid-State Circuits Conference (ISSCC)*. IEEE, 2007, 98–589.

9. U. Ogras and R. Marculescu, "Application-specific network-on-chip architecture customization via long-range link insertion," In *Proc. of ICCAD* (6–10 Nov.) 2005, 246–253.

10. E. Bolotin, I. Cidon, R. Ginosar, and A. Kolodny, "Cost considerations in network on chip," *Integr. VLSI J.* 38 (2004) (1): 19–42.

11. E. Beyne, "3D system integration technologies," In *International Symposium on VLSI Technology, Systems, and Applications*, Hsinchu, Taiwan, April 2006, 1–9.

12. ——, "The rise of the 3rd dimension for system integration," In *Proc. of International Interconnect Technology Conference*, Burlingame, CA 5–7 June, 2006, 1–5.

13. J. Joyner, R. Venkatesan, P. Zarkesh-Ha, J. Davis, and J. Meindl, "Impact of three-dimensional architectures on interconnects in gigascale integration," *IEEE Transactions on Very Large Scale Integration (VLSI) Systems*, 9 (Dec. 2001) (6): 922–928.

14. R. Marculescu, U. Y. Ogras, and N. H. Zamora, "Computation and communication refinement for multiprocessor SoC design: A system-level perspective," In *Proc. of DAC*. New York: ACM Press, 2004, 564–592.

15. J. Duato, S. Yalamanchili, and N. Lionel, *Interconnection Networks: An Engineering Approach*. San Francisco, CA: Morgan Kaufmann Publishers Inc., 2002.

16. W. Dally and B. Towles, *Principles and Practices of Interconnection Networks*. San Francisco, CA: Morgan Kaufmann Publishers Inc., 2003.

17. H. G. Lee, N. Chang, U. Y. Ogras, and R. Marculescu, "On-chip communication architecture exploration: A quantitative evaluation of point-to-point, bus, and network-on-chip approaches," *ACM Trans. Des. Autom. Electron. Syst.*, 12 (2007) (3): 23.

18. Z. Lu, R. Thid, M. Millberg, E. Nilsson, and A. Jantsch, "NNSE: Nostrum network-on-chip simulation environment," In *Proc. of SSoCC*, April 2005.

19. V. Soteriou, N. Eisley, H. Wang, B. Li, and L.-S. Peh, "Polaris: A system-level roadmap for on-chip interconnection networks," In *Proc. of ICCD*, October 2006. [Online]. Available: http://www.gigascale.org/pubs/930.html.

20. M. Dall'Osso, G. Biccari, L. Giovannini, D. Bertozzi, and L. Benini, "xPipes: a latency insensitive parameterized network-on-chip architecture for multi-processor SoCs," In *Proc. of ICCD*. IEEE Computer Society, 2003.

21. Open SystemC Initiative, *IEEE Std 1666-2005: IEEE Standard SystemC Language Reference Manual*. IEEE Computer Society, March 2006.

22. V. Puente, J. Gregorio, and R. Beivide, "SICOSYS: An integrated framework for studying interconnection network performance in multiprocessor systems," In *Proc. of 10th Euromicro Workshop on Parallel, Distributed and Network-Based Processing*, 2002, 15–22.

23. V. Soteriou, H. Wang, and L.-S. Peh, "A statistical traffic model for on-chip inter-connection networks," In *Proc. of MASCOTS*. Washington, DC: IEEE Computer Society, 2006, 104–116.

24. W. Heirman, J. Dambre, and J. V. Campenhout, "Synthetic traffic generation as a tool for dynamic interconnect evaluation," In *Proc. of SLIP*. New York: ACM Press, 2007, 65–72.

25. F. Ridruejo and J. Miguel-Alonso, "INSEE: An interconnection network sim-ulation and evaluation environment," In *Proc. of Euro-Par Parallel Processing*, 3648/2005. Berlin: Springer, 2005, 1014–1023.

26. U. Y. Ogras, J. Hu, and R. Marculescu, "Key research problems in NoC design: A holistic perspective," In *Proc. of CODES+ISSS*, 2005, 69–74.

27. F. Li, C. Nicopoulos, T. Richardson, Y. Xie, V. Narayanan, and M. Kandemir, "Design and management of 3D chip multiprocessors using network-in-memory," In *Proc. of ISCA*. Washington, DC: IEEE Computer Society, 2006, 130–141.

28. M. Koyanagi, H. Kurino, K. W. Lee, K. Sakuma, N. Miyakawa, and H. Itani, "Future system-on-silicon lsi chips," *IEEE Micro* 18 (1998) (4): 17–22.

29. K. Lee, T. Nakamura, T. Ono, Y. Yamada, T. Mizukusa, H. Hashimoto, K. Park, H. Kurino, and M. Koyanagi, "Three-dimensional shared memory fabricated using wafer stacking technology," *IEDM Technical Digest*, Electron Devices Meeting (2000) 165–168.

30. A. Iwata, M. Sasaki, T. Kikkawa, S. Kameda, H. Ando, K. Kimoto, D. Arizono, and H. Sunami, "A 3D integration scheme utilizing wireless interconnections for implementing hyper brains," 2005.

31. J. Meindl, "Interconnect opportunities for gigascale integration," *IEEE Micro* 23 (IEEE Computer Society Press, May/June 2003) (3): 28–35.

32. J. Joyner, P. Zarkesh-Ha, J. Davis, and J. Meindl, "A three-dimensional stochastic wire-length distribution for variable separation of strata," In *Proc. of the IEEE 2000 International Interconnect Technology Conference*. IEEE, 2000, 126–128.

33. J. Joyner and J. Meindl, "Opportunities for reduced power dissipation using three-dimensional integration," In *Proc. of the IEEE 2002 International Interconnect Technology Conference*. IEEE, 2002, 148–150.

34. P. Benkart, A. Kaiser, A. Munding, M. Bschorr, H.-J. Pfleiderer, E. Kohn, A. Heittmann, H. Huebner, and U. Ramacher, "3D chip stack technology using through-chip interconnects," *IEEE Des. Test* 22 (2005) (6): 512–518.

35. W. R. Davis, J. Wilson, S. Mick, J. Xu, H. Hua, C. Mineo, A. M. Sule, M. Steer, and P. D. Franzon, "Demystifying 3D ICs: The pros and cons of going vertical," *IEEE Des. Test* 22 (2005) (6): 498–510.

36. C. Ababei, Y. Feng, B. Goplen, H. Mogal, T. Zhang, K. Bazargan, and S. Sapatnekar, "Placement and routing in 3D integrated circuits," *IEEE Des. Test* 22 (2005) (6): 520–531.

37. S. K. Lim, "Physical design for 3D system on package," *IEEE Des. Test* 22 (2005) (6): 532–539.

38. H. Hua, C. Mineo, K. Schoenfliess, A. Sule, S. Melamed, R. Jenkal, and W. R. Davis, "Exploring compromises among timing, power and temperature in three-dimensional integrated circuits," In *Proc. of the 43rd Annual Conference on Design Automation.* New York: ACM, 2006, 997–1002.

39. S. Im and K. Banerjee, "Full chip thermal analysis of planar (2-D) and vertically integrated (3-D) high performance ICs," In *International Electron Devices Meeting, IEDM Technical Digest.*, 2000, 727–730.

40. T.-Y. Chiang, S. Souri, C. O. Chui, and K. Saraswat, "Thermal analysis of heterogeneous 3D ICs with various integration scenarios," In *Proc. of International Electron Devices Meeting*, 2001.

41. K. Puttaswamy and G. H. Loh, "Thermal analysis of a 3D die-stacked high-performance microprocessor," In *Proc. of the 16th ACM Great Lakes Symposium on VLSI.* New York: ACM, 2006, 19–24.

42. C. Addo-Quaye, "Thermal-aware mapping and placement for 3-D NoC designs," In *Proc. of IEEE SOC*, 2005, 25–28.

43. B. Goplen and S. Sapatnekar, "Thermal via placement in 3D ICs," *In Proc. of the 2005 International Symposium on Physical Design.* ACM, 2005, 167–174.

44. J. Cong and Y. Zhang, "Thermal via planning for 3-D ICs," In *Proc. of the 2005 IEEE/ACM International Conference on Computer-Aided Design.* Washington, DC: IEEE Computer Society, 2005, 745–752.

45. U. Y. Ogras and R. Marculescu, "Analytical router modeling for networks-on-chip performance analysis," In *Proc. of the Conference on Design, Automation and Test in Europe.* EDA Consortium, 2007, 1096–1101.

46. P. P. Pande, C. Grecu, M. Jones, A. Ivanov, and R. Saleh, "Performance evaluation and design trade-offs for networks-on-chip interconnect architectures," *IEEE Trans. on Comp.*, 54 (Aug. 2005) (8): 1025–1040.

47. B. Feero and P. P. Pande, "Performance evaluation for three-dimensional networks-on-chip," In *Proc. of ISVLSI*, 2007, 305–310.

48. V. F. Pavlidis and E. G. Friedman, "3-D topologies for networks-on-chip," *IEEE Trans. on VLSI Sys.*, 15 (2007) (10): 1081–1090.

49. J. Kim, C. Nicopoulos, D. Park, R. Das, Y. Xie, V. Narayanan, M. S. Yousif, and C. R. Das, "A novel dimensionally-decomposed router for on-chip communication in 3D architectures," In *Proc. of ISCA.* ACM Press, 2007, 138–149.

50. K. Siozios, K. Sotiriadis, V. F. Pavlidis, and D. Soudris, "Exploring alternative 3D FPGA architectures: Design methodology and CAD tool support," In *Proc. of FPL*, 2007.

51. L. M. Ni and P. K. McKinley, "A survey of wormhole routing techniques in direct networks," *Computer* 26 (1993) (2): 62–76.

52. T. Ye, L. Benini, and G. De Micheli, "Analysis of power consumption on switch fabrics in network routers," In *Proc. of DAC* (10–14 June) 2002, 524–529.
53. R. Reif, A. Fan, K.-N. Chen, and S. Das, "Fabrication technologies for three-dimensional integrated circuits," In *Proc. of International Symposium on Quality Electronic Design* (18–21 March) 2002, 33–37.
54. MIT Lincoln Labs, *Mitll Low-Power FDSOI CMOS Process Design Guide*, September 2006.
55. A. W. Topol, J. D. C. La Tulipe, L. Shi, D. J. Frank, K. Bernstein, S. E. Steen, A. Kumar, et al., "Three-dimensional integrated circuits," *IBM J. Res. Dev.* 50 (2006) (4/5): 491–506.
56. A. W. Topol, J. D. C. La Tulipe, L. Shi, D. J. Frank, K. Bernstein, S. E. Steen, A. Kumar, "Techniques for producing 3D ICs with high-density interconnect," In *VLSI Multi-Level Interconnection Conference*, 2004.
57. S. M. Alam, R. E. Jones, S. Rauf, and R. Chatterjee, "Inter-strata connection characteristics and signal transmission in three-dimensional (3D) integration technology," In *ISQED '07: Proceedings of the 8th International Symposium on Quality Electronic Design*. Washington, DC: IEEE Computer Society, 2007, 580–585.
58. G.-M. Chiu, "The odd-even turn model for adaptive routing," *IEEE Trans. Parallel Distrib. Syst.* 11 (2000) (7): 729–738.

2

Resource Allocation for QoS On-Chip Communication

Axel Jantsch and Zhonghai Lu

CONTENTS

2.1   Introduction

The provision of communication services with well-defined performance characteristics has received significant attention in the NoC community because for many applications it is not sufficient or adequate to simply maximize average performance. It is envisioned that complex NoC-based architectures will host complex, heterogeneous sets of applications. In a scenario where many applications compete for shared resources, a fair allocation policy that gives each application sufficient resources to meet its delay, jitter, and throughput requirements is critical. Each application, or each part of an application, should obtain exactly those resources needed to accomplish its task, not more nor less. If an application gets too small a share of the resources, it will either fail completely, because of a critical deadline miss, or its utility will be degraded, for example, due to bad video or audio quality. If an application gets more of a resource than needed, the system is over-dimensioned and not cost effective. Moreover, well-defined performance characteristics are a prerequisite for efficient composition of components and subsystems into systems [1]. If all subsystems come with QoS properties the system performance can be statically analyzed and, most importantly, the impact of the composition on the performance of individual subsystems can be understood and limited. In the absence of QoS characteristics, all subsystems have to be reverified because the interference with other subsystems may severely affect a subsystem's performance and even render it faulty. Thus, QoS is an enabling feature for compositionality.

This chapter discusses resource allocation schemes that provide the shared NoC communication resources with well-defined Quality of Service (QoS) characteristics. We exclusively deal with the performance characteristics delay, throughput, and, to a lesser extent, delay variations (jitter).

We group the resource allocation techniques into three main categories. Circuit switching* allocates all necessary resources during the entire lifetime of a connection. Figure 2.1(b) illustrates this scheme. In every switch there is a table that defines the connections between input ports and output ports. The output port is exclusively reserved for packets from that particular input port. In this way all the necessary buffers and links are allocated for a connection between a specific source and destination. Before a data packet can be sent, the complete connection has to be set up; and once it is done, the communication is very fast because all contention and stalling is avoided. The table can be implemented as an optimized hardware structure leading to a very compact and fast switch. However, setting up a new connection has a relatively high delay. Moreover, the setup delay is unpredictable because it is not guaranteed that a new connection can be set up at all. Circuit switching is justified only if a connection is stable over a long time and utilizes the resources to a very high degree. With few exceptions such as SoCBUS [2] and Crossroad [3], circuit switching has not been widely used in NoCs because only few applications justify the exclusive assignment of resources to individual connections. Also, the problem of predictable communication is not avoided but only moved from data communication time to circuit setup time. Furthermore, the achievable load of the network as a whole is limited in practice because a given set of circuits blocks the setup

---

* Note that some authors categorize time division multiplexing (TDM) techniques as a circuit switching scheme. In this chapter we reserve the term *circuit switching* for the case when resources are allocated exclusively during the entire lifetime of a connection.

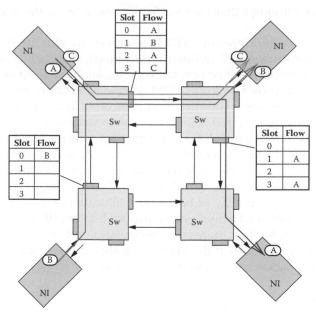

(a) TDM based allocation of likes with slot allocation tables.

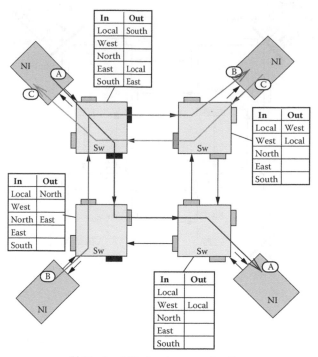

(b) Circuit switching based resource allocation

FIGURE 2.1
Resource allocation schemes based on TDM and circuit switching. (Sw = switch; NI = network interface; A, B, C = traffic flows).

of new circuits although there are sufficient resources in the network as a whole.

In time division multiplexing (TDM) resources are allocated exclusively to a specific user during well defined time periods. If a clock is available as a common time reference, clock cycles, or slots, are often used as allocation units. Figure 2.1(a) shows a typical TDM scheme where links are allocated to flows in specific time slots. The allocation is encoded in a slot allocation table with one table for each shared resource, a link in this example. The example assumes four different time slots. The tables are synchronized such that a flow, which has slot k in one switch, gets slot $(k + 1)$ mod 4 in the following switch, assuming it takes one cycle for a packet to traverse a switch.

The example illustrates two drawbacks of TDM schemes. First, there is a trade-off between granularity of bandwidth allocation and table size. If we have more flows and need a finer granularity for bandwidth allocation, larger tables are required. Second, there is a direct relation between allocated bandwidth and maximum delay. If the bandwidth allocated is k/n with k out of n slots allocated, a packet has to wait $n/k - 1$ cycles in the worst case for the next

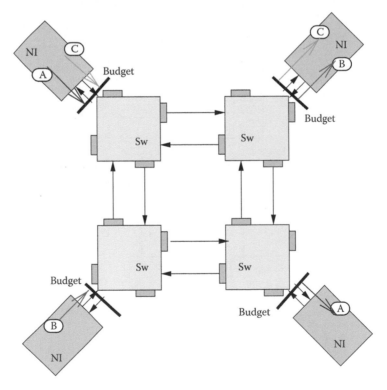

Aggregate resource allocation

FIGURE 2.2
Aggregate resource allocation.

slot to appear. This is a problem for low delay, low throughput traffic because it either gets much more bandwidth than needed or its delay is very high.

Aggregate resource allocation is a course grained and flexible allocation scheme. Figure 2.2 shows that each resource is assigned a traffic budget for both sent and received traffic. The reason for this is that if all resources comply with their budget bounds, the network is not overloaded and can guarantee minimum bandwidth and maximum delay properties for all the flows. Traffic budgets can be defined per resource or per flow, and they have to take into account the communication distance to correctly reflect the load in the network. Aggregate allocation schemes are flexible but provide looser delay bounds and require larger buffers than more fine grained control mechanisms such as TDM and circuit switching. This approach has been elaborated by Jantsch [1] and suitable analysis techniques can be adapted from flow regulation theories in communication networks such as network calculus [4].

In the following sections we will discuss these three main groups of resource allocation in more detail. In Section 2.5 we take up dynamic setup of connections, and in Section 2.6 we elaborate some aspects of priority schemes and fairness of resource allocation. Finally we give an example of how to use a TDM resource allocation scheme in a complex telecom system.

2.2   Circuit Switching

Circuit switching means that all the necessary resources between a source node and a destination node are exclusively allocated to a particular connection during its entire lifetime. Thus, no arbitration is needed and packets never stall on the way. Consequently, circuit switching allows for very fast communication and small, low-power switches, *if the application is suitable*. In an established connection, each packet experiences 1 cycle delay per switch in SoCBUS [2] and 3.48 ns in a 180 nm implementation of a crossroad switch [3]. Hence, the communication delay in established connections is both low and predictable. However, setting up a connection takes more time and is unpredictable. For SoCBUS, the authors report a setup time of at least 5 cycles per switch.

Figure 2.3 shows many, but not all, resources needed for a circuit switched connection. We have the access port in the network interface (NI), up- and downstream buffers in the NI, buffers and crossbars in the switches, and the links between the NIs and the switches. If the entire resource chain from the NI input port across the network to the NI output port is reserved exclusively for a specific connection, strong limitations are imposed on setting up other connections. For instance, in this scenario, one source node can only have one single sending and one receiving connection at the same time.

Figure 2.4 illustrates another aspect of the inflexibility of circuit switching. If the links attributed with "in use" labels are allocated to connections, no new

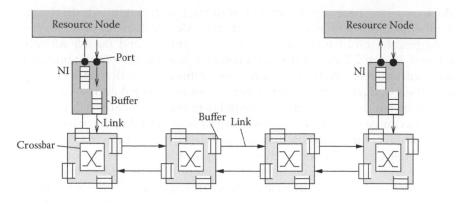

FIGURE 2.3
All the resources used for communication between a source and a destination can be allocated in different ways.

communication from the entire left half of the network to the right half is possible. If these four connections live for a long time, they will completely block a large set of new connections, independent of the routing policy employed, even if they utilize only a tiny fraction of the network or link bandwidth. If restrictive routing algorithms such as deterministic dimension order routing

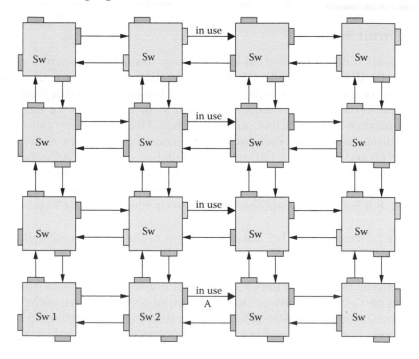

FIGURE 2.4
A few active connections may inhibit the setup of new connections although communication bandwidth is available.

are used, a few allocated links can already stall completely the communication between large parts of the system. For instance, if only one link, that is, link A in Figure 2.4, is used in both nodes connected to Sw 1, then Sw 2 will not be able to communicate to any of the nodes in the right half of the system under X–Y dimension order routing.

Consequently, neither the setup delay nor the unpredictability of the setup time is the most severe disadvantage of circuit switching when compared to other resource allocation schemes, because the connection setup problem is very similar in TDM-based techniques (see Section 2.5 for a discussion on circuit setup). The major drawback of circuit switching is its inflexibility and, from a QoS point of view, the limited options for selecting a particular QoS level. For a given source–destination pair the only choice is to set up a circuit switched connection which, once established, gives the minimal delay (1 cycle per hop × the number of hops in SoCBUS) and the full bandwidth. If an application requires many overlapping connections with moderate bandwidth demands and varying delay requirements, a circuit switched network has little to offer. Thus, a pure circuit switching allocation scheme can be used with benefit in the following two scenarios:

1. If the application exhibits a well-understood, fairly static communication pattern with a relatively small number of traffic streams with very high bandwidth requirements and long lifetime, these streams can be mapped on circuit switched connections in a cost- and power-efficient and low-delay implementation, as demonstrated in a study by Chung et al. [3].

2. For networks with a small number of hops (up to two), connections can be quickly built up and torn down. The setup overhead may be compensated by efficient data traversal even if the packet length is only a few words. Several proposals argue for circuit switching implementations based implicitly on this assumption [2,5,6]. But even for small-sized networks we have the apparent trade-off between packet size and blocking time of resources. Longer packets decrease the relative overhead of connection setup but block the establishment of other connections for a longer time.

For large networks and applications with communications having different QoS requirements that demand more flexibility in allocating resources, circuit switched techniques are only part of the solution at best.

This inflexibility of circuit switching can be addressed by duplicating some bottleneck resources. For instance, if the resources in the NI are duplicated, as shown in Figure 2.5(a), each node can entertain two concurrent connections in each direction, which increases the overall utilization of the network.

A study by Millberg et al. [7] has demonstrated that by duplicating the outgoing link capacity of the network, called dual packet exit [Figure 2.5(b)], the average delay is reduced by 30% and the worst case delay by 50%. Even though that study was not concerned with circuit switching, similar or higher gains are expected in circuit switched approaches. Leroy et al. [8] essentially

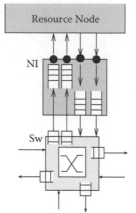
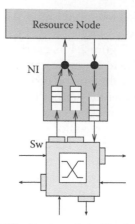

(a) Duplication of selected resources can increase the overall utilization of the network.

(b) Dual packet exit doubles the upstream buffers in the NI [7].

FIGURE 2.5
Duplication of NI resources.

propose to duplicate the links between switches in an approach they call spatial division multiplexing (SDM). As illustrated in Figure 2.6(a), parts of the link, that is, subsets of its wires, can be allocated to different connections, thus relaxing the exclusivity of link allocation in other circuit switched schemes. The switch then becomes a sophisticated multiplexer structure that allows the routing of input wires to output wires in a highly flexible and configurable manner. Leroy et al. compare this method to a TDM approach and report 8% less energy consumption, 31% less area, and 37% higher delay for their implementation of an SDM switch.

Circuit switching can be combined with time sharing by exclusively reserving some resources while sharing others. Those resources that are exclusively allocated can be duplicated to combine maximum flexibility with short delays and efficient implementation. A good example for a mixed approach is the Mango NoC [9], which exclusively allocates buffers in a switch but allows sharing of the links between switches. In Figure 2.6(b) the four buffers A–D are allocated exclusively to a connection but the link between the two switches is shared among them. Flits from the output buffers in the left-hand switch access the link based on a mix of round robin and priority arbitration that allows calculation of predictable end-to-end delays for each flit.

This clever scheme decouples to some extent delay properties of a connection from throughput properties. The maximum delay of a flit is controlled by assigning priorities. The higher the priority of a packet, the lower the maximum waiting time in the VC buffer for accessing the link. But even low priority flits have a bounded waiting time because they can be stalled by at most one flit of each higher priority connection. Hence, a priority Q flit has to wait at most $Q \cdot F$, where F is the time for one flit to access the link and there are $Q - 1$ higher priority classes. Because there is no other arbitration

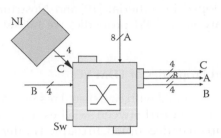

(a) Spatial division multiplexing (SDM) assigns different
wires to different connections on the links [8].

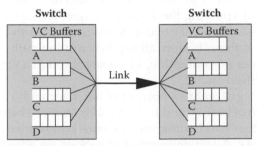

(b) A mixed circut switched and time shared allocation
scheme of the Mango NoC [9].

FIGURE 2.6
Duplication of switch resources.

in the switch, the end-to-end delay (not considering the network interfaces)
is bounded by $(Q \cdot F + \Delta) \cdot h$, where Δ is the constant delay in the crossbar
and the input buffer of the switch, and h is the number of hops.

The number of VCs determine the granularity of bandwidth allocation,
and the bandwidth allocated to a connection can be increased by assigning
more VCs.

One drawback of this method is that a connection exclusively uses a resource, a VC buffer, and to support many concurrent connections, many VCs
are required. This drawback is inherited from the exclusive resource allocation
of circuit switching, but it is a limited problem here because it is confined to the
VC buffers. Also, there is a trade-off between high granularity of bandwidth
allocation and the number of VCs. But this example demonstrates clearly that
the combination of different allocation schemes can offer significant benefits
in terms of increased flexibility and QoS control at limited costs.

2.3 Time Division Multiplexing Virtual Circuits

Next we discuss a less strict reservation method which exclusively allocates
a resource for specific connections only in individual time slots. In different
time slots the resource is used by different connections. We use the TDM

VC techniques developed in Ætherial [10] and Nostrum [11] as examples. For a systematic analysis of TDM properties we follow the theory of logic networks [12,13].

2.3.1   Operation and Properties of TDM VCs

In a network, we are concerned with two shared resources: buffers in switches and links (thus link bandwidth) between switches. The allocation for the two resources may be coupled or decoupled. In coupled buffer and link allocation, the allocation of buffers leads to associated link allocation. In decoupled buffer and link allocation, the allocation of buffers and that of links is independent. In this section, we consider the coupled buffer and link allocation using TDM. The consequence of applying the TDM technique to coupled buffer and link allocation is the reservation of exclusive slots in using both buffers and links. When packets pass these buffers along their routing path in reserved time slots, they encounter no contention, like going through a virtually dedicated circuit, called a virtual circuit.

On one hand, to guarantee a portion of link bandwidth for a traffic flow, the exclusive share of buffers and links must be reserved before actual packet delivery. On the other hand, traffic on a VC must be admitted into the network in a disciplined way. A certain number of packets are admitted in precalculated slots within a given window. This forms an *admission pattern* that is repeated without change throughout the lifetime of the VC. We call the window *admission cycle*. In the network, VC packets synchronously advance one hop per time slot. Because VC packets encounter no contention, they never stall, using consecutive slots in consecutive switches. As illustrated in Figure 2.7, VC v passes switches sw_1, sw_2, and sw_3 through $\{b_1 \rightarrow b_2 \rightarrow b_3\}$. On v, two packets are injected into the network every six slots (we say that the window size is 6). Initially, the slots of buffer b_1 at the first switch sw_1 occupied by the packet flow are 0 and 2. Afterward, this pattern repeats, b_1's slots 6 and 8, 12 and 14, and so on are taken. In the second switch sw_2, the packets occupy b_2's slots 1 and 3, 7 and 9, 13 and 15, and so on. In switch sw_3, they occupy b_3's slots 2 and 4, 8 and 10, 14 and 16, and so on.

As such, TDM VC makes two assumptions: (1) Network switches share the same notion of time. They have the same clock frequency but may allow phase

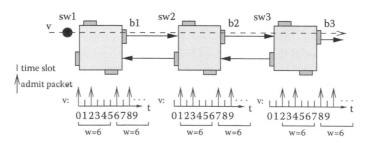

FIGURE 2.7
An example of packet delivery on a VC.

difference [14]. (2) Buffer and link allocations are coupled, as stated previously. Because packets are transmitted over these shared resources without stall and in a time-division fashion, we need only one buffer per link. This buffer may be situated at the input or output of a switch. As can be observed in Figure 2.7, we assumed that the buffer is located at the output. In terms of QoS, TDM VC provides strict guarantees in delay and bandwidth with low cost. Compared with circuit switching, it utilizes resources in a shared fashion (but with exclusive time slots), thus is more efficient. As with circuit switching, it must be established before communication can start. The establishment can be accomplished through configuring a routing table in switches. Routing for VC packets is performed by looking up these tables to find the output port along the VC path.

Before discussing VC configuration, we introduce two representative TDM VCs proposed for on-chip networks, the Æthereal VC [10] and the Nostrum VC [11].

2.3.2 On-Chip TDM VCs

Figure 2.8 shows two VCs, v_1 and v_2, and the respective routing tables for the switches. The output links of a switch are associated with a buffer or register. A routing table (t, in, out) is equivalent to a routing or slot allocation function $\mathcal{R}(t, in) = out$, where t is time slot, in an input link, and out an output link. v_1 passes switches sw_1 and sw_2 through $\{b_1 \rightarrow b_2\}$; v_2 passes switches sw_3 and sw_2 through $\{b_3 \rightarrow b_2\}$. The Æthereal NoC [10] proposes this type of VC for QoS. Because the path of such a VC is not a loop, we call it an open-ended VC.

The Nostrum NoC [11] also suggests TDM VC for QoS. However, a Nostrum VC has a cyclic path, that is, a closed loop. On the loop, at least one *container* is rotated. A container is a *special packet* used to carry data packets, like a vehicle carrying passengers. The reason to have a loop is due to the fact that Nostrum uses deflection routing [15] whereas switches have no buffer queues. If a packet arrives at a switch, it must be switched out using one output port the next cycle. A Nostrum switch has $k + 1$ inports/outports, k of which are network ports connected to other switches and one of which is a

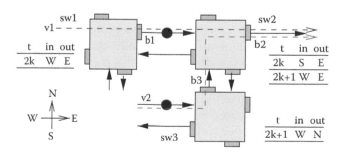

FIGURE 2.8
Open-ended virtual circuits.

local duplex port for admitting/sinking packets into/from the network. If k network packets arrive at a switch but none of them has reached its destination, none of the packets will be sunk and will occupy all the k network output ports the next cycle. This situation makes any packet admission at this time impossible because there is no output port available. This problem is solved by a looped container. The looped container ensures that there is always an output port or link available for locally admitting a VC packet into the container and thus the network. VC packets are loaded into the container from a source and copied (for multicast) or unloaded at the destination, bypassing other switches. Similarly to open-ended VCs, containers as VC packet carriers enjoy higher priority than best-effort packets and must not contend with each other.

2.3.3   TDM VC Configuration

As introduced in Section 2.3.2, TDM VC requires an establishment phase to configure routing tables. The entries for the routing tables are globally orchestrated such that no simultaneous use of shared buffers and links is possible, that means the network is free from contention. This process is called TDM VC configuration. We can loosely formulate the problem as follows: *Given a specification set of n VCs, each with a set of source and destination nodes and minimum bandwidth requirement, determine visiting nodes in sequence for each VC and exact time slots when VC packets visit each node.* Note that here we only use bandwidth as a constraint in the formulation, but apparently, other design constraints such as delay, jitter, and power can be added into the formulation, if needed. Also, we do not include a cost function as an optimization criterion.

VC configuration is a complex problem that can be elaborated as two sequential but orthogonal subproblems:

1. Path selection: Because a network is rich in connectivity, given a set of source and destination nodes,* there exist diverse ways to traverse all nodes in the set. Minimal routes are typically preferable. However, in some scenarios, nonminimal routes are also useful to balance traffic and have the potential to enable more VCs to be configurable. Allowing nonminimal routes further complicates the problem. In both cases, we need to explore the network path diversity. This involves an exponentially increased search space. Suppose each VC has m alternative paths, configuring n VC routes has a search space of m^n.

2. Slot allocation: Because VC packets cannot contend with each other, VCs must be configured such that an output link of a switch is allocated to one VC per slot. Again, finding optimal slot allocation, that is, reserving sufficient but not more than necessary bandwidth, requires exploring an exponentially increased design space.

---

* We allow that a VC may comprise more than one source and one destination node.

VC configuration is typically an iterative process, starting with the path selection, slot allocation, and then repeating the two sequential steps until a termination condition is reached such that solutions are found, or solutions cannot be found within a time threshold. VC configuration is a combinatorial optimization problem. Depending on the size of the problem, one can use different techniques to solve it, discovering whether there exist any optimal or suboptimal solutions. If the problem size is small, standard search techniques such as branch-and-bound backtracking may be used to constructively search the solution space, finding optimal or suboptimal solutions. If the size of the problem is large, heuristics such as dynamic programming, randomized algorithms, simulated annealing and genetic algorithms may be employed to find suboptimal solutions within reasonable time.

No matter what search techniques we use, any solution must satisfy three conditions: (1) All VCs are contention free. (2) All VCs are allocated sufficient slots in a time wheel, thus guaranteeing sufficient bandwidth. (3) The network must be deadlock-free and livelock-free. Condition (3) is important for networks with both best-effort and TDM VC traffic. The preallocated bandwidth must leave room to route best-effort traffic without deadlock and livelock. For example, a critical link should not have all its bandwidth reserved, making it unusable for best-effort packets. In the following sections, we focus on conditions (1) and (2), discussing how to guarantee contention-free VCs and how to allocate sufficient bandwidth. This discussion is based on the logical network theory [12,13]. The theory is equally suited for open and closed VCs; in the following sections we use closed VCs to illustrate the concepts.

2.3.4 Theory of Logical Network for TDM VCs

One key of success for synchronous hardware design is that we are able to reason out the logic behavior for each signal at each and every cycle. In this way, the logic design is fully deterministic. Traffic flows delivered on TDM VCs exhibit well-defined synchronous behavior, fully pipelined and moving one hop per cycle through preallocated resources. To precisely and collectively express the resources used or reserved by a guaranteed service flow, we define a logical network (LN) as an infinite set of associated (time slot, buffer) pairs with respect to a buffer on the flow path. This buffer is called the *reference buffer*. When VCs overlap, we use a shared buffer as the reference buffer because it is the point of interest to avoid contention. Because LNs use exclusive sets of resources, VCs allocated to different LNs are free from conflict.

LNs may be constructed in two steps: *slot partitioning* and *slot mapping*. Slot partitioning is performed in the time domain and slot mapping is performed in the space domain. We exemplify the LN construction using Figure 2.9, where two VCs v_1 and v_2 are to be configured. The loop length of v_1 is 4 and a container revisits the same buffer every 4 cycles. Assuming uniform link bandwidth 1 packet/cycle, the bandwidth granularity of v_1 is 1/4 packet/cycle and the packet admission cycle on v_1 is 4. Because two containers are launched, v_1 offers a bandwidth of 1/2 packet/cycle. Similarly, v_2 with one container

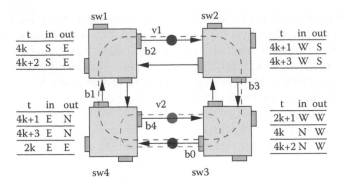

FIGURE 2.9
Closed-loop virtual circuits.

supports bandwidth of 1/2 packet/cycle, and the packet flow on v_2 has an admission cycle of 2.

1. Slot partitioning: As b_0 is the only shared buffer of v_1 and v_2, $v_1 \cap v_2 = \{b_0\}$, we use b_0 as the reference buffer, denoted as $Ref(v_1, v_2) = b_0$. Because v_1 and v_2 use b_0 once every two slots, their bandwidth equals $1/2$. Thus, we partition the slots of b_0 into two sets, an even set $s_0^2(b_0)$ for $t = 2k$ and an odd set $s_1^2(b_0)$ for $t = 2k + 1$, as highlighted in Figure 2.10 by the underlined number set $\{0, 2, 4, 6, 8, \cdots\}$ and $\{1, 3, 5, 7, 9, \cdots\}$, respectively. The notation $s_\tau^T(b)$ represents pairs $(\tau + kT, b)$, which is the τth slot set of the total T slot sets, $\forall k \in \mathbb{N}$, $\tau \in [0, T)$ and $T \in \mathbb{N}$. The pair (t, b) refers to the slot of b at time instant t. Notation $s_{\tau_1, \tau_2, \cdots, \tau_n}^T(b)$ collectively represents a set of pair sets $\{(\tau_1 + kT, b), (\tau_2 + kT, b), \cdots, (\tau_n + kT, b)\}$.

2. Slot mapping: The partitioned slot sets can be mapped to slot sets of other buffers on a VC regularly and unambiguously because a VC

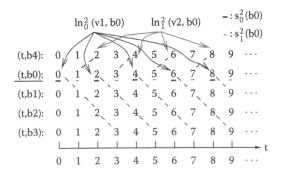

FIGURE 2.10
LN construction by partitioning and mapping slots in the time and space domain.

packet or container advances one hop along its path each and every slot. For example, v_1 packets holding slot t at buffer b_0, that is, pair (t, b_0), will consecutively take slot $t+1$ at b_1 (pair $(t+1, b_1)$), slot $t+2$ at b_2 (pair $(t+2, b_2)$), and slot $t+3$ at b_3 (pair $(t+3, b_3)$). In this way, the slot partitionings are propagated to other buffers on the VC. In Figure 2.10, after mapping the slot set $s_0^2(b_0)$ on v_1 and $s_1^2(b_0)$ on v_2, we obtain two sets of slot sets $\{s_0^2(b_0), s_1^2(b_1), s_0^2(b_2), s_1^2(b_3)\}$ and $\{s_1^2(b_0), s_0^2(b_4)\}$, as marked by the dashed diagonal lines. We refer to the logically networked slot sets in a set of buffers of a VC as an LN. Thus an LN is a composition of associated (time slot, buffer) pairs on a VC with respect to a buffer. We denote the two LNs as $ln_0^2(v_1, b_0)$ and $ln_1^2(v_2, b_0)$, respectively. The notation $ln_\tau^T(v, b)$ represents the τth LN of the total T LNs on v with respect to b. Figure 2.10 illustrates the mapped slot sets for $s_0^2(b_0)$ and $s_1^2(b_0)$ and the resulting LNs. We may also observe that slot mapping is a process of assigning VCs to LNs. LNs can be viewed as the result of VC assignment to slot sets, and an LN is a function of a VC. In our case, v_1 subscribes to $ln_0^2(v_1, b_0)$ and v_2 to $ln_1^2(v_2, b_0)$.

As $ln_0^2(v_1, b_0) \cap ln_1^2(v_2, b_0) = \emptyset$, v_1 and v_2 are conflict-free. In addition, the bandwidth supply of both VCs equals 1/2 packet/cycle, $BW(ln_0^2(v_1, b_0)) = BW(ln_1^2(v_2, b_0)) = 1/2$ packet/cycle.

Suppose that v_1 and v_2 are two overlapping VCs with D_1 and D_2 being their admission cycles, respectively, we have proved a set of important theorems, which we summarize as follows:

- The maximum number N_{ln} of LNs, which v_1 and v_2 can subscribe to without conflict, equals $GCD(D_1, D_2)$, the greatest common divisor (GCD) of D_1 and D_2. The bandwidth that an LN possesses equals $1/N_{ln}$ packet/cycle.

- Assigning v_1 and v_2 to different LNs is the sufficient and necessary condition to avoid conflict between them.

- If v_1 and v_2 have multiple (more than one) shared buffers, these buffers must satisfy *reference consistency* to be free from conflict. If so, any of the shared buffers can be used as the reference buffer to construct LNs. Two shared buffers b_1 and b_2 are termed *consistent* if it is true that, "v_1 and v_2 packets do not conflict in buffer b_1" if and only if "v_1 and v_2 packets do not conflict in buffer b_2." The sufficient and necessary condition for them to be consistent is that the distances of b_1 and b_2 along the two VCs, denoted $d_{b_1 \bar{b}_2}(v_1)$ and $d_{b_1 \bar{b}_2}(v_2)$, respectively, satisfy $d_{b_1 \bar{b}_2}(v_1) - d_{b_1 \bar{b}_2}(v_2) = kN_{ln}$, $k \in \mathbb{Z}$. Furthermore, instead of pair-wise checking, the reference consistency can be linearly checked.

2.3.5   Application of the Logical Network Theory for TDM VCs

Guided by the LN theory, slot allocation is a procedure of computing and consuming LNs, by which VCs are assigned to different LNs. To begin this procedure, we need to know the admission cycle and bandwidth demand of VCs. We draw a diagram for allocating slots for two VCs, as shown in Figure 2.11. It has three main steps:

Step 1: Reference consistency check. If two VCs have more than one shared buffer, this step checks whether they are consistent.

Step 2: Compute available LNs. The available LNs (thus bandwidth) are computed to check if they can satisfy the VC's bandwidth requirement.

Step 3: Consume LNs. This allocates and claims the exclusive slots (slot sets).

As an example, in Figure 2.12, two VCs v_1 and v_2 comprise a buffer set, $\{b_1, b_2, b_5, b_6\}$ and $\{b_2, b_3, b_4, b_6\}$, respectively, and their bandwidth demand is $BW(v_1) = 3/8$ and $BW(v_2) = 7/16$. Following the procedure, the slot allocation is conducted as follows:

Step 1: The number N_{ln} of LNs for v_1 and v_2 equals $N_{ln}(v_1, v_2) = GCD(8, 16) = 8$. Since $v_1 \cap v_2 = \{b_2, b_6\}$, the distance between b_2 and b_6 along v_1 is $d_{b_2 b_6}(v_1) = 2$. Similarly, the distance between b_2 and

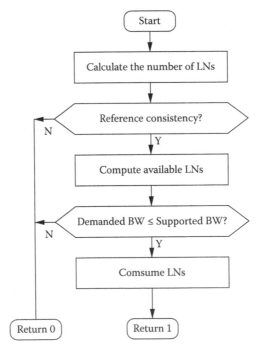

FIGURE 2.11
LN-oriented slot allocation.

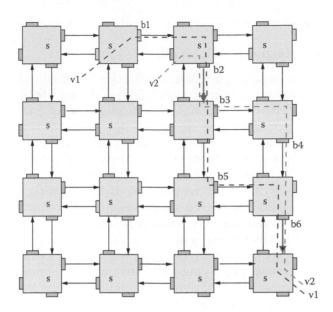

FIGURE 2.12
An example of LN-oriented slot allocation.

b_6 along v_2 is $d_{b_2 b_6}(v_2) = 2$. Thus the difference between the two distances is $d_{b_2 b_6}(v_1, v_2) = d_{b_2 b_6}(v_1) - d_{b_2 b_6}(v_2) = 0$. Therefore, b_2 and b_6 satisfy reference consistency. We can select either of them as the reference buffer, say, b_2.

Steps 2 and 3: We can start with either of the two VCs, say v_1. Referring to b_2, the slot set to consider is $\{0, 1, 2, 3, 4, 5, 6, 7\}$. Each of the elements in the set represents an LN with bandwidth supply 1/8 packet/cycle. Initially all slots are available. Because BW$(v_1) = 3/8$, we allocate slots $\{0, 2, 4\}$ to v_1. The remaining slot set is $\{1, 3, 5, 6, 7\}$, providing a bandwidth of 5/8. Apparently, this can satisfy the bandwidth requirement of v_2. We allocate slots $\{1, 3, 5, 7\}$ to v_2. The resulting slot allocation meets both VCs' bandwidth demand and packets on both VCs are contention-free.

The LN-based theory provides us a formal method to conduct slot allocation. The application of this theory to open-ended VCs is straightforward, as we have shown. Applying this theory to closed-loop VCs is also simple. In this case, the admission cycle for VCs is predetermined by the length of VC loops. Although this approach opens a new angle for slot allocation avoiding ad hoc treatment of this problem, the complexity of optimal slot allocation still remains. In this regard, the key questions include how to make the right bandwidth granularity by scaling admission cycles without jeopardizing application requirements, and how to consume LNs to leave room for VCs that remain to be configured.

2.4   Aggregate Resource Allocation

In TDM and circuit switching approaches resources are allocated exclusively to connections either for short time slots (TDM) or during the entire lifetime of the connection (circuit switching). This results in precise knowledge of delay and throughput of a connection independent of other activities in the network. Once a connection is established, data communication in other connections cannot interfere or influence the performance. This exclusive allocation of resources results in strong isolation properties, which is ideal for QoS guarantees. The disadvantage is the potential underutilization of resources. A resource (buffer, link) that is exclusively allocated to a connection, cannot be used by other connections even when it is idle. Dynamic and adaptive reallocation of resources for the benefit of overall performance is not possible. Thus, both TDM and circuit switching schemes are ideally suited for the well-known, regular, long-living traffic streams that require strong QoS guarantees. They are wasteful for dynamic, rapidly changing traffic patterns for which the occasional miss of a deadline is tolerable.

Resource planning with TDM and circuit switching schemes will optimize for the worst case situation and will provide strong upper bounds for delay and lower bounds for throughput. The average case delay and throughput is typically very close or even identical to the worst case. If a connection is set up, based on circuit switching, the delay of every packet through the network is the same and the worst case delay is the same as the average case delay. For TDM connections, the worst case delay occurs when a packet has just missed its time slot and has to wait one full period for the next time slot. On average this waiting time will be only half the worst case waiting time, but the delivery time through the network is identical for all packets. Hence, the average case delay will be lower but close to the worst case.

We can relax the assumption of exclusive resource ownership to allow for dynamic and more adaptive allocation of resources while still being able to provide QoS guarantees. *Aggregate resource allocation* assigns a resource to a group of users (connections or processing elements) and dynamically arbitrates the requests for resource usage. By constraining the volume of traffic injected by users and by employing proper arbitration policies, we can guarantee worst case bounds on delay and throughput while maximizing resource utilization and, thus, average performance.

2.4.1   Aggregate Allocation of a Channel

Consider the situation in Figure 2.13(a). Two flows share a single channel. We know the capacity of the channel (32 Mb/s) and the delay (2 μs for 32 bits that are transferred in parallel). However, to give bounds on delay and throughput for the two individual flows we need to know the following:

1. Characteristics of the flows
2. Arbitration policy for channel access

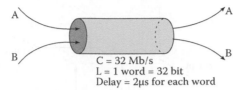

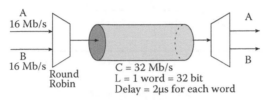

C = 32 Mb/s
L = 1 word = 32 bit
Delay = 2μs for each word

(a) One channel is allocated to two flows.

C = 32 Mb/s
L = 1 word = 32 bit
Delay = 2μs for each word

(b) The channel access is arbitrated with a round-robin policy.

FIGURE 2.13
Shared channel.

The flows have to be characterized, for example, in terms of their average traffic rate and their burstiness. The latter is important because a flow with low average rate and unlimited burst size can incur an unlimited delay on its own packets and, depending on the isolation properties of the arbiter, on the other flow as well. The arbitration policy has to be known because it determines how much the two flows influence each other. Figure 2.13(b) shows the situation where each flow has an average rate of 16 Mb/s and the channel access is controlled by a round-robin arbiter. Assuming a fixed word length of L in both flows, round-robin arbitration means that each flow gets at least 50% of the channel bandwidth, which is 16 Mb/s. A flow may get more if the other flow uses less, but we now know a worst case lower bound on the bandwidth. Round-robin arbitration has good isolation properties because the minimum bandwidth for each flow does not depend on the properties of the other flow.

To derive an upper bound on the delay for each flow, we have to know the maximum burst size. There are many ways to characterize burstiness of flows. We use a simple, yet powerful, traffic volume-based flow model from network calculus [4,16]. In this model a traffic flow can be characterized by a pair of numbers (σ, ρ), where σ is the burstiness constraint and ρ is the average bit rate. Call $F(t)$ the total volume of traffic in bits on the flow in the period $[0, t]$. Then a flow is (σ, ρ)-regulated if

$$F(t_2) - F(t_1) \leq \sigma + \rho(t_2 - t_1)$$

for all time intervals $[t_1, t_2]$ with $0 \leq t_1 \leq t_2$. Hence, in any period the number of bits moving in the flow cannot exceed the average bit rate by more than σ. This concept is illustrated in Figure 2.14 where the solid line shows a flow that is constrained by the function $\sigma + \rho t$.

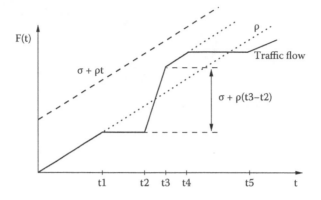

FIGURE 2.14
A (σ, ρ)-regulated flow.

We use this notation in our shared channel example and model a round-robin arbiter as a *rate-latency server* [4] that serves each input flow with a minimum rate of $C/2$ after a maximum initial delay of L/C, assuming a constant word length of L in both flows. Then, based on the network calculus theory, we can compute the maximum delay and backlog on flow A ($\bar{D}_A$, $\bar{B}_A$) and flow B ($\bar{D}_B$, $\bar{B}_B$), and the characteristics of the output flows A^* and B^*, as shown in Figure 2.15.

We cannot derive these formulas here (see Le Boudec [4] for detailed derivation and motivation) due to the limited space, but we can make several observations. The delay in each flow consists of three components. The first two are due to arbitration and the last one, 2 μs, is the channel delay. The term L/C is the worst case, as the time it takes for a word in one flow to get access to the channel if there are no other words in the same flow queued up before the arbiter. The second term, $2\sigma/C$, is the delay of a worst case burst. The formula for the maximum backlog also consists of two terms: one due to the worst

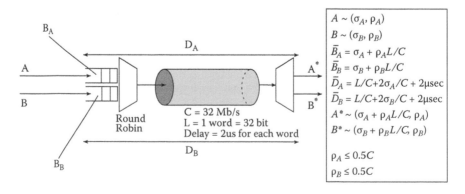

FIGURE 2.15
The shared channel serves two regulated flows *A* and *B* with round-robin arbitration.

TABLE 2.1

Maximum Delay, Backlog, and Output Flow Characteristics for a Round-Robin Arbitration. Delays Are in μs, Rates Are in Mb/s, and Backlog and Delay Values Are Rounded Up to Full 32-Bit Words.

A	B	A			B		
(σ_A, ρ_A)	(σ_B, ρ_B)	$\bar{B}_A$	$\bar{D}_A$	$(\sigma_{A^*}, \rho_{A^*})$	$\bar{B}_B$	$\bar{D}_B$	$(\sigma_{B^*}, \rho_{B^*})$
(0, 16.00)	(0, 16.00)	32	3	(32, 16.00)	32	3	(32, 16.00)
(0, 12.80)	(0, 12.80)	32	3	(32, 12.80)	32	3	(32, 12.80)
(0, 9.60)	(0, 16.00)	32	3	(32, 9.60)	32	3	(32, 16.00)
(0, 6.40)	(0, 16.00)	32	3	(32, 6.40)	32	3	(32, 16.00)
(0, 3.20)	(0, 16.00)	32	3	(32, 3.20)	32	3	(32, 16.00)
(0, 16.00)	(0, 16.00)	32	3	(32, 16.00)	32	3	(32, 16.00)
(32, 16.00)	(0, 16.00)	64	5	(64, 16.00)	32	3	(32, 16.00)
(64, 16.00)	(0, 16.00)	96	7	(96, 16.00)	32	3	(32, 16.00)
(128, 16.00)	(0, 16.00)	160	11	(160, 16.00)	32	3	(32, 16.00)
(256, 16.00)	(0, 16.00)	288	19	(288, 16.00)	32	3	(32, 16.00)

case arbitration time $(\rho L/C)$ and the other due to bursts (σ). The rates of the output flows are unchanged, as is expected, but the burstiness increases due to the variable channel access delay in the arbiter. It can be seen in the formulas that delay and backlog bounds and the output flow characteristics of each flow do not depend on the characteristics of the other flow. This demonstrates the strong isolation of the round-robin arbiter that in the worst case always offers half the channel bandwidth to each flow. However, the average delay and backlog values of one flow do depend on the actual behavior of the other flow, because if one flow does not use its maximum share of the channel bandwidth (0.5C), the arbiter allows the other flow to use it. This dynamic reallocation of bandwidth will increase average performance and channel utilization. However, note that these formulas are only valid under the given assumptions, that is, the average flows of both rates must be lower than 50% of the channel bandwidth. If one flow has a higher average rate, its worst case backlog and delay are unbounded.

Table 2.1 shows how the delay and backlog bounds depend on input rates and burstiness. In the upper half of the table, both flows have no burstiness but the rate of flow *A* is varying. It can be seen that flow *B* is not influenced at all and for flow *A* only the output rate changes but delay and backlog bounds are not affected. This is because as long as the flow does not request more than 50% of the channel bandwidth (16 Mb/s), both backlog and delay in the arbiter are only caused by the arbitration granularity of one word. In the lower part of the table, the burstiness of flow *A* is steadily increased. This affects the backlog bound, the delay bound, and the output flow characteristics of *A*. However, flow *B* is not affected at all, which underscores the isolation property of round-robin arbitration under the given constraints.

To illustrate the importance of the arbitration policy on QoS parameters and the isolation of flows, we present priority-based arbitration as another example. Figure 2.16 shows the same situation, but the arbiter gives higher

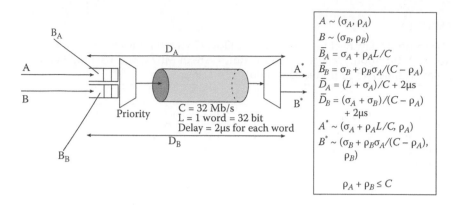

$$A \sim (\sigma_A, \rho_A)$$
$$B \sim (\sigma_B, \rho_B)$$
$$\bar{B}_A = \sigma_A + \rho_A L/C$$
$$\bar{B}_B = \sigma_B + \rho_B \sigma_A/(C - \rho_A)$$
$$\bar{D}_A = (L + \sigma_A)/C + 2\mu s$$
$$\bar{D}_B = (\sigma_A + \sigma_B)/(C - \rho_A) + 2\mu s$$
$$A^* \sim (\sigma_A + \rho_A L/C, \rho_A)$$
$$B^* \sim (\sigma_B + \rho_B \sigma_A/(C - \rho_A), \rho_B)$$
$$\rho_A + \rho_B \leq C$$

FIGURE 2.16
The shared channel serves two regulated flows A and B with a priority arbitration.

priority to flow A and flow B word cannot be preempted. If a flow B word has obtained access to the channel, an arriving flow A word has to wait until the complete flow B word is emitted into the channel. As can be seen from the formulas in Figure 2.16, flow A is served at full channel bandwidth C, but the service rate of flow B is dependent on flow A, which is $C - \rho_A$. The bounds and output characteristics of flow A are entirely independent of flow B. In fact, flow B is almost invisible to flow A because the full channel capacity is available to flow A if requested. The only delay incurred by flow B is L/C if a flow B word has been granted access and is allowed to complete transmission. On the other hand, the bounds and output characteristics of flow B depend heavily on σ_A and ρ_A. This impact is shown quantitatively in Table 2.2.

To summarize this example, the round-robin policy offers a fair access to the channel and provides very good isolation properties such that the

TABLE 2.2

Maximum Delay, Backlog, and Output Flow Characteristics for an Arbitration Giving a Higher Priority. Delays Are in μs, Rates Are in Mb/s, and Backlog and Delay Values Are Rounded Up to Full 32-Bit Words.

A	B	A			B		
(σ_A, ρ_A)	(σ_B, ρ_B)	$\bar{B}_A$	$\bar{D}_A$	$(\sigma_{A^*}, \rho_{A^*})$	$\bar{B}_B$	$\bar{D}_B$	$(\sigma_{B^*}, \rho_{B^*})$
(0, 16.00)	(0, 16.00)	32	3	(32, 16.00)	32	4	(32, 16.00)
(0, 12.80)	(0, 12.80)	32	3	(32, 12.80)	32	4	(32, 12.80)
(0, 9.60)	(0, 16.00)	32	3	(32, 9.60)	32	3	(32, 16.00)
(0, 6.40)	(0, 16.00)	32	3	(32, 6.40)	32	3	(32, 16.00)
(0, 3.20)	(0, 16.00)	32	3	(32, 3.20)	32	3	(32, 16.00)
(0, 16.00)	(0, 16.00)	32	3	(32, 16.00)	32	4	(32, 16.00)
(32, 16.00)	(0, 16.00)	64	4	(64, 16.00)	32	4	(32, 16.00)
(64, 16.00)	(0, 16.00)	96	5	(96, 16.00)	64	6	(64, 16.00)
(128, 16.00)	(0, 16.00)	160	7	(160, 16.00)	128	10	(128, 16.00)
(256, 16.00)	(0, 16.00)	288	11	(288, 16.00)	256	18	(256, 16.00)

performance of one flow can be analyzed independent of the behavior of the other flow. Priority-based arbitration offers better QoS figures for one flow at the expense of the other. Its isolation properties are weaker because the performance of the low priority flow can only be analyzed if the characteristics of the high priority flow are known. An extensive analysis and comparison of different service disciplines is presented by Hui Zhang [17]. An analysis of some arbitration policies in the network calculus framework is elaborated by LeBoudec [4].

2.4.2 Aggregate Allocation of a Network

In a network each connection needs a chain of resources. To perform aggregate resource allocation for the entire connection, we can first do it for each individual resource and then compute performance properties for the entire connection. Network calculus is a suitable framework for this approach because it allows us to derive tighter bounds for sequences of resources than what is possible by simply adding up the worst cases of individual resources. This feature is known as *pay bursts only once*. Although this is a feasible and promising approach, we illustrate here an alternative technique that views the entire network as a resource to derive QoS properties. It has been elaborated in the context of the Nostrum NoC [1], which we take as an example in the following.

Each processing element in the network is assigned a traffic budget for both incoming and outgoing traffic. The amount of traffic in the network is bounded by these node budgets. As a consequence, the network exhibits predictable performance that can be used to compute QoS characteristics for each connection.

The Nostrum NoC has mesh topology and a deflection routing algorithm, which means that packets that lose competition for a resource are not buffered but deflected to a nonideal direction. Hence, packets that are deflected take nonminimal paths through the network. A connection h between a sender **A** and a receiver **B** loads the network with

$$E_h = n_h d_h \delta \tag{2.1}$$

where n_h is the number of packets **A** injects into the network during a given window, W, d_h is the shortest distance between **A** and **B**, and δ is the average deflection factor. It expresses the average amount of deflections a packet experiences and is defined as

$$\delta = \frac{\text{sum of traveling time of all packets in cycles}}{\text{sum of shortest path of all packets in cycles}}$$

δ is load dependent and, as we will see in the following equations, the network load has to be limited in order to bound δ. Call H_r^o and H_r^i the sets of all outgoing and incoming connections of node r, respectively. We assign traffic budgets for each node as follows:

$$\sum_{h \in H_r^o} E_h \leq B_r^o \tag{2.2}$$

$$\sum_{h \in H_r^i} E_h \leq B_r^i \tag{2.3}$$

$$\sum_r B_r^o = \sum_r B_r^i \leq \kappa C_{\text{Net}} \tag{2.4}$$

B_r^o and B_r^i constitute the traffic budgets for each node r and C_{Net} is the total communication capacity of the network during the time window $W.\kappa$, with $0 \leq \kappa \leq 1$, which is called the *traffic ceiling*. It is an empirical constant that has to be set properly to bound the deflection factor δ. A node is allowed to set up a new connection as long as the constraints shown in Equations (2.2) and (2.3) are met. In return, every connection is characterized by the following bandwidth, average delay, and maximum delay bounds [1],

$$\text{BW}_h = \frac{n_h}{W} \tag{2.5}$$

$$\text{maxLat}_h = 5DN \tag{2.6}$$

$$\text{avgLat}_h = d_h \delta \tag{2.7}$$

where D is the diameter of the network and N the number of nodes.

Thus, to summarize, by constraining the traffic for the whole network (by κ), for each resource node (B_r^i, B_r^o) and for each connection (n_h/W), QoS characteristics Equations (2.5), (2.6), and (2.7) are obtained. But note the dependency of the deflection factor δ on the traffic ceiling κ, for which a closed analytic formula is not known for a deflective routing network with its complex, adaptive behavior. In the paper by Jantsch [1], D_1 is suggested as an upper bound for δ. D_1 is the delay bound for 90% of the packets under uniformly distributed traffic. It can be determined empirically and has been found to be fairly tight bound when the network is only lightly loaded. When the network is operated close to its saturation point, the bound is much looser. However, the important point is that D_1 has been found to be an upper bound for δ even for a large number of different traffic patterns. Hence, it can serve as an empirical upper bound under a wide range of conditions. Table 2.3 shows a given traffic ceiling κ and the measured corresponding D_1 for a range of network sizes. The *sat.* entries mean that the network is saturated and delays are unbounded.

This approach of using the entire network as an aggregate resource that is managed by controlling the incoming traffic gives very loose worst case bounds. The worst case bounds on maximum delay and minimum bandwidth are conservative and are always honored. In contrast, the given average delay may be violated. It is an upper bound in the sense that in most cases the observed average delay is below, but it is not guaranteed because it is possible to construct traffic patterns that violate this bound.

TABLE 2.3

(κ, D_1) Pairs for Various Network Sizes N and Emission Budgets per Cycle B_r^o / W

B_r^o / W	16 (κ, D_1)	30 (κ, D_1)	50 (κ, D_1)	70 (κ, D_1)	100 (κ, D_1)
0.05	(0.04, 1.12)	(0.06, 1.12)	(0.07, 1.15)	(0.08, 1.16)	(0.09, 1.11)
0.10	(0.09, 1.12)	(0.11, 1.15)	(0.14, 1.23)	(0.16, 1.23)	(0.19, 1.23)
0.15	(0.13, 1.12)	(0.17, 1.30)	(0.21, 1.41)	(0.24, 1.35)	(0.28, 1.35)
0.20	(0.18, 1.36)	(0.22, 1.40)	(0.27, 1.46)	(0.32, 1.46)	(0.37, 1.55)
0.25	(0.22, 1.44)	(0.28, 1.45)	(0.34, 1.64)	(0.40, 1.80)	(0.46, *sat.*)
0.30	(0.27, 1.44)	(0.34, 1.61)	(0.41, 4.65)	(0.48, *sat.*)	(0.56, *sat.*)
0.35	(0.31, 1.60)	(0.39, 1.72)	(0.48, *sat.*)	(0.55, *sat.*)	(0.65, *sat.*)
0.40	(0.36, 1.60)	(0.45, 6.10)	(0.55, *sat.*)	(0.63, *sat.*)	(0.74, *sat.*)
0.45	(0.40, 1.80)	(0.50, *sat.*)	(0.62, *sat.*)	(0.71, *sat.*)	(0.83, *sat.*)
0.50	(0.44, 6.17)	(0.56, *sat.*)	(0.69, *sat.*)	(0.79, *sat.*)	(0.93, *sat.*)

This approach, while giving loose worst case bounds, optimizes the average performance, because all network resources are adaptively allocated to the traffic that needs them. It is also cost effective, because in the network there is overhead for reserving resources and no sophisticated scheduling algorithm for setting up connections is required. The budget regulation at the network entry can be implemented cost efficiently and the decision for setting up a new connection can be taken quickly, based on locally available information. However, to check if the receiving node has sufficient incoming traffic capacity is more time consuming because it requires communication and acknowledgment across the network.

2.5 Dynamic Connection Setup

The reservation of resources to provide performance guarantees poses a dilemma. Once all resources are allocated, it is straightforward to calculate the delay of a packet from a sender to a receiver. However, the setup time to establish a new connection may be subjected to arbitrary delay and unbounded. Hence, the emission time of the first packet in a connection cannot be part of the QoS guarantees. This feature of pushing the uncertainty of delays from the communication of an individual packet to the setup of a connection is in common to all the three resource allocation classes discussed in Section 2.4. TDM, circuit switching, and aggregate resource allocation schemes all have to set up a connection first, to be able to provide QoS guarantees for established connections. We have three possibilities to deal with this problem:

1. Setup of static connections at design time
2. Limit the duration of connections to bound the setup time for a statically defined set of connections
3. Accept unbounded setup time

Alternative (1), to allow only statically defined connections, is acceptable only for a certain class of applications that exhibit a well-known and static traffic pattern. For more dynamic applications, this option is either too limiting or too wasteful.

Alternative (2) is a compromise between (1) and (3). It defines statically at design time a set of connections. Each connection is characterized and assigned a maximum traffic volume and lifetime. Because the maximum lifetime of all connections is known, the setup time for a new connection is bounded and becomes part of the QoS characteristics of a connection. Only a small subset of all connections is active concurrently at any time and each connection competes for getting access to the network. For this process, we can use many of the same techniques such as scheduling, arbitration, priority schemes, preemption, etc. as we use for individual packet transmission. Thus the QoS parameters would describe a two-level hierarchy: (a) the worst case setup time, the minimum frequency of network access, and minimum lifetime of a connection, (b) the worst case delay and minimum bandwidth for packets belonging to established connections.

Alternative (3) is acceptable for many applications. For instance, a multimedia supporting user device or a telecom switch may simply refuse new requests if the system is already overloaded. Every finite resource system can only handle a finite number of applications and tasks; thus, it is only natural and unavoidable to sometimes reject new requests.

In the following we briefly discuss connection setup in the context of circuit switching. TDM-based connections are established in essentially the same way. In aggregate resource allocation schemes the setup may work in a similar way as well but, depending on the schemes, some problems may not appear or may be posed differently.

If circuit switched connections are configured dynamically, as in SoCBUS [2] and Crossroad [3], the connection is set up by transmitting a special request signal or packet from the source node to the destination node along the intended route of the connection. This is illustrated in Figure 2.17 for a connection extending over two intermediate switches. If the connection is built successfully, the destination node responds by returning an acknowledgment packet (Ack) to the source. Once the source node has received the acknowledgment, it sends the data packets. Because the delivery of data packets along the active connection is guaranteed, no acknowledgment is required and data transmission can proceed at a very high speed. When the last data packet is emitted by the source node, it tears down the connection by sending a cancel packet that releases all reserved resources of the connection.

Figure 2.17(b) illustrates the case when a requested resource is not available because it is in use by another connection, for example, the link between switch 2 and the destination node. In this situation there are two main possibilities. First, the request packet waits in switch 2 until the requested resource (the link) is free. When the link becomes available, the request packet proceeds further while building the connection until it arrives at the destination node. The second alternative, which is shown in the figure, is to

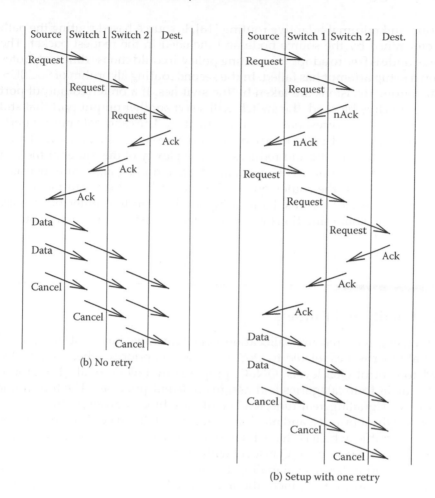

(b) No retry

(b) Setup with one retry

FIGURE 2.17

The three phases of circuit switched communication in SoCBUS [2].

tear down the partially built-up connection with a negative acknowledgment (nAck) packet. The main disadvantage of the first approach is the occurrence of deadlocks. Two or more partially built-up connections could end up in a cyclic dependency, each of them waiting indefinitely for a resource that one of the others blocks. The second disadvantage of the waiting approach is that a partially built-up connection may potentially block resources for a very long time without using them and prohibiting other connections to use them as well. Consequently, both SoCBUS and Crossroad tear down partially built-up connections when a requested resource is not available. After some delay the source node makes a new attempt.

In principle, the connection setup can use any routing policy, deterministic or adaptive, minimal or nonminimal path routing. For instance, the SoCBUS designers have implemented two different routing algorithms: source-based routing and minimal-adaptive routing [18]. In source-based routing the path is determined by the source node and included in the request packet. The source node is free to adopt any routing policy; it could choose different routes when a setup attempt has failed. In the second routing algorithm of SoCBUS, local routing decisions are taken by the switches. If a preferred output port of the switch is blocked, the switch will select another output port that still has to be on the shortest path to the destination. The candidate output ports lying on a minimal path are tried in a round-robin fashion. Source-based routing is deterministic and reduces the complexity of the router at the cost of the network interface. The minimal adaptive routing algorithm increases the complexity of the router but is able to use local load information for the routing decision. It results in a higher delay in the router but it may find a path in some situations where the deterministic source-based routing fails.

2.6   Priority and Fairness

By focusing on a resource allocation perspective, we have not illuminated several other issues related to QoS. For instance, priority schemes are often used to control QoS levels. QNoC, proposed by Bolotin et al. [19], groups all traffic in four categories and assigns different priorities. The four traffic classes are signaling, real-time, read/write, and block-transfer, with signaling having highest priority, and block-transfer lowest. It can be observed that the signaling traffic, which is characterized by low bandwidth and delay requirements, enjoys a very good QoS level without having a strong, adverse impact on other traffic. Because signaling traffic is rare, its preferential treatment does not diminish, too much, the average delay of other high throughput traffic.

In general, a priority scheme allows control of the access to a resource, which is not exclusively reserved. Hence, it is an arbitration technique in aggregate allocation schemes. Its effect is to decrease the delay of one traffic class at the expense of the other traffic, and it makes all low priority traffic invisible to high priority traffic. To compute the delay and throughput bounds of a traffic class we only have to consider traffic of the same and higher priority.* We have seen this phenomenon in Section 2.4, Figure 2.15, where the high priority flow could command the entire channel capacity. However, if we know that the high priority flow A uses only a small fraction of the channel bandwidth, say $\rho_A \leq 0.05C$, even flow B will be served very well. This knowledge of application traffic characteristics is utilized in QNoC and most other priority schemes leading to cost-efficient implementations with good QoS levels.

---

* For the sake of simplicity we ignore priority inversion. *Priority inversion* is a time period when a high priority packet waits for a low priority packet. This period is typically limited and known as L/C in Figure 2.15.

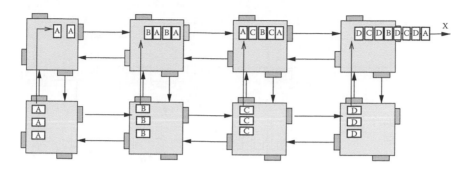

FIGURE 2.18
Local fairness may be very unfair globally.

In summary, we note that hard bounds on delay and bandwidth can only be given if the rate and burstiness of all higher priority traffic is constrained and known. Priority schemes work best with a relatively small number of priority levels (2–8) and, if well characterized, low throughput traffic is assigned to high priority levels.

All arbitration policies should feature a certain fairness of access to a shared resource. But which notion of fairness to apply is, however, less obvious. Local versus global fairness is a case in point, illustrated in Figure 2.18. Packets of connection A are subject to three arbitration points. At each point a round-robin arbiter is fair to both connections. However, at channel **X** connection D occupies four times the bandwidth and experiences 1/4 of the delay as connection A. This example shows that if only local fairness is considered, the number of arbitration points that a connection meets has a big impact on its performance because its assigned bandwidth drops by a factor two at each arbitration point. Consequently, in multistage networks often age-based fairness or priority schemes are used. This can be implemented with a counter in the packet header that is set to zero when the packet enters the network and is incremented in every cycle. For instance, Nostrum uses an age-based arbitration scheme to guarantee the maximum delay bound given in Section 2.4, Equation (2.6).

Another potential negative effect of ill-conceived fairness is shown in Figure 2.19. Assume we have two messages A and B, each consisting of 10 packets each. Assume further that the delay of the message is determined by the delay of the last packet. Packets of messages A and B compete for channel **X**. If they are arbitrated fairly in a round-robin fashion, they occupy the channel alternatively. Assume it takes one cycle to cross channel **X**. If a packet A gets access first, the last A packet will have crossed the channel after 19 cycles, and the last B packet after 20 cycles. If we opt for an alternative strategy and assign channel **X** exclusively to message A, all A packets will have crossed the channel after 10 cycles although all B packets will still need 20 cycles. Thus, a winner-takes-it-all arbitration policy would decrease the delay of message A by half without adversely affecting the delay of message B. Moreover, if the buffers are exclusively reserved for a message, both

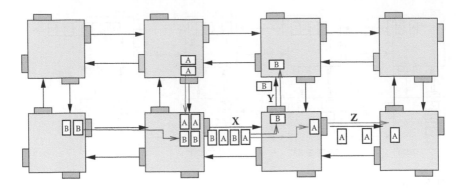

FIGURE 2.19
Local fairness may lead to lower performance.

messages will block their buffers for a shorter time period compared to the fair round-robin arbitration.

These examples illustrate that fairness issues require attention, and the effects of arbitration policies on global fairness and performance are not always obvious. For a complete trade-off analysis the cost of implementation has to be also taken into account. For a discussion on implementation of circuits, their size, and delay that realize different arbitration policies see Dally and Towles [20, Chapter 18].

2.7   QoS in a Telecom Application

To illustrate the usage of QoS communication, we present a case study on applying TDM VCs to an industrial application provided by Ericsson Radio Systems [13].

2.7.1   Industrial Application

As mapped onto a 4×4 mesh NoC in Figure 2.20, an industrial application is a radio system consisting of 16 IPs. Specifically, n_2, n_3, n_6, n_9, n_{10}, and n_{11} are ASICs; n_4, n_7, n_{12}, n_{13}, n_{14}, and n_{15} are DSPs; n_5, n_8, and n_{16} are FPGAs; n_1 is a device processor that loads all nodes with program and parameters at start-up, sets up and controls resources in normal operation. Traffic to/from n_1 is for the system's initial configuration and no longer used afterward. The mesh has 48 duplex links with uniform link capacity. There are 26 node-to-node traffic flows that are categorized into 11 types of traffic flows {a, b, c, d, e, f, g, h, i, j, k}, as marked in the figure. The traffic flows are associated with a bandwidth requirement. In the example, a and h are multicast traffic,

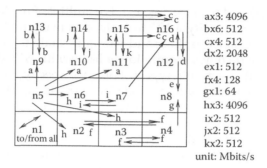

FIGURE 2.20
Node-to-node traffic flows for a radio system.

and others are unicast traffic. As the application requires strict bandwidth guarantees for processing traffic streams, we use TDM VCs to serve the traffic flows. In this case study, we use closed-loop VCs.

The case study comprises two phases: VC specification and VC configuration. The VC specification phase defines a set of source and destination (sink) nodes, and normalized bandwidth demand for each VC. The VC configuration phase constructs VC implementations satisfying the VCs' specification requirement—one VC implementation for one VC specification. In this case study, a VC implementation is a looped TDM VC. Note that a VC specification only consists of source and destination nodes, although its corresponding VC implementation consists of the source and destination nodes plus intermediate visiting nodes.

2.7.2 VC Specification

The VC specification phase consists of three steps: determining link capacity, merging traffic flows, and normalizing VC bandwidth demand.

We first determine the minimum required link capacity by identifying a critical (heaviest loaded) link. The most heavily loaded link may be the link directing from n_5 to n_9. The **a**-type traffic passes it and $BW_a = 4096$ Mbits/s. To support bw_a, link bandwidth bw_{link} must be not less than 4096 Mbits/s. We choose the minimum 4096 Mbits/s for BW_{link}. This is an initial estimation and subject to adjustment and optimization, if necessary.

Because the VC path search space increases exponentially with the number of VCs, reducing the number of VCs when building a VC specification set is crucial. In our case, we intend to define 11 VCs for the 11 types of traffic. To this end, we merge traffic flows by taking advantage of the fact that the VC loop allows multiple source and destination nodes (multinode VCs) on it, functioning as a virtual bus supporting arbitrary communication patterns [13]. Specifically, this merging can be done for multicast, multiple-flow low-bandwidth, and round-trip (bidirectional) traffic. In the example, for the two multicast traffic **a** and **h**, we specify two multinode VCs for them as $\bar{v}_a(n_5, n_9, n_{10}, n_{11})$ and $\bar{v}_h(n_5, n_6, n_2, n_3)$. For multiple-flow low-bandwidth type of traffic, we can specify a VC to include as many nodes as a type of traffic spreads. For instance, traffic **c** and **f** include 4 node-to-node flows each, and their node-to-node flows require lower bandwidth, 512 Mbits for traffic type **c** and 128 Mbits for traffic type **f**. For **c**, we specify a a five-node VC $\bar{v}_c(n_{13}, n_{14}, n_{15}, n_{16}, n_7)$; for **f**, a three-node VC $\bar{v}_f(n_2, n_3, n_4)$. Furthermore, as we use a closed-loop VC, two simplex traffic flows can be merged into one duplex flow. For instance, for two **i** flows, we specify only one VC $\bar{v}_i(n_6, n_7)$. This also applies to traffic **b, d, j,** and **k**.

Based on results from the last two steps, we compute normalized bandwidth demand for each VC specification. Suppose link capacity $bw_{link} = 4096$ Mbits/s, 512 Mbits/s is equivalent to $1/8$ bw_{link}. While calculating this, we need to be careful of duplex traffic. Because the VC implementation is a loop, a container on it offers equal bandwidth in a round trip. Therefore, duplex traffic can exploit this by utilizing bandwidth in either direction. For example, traffic **d** has two flows, one from n_{16} to n_{12}, the other from n_{12} to n_{16}, requiring $1/2$ bandwidth in each direction. By using a looped VC, the actual bandwidth demand on the VC is still $1/2$ (not $2 \times 1/2$). Because of this, the bandwidth requirements on VCs for traffic **b, d, f, i, j,** and **k** are $1/8$, $1/2$, $1/16$, $1/8$, $1/8$, and $1/8$, respectively.

With the steps mentioned above, we obtain a set of VC specifications as listed in Table 2.4.

2.7.3 Looped VC Implementation

In the VC implementation phase, we find a route for each VC and then compute the number n_c of the required containers to support the required bandwidth. This is calculated by $n_c \geq \overline{bw} \cdot |v|$, where $\overline{bw}$ is the normalized

TABLE 2.4

VC Specification for Traffic Flows

VC Spec.	Traffic	BW (Mbits/s)	Number of Node-to-Node Flows	Source and Sink Nodes	BW Demand
1	a	4096	3	n5, n9, n10, n11	1
2	b	512	2	n9, n13	1/8
3	c	512	4	n7, n13, n14, n15, n16	1/2
4	d	2048	2	n12, n16	1/2
5	e	512	1	n8, n12	1/8
6	f	128	4	n2, n3, n4	1/16
7	g	64	1	n4, n8, n4	1/64
8	h	4096	3	n5, n6, n2, n3	1
9	i	512	2	n6, n7	1/8
10	j	512	2	n10, n14	1/8
11	k	512	2	n11, n15	1/8

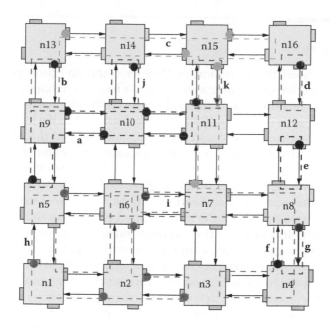

FIGURE 2.21
One solution of looped VC implementations with a snapshot of containers on VCs.

bandwidth demand of $\bar{v}$ and $|v|$ is the loop length of the VC implementation v. After configuration, we obtain TDM VC implementations for all traffic flows, one for each VC specification. One feasible solution when link capacity is set to be 4096 Mbits/s is shown in Figure 2.21.

The VC implementation details are listed in Table 2.5. In total, there are 25 containers launched on 11 VCs. The network has a utilization of 52%.

TABLE 2.5

Looped TDM VC Implementations for Traffic Flows

VC Impl.	Traffic	Visiting Nodes	Loop Length	Containers	BW Supply
1	a	n5, n9, n10, n11, n10, n9, n5	6	6	1
2	b	n9, n13, n9	2	1	1/2
3	c	n7, n11, n15, n14, n13, n14, n15, n16, n15, n11, n7	10	5	1/2
4	d	n12, n16, n12	2	1	1/2
5	e	n8, n12, n8	2	1	1/2
6	f	n3, n4, n8, n7, n6, n5, n1, n2, n6, n7, n8, n4, n3	12	1	1/12
7	g	n4, n8, n4	2	1	1/2
8	h	n1, n5, n6, n2, n3, n2, n1	6	6	1
9	i	n6, n7, n6	2	1	1/2
10	j	n10, n14, n10	2	1	1/2
11	k	n11, n15, n11	2	1	1/2

2.8 Summary

We have addressed the provision of QoS for communication performance from the perspective of resource allocation. We have seen that we can reserve communication resources exclusively throughout the lifetime of a connection (circuit switching) or during individual time slots (TDM). We have discussed nonexclusive usage of resources in Section 2.4 and noticed that QoS guarantees can be provided by analyzing the worst case interaction of all involved connections. We have observed a general trade-off between the utilization of resources and the tightness of bounds. If we exclusively allocate resources to a single connection, their utilization may be very low because no other connection can use them. But the delay of packets is accurately known and the worst case is the same as the average and the best cases. In the other extreme we have aggregate allocation of the entire network to a set of connections. The utilization of resources is potentially very high because they are adaptively assigned to packets in need. However, the worst case delay can be several times the average case delay because many connections may compete for the same resource simultaneously. Which solution to select depends on the application's traffic patterns, on the real-time requirements, and on what constitutes an acceptable cost.

In practice all the presented techniques of resource allocation and arbitration can be mixed. By using different techniques for managing the various resources such as links, buffers, crossbars, and NIs, a network can be optimized for a given set of objectives while exploiting knowledge of application features and requirements.

References

[1] A. Jantsch, "Models of computation for networks on chip." In *Proc. of Sixth International Conference on Application of Concurrency to System Design*, June 2006, invited paper.

[2] D. Wiklund and D. Liu, "SoCBUS: Switched network on chip for real time embedded systems." In *Proc. of Parallel and Distributed Processing Symposium*, Apr. 2003.

[3] K.-C. Chang, J.-S. Shen, and T.-F. Chen, "Evaluation and design trade-offs between circuit-switched and packet-switched NOCs for application-specific SOCs." In *Proc. of 43rd Annual Conference on Design Automation*, 2006, 143–148.

[4] J.-Y. LeBoudec, *Network Calculus*. Lecture Notes in Computer Science, no. 2050. Berlin: Springer Verlag, 2001.

[5] C. Hilton and B. Nelson, "A flexible circuit switched NOC for FPGA based systems." In *Proc. of Conference on Field Programmable Logic (FPL)*, Aug. 2005, 24–26.

[6] A. Lines, "Asynchronous interconnect for synchronous SoC design," *IEEE Micro* 24(1) (Jan-Feb 2004): 32–41.

[7] M. Millberg and A. Jantsch, "Increasing NoC performance and utilisation using a dualpacket exit strategy." *In 10th Euromicro Conference on Digital System Design*, Lubeck, Germany, Aug. 2007.

[8] A. Leroy, P. Marchal, A. Shickova, F. Catthoor, F. Robert, and D. Verkest, "Spatial division multiplexing: A novel approach for guaranteed throughput on NoCs." In *Proc. of International Conference on Hardware/Software Codesign and System Synthesis*, Sept. 2005, 81–86.

[9] T. Bjerregaard and J. Sparso, "A router architecture for connection-oriented service guarantees in the MANGO clockless network-on-chip." In *Proc. of Conference on Design, Automation and Test in Europe—Volume 2*, Mar. 2005, 1226–1231.

[10] K. Goossens, J. Dielissen, and A. Rădulescu, "The Æthereal network on chip: Concepts, architectures, and implementations," *IEEE Design and Test of Computers* 22(5), (Sept-Oct 2005): 21–31.

[11] M. Millberg, E. Nilsson, R. Thid, and A. Jantsch, "Guaranteed bandwidth using looped containers in temporally disjoint networks within the Nostrum network on chip." In *Proc. of Design Automation and Test in Europe Conference*, Paris, France, Feb. 2004.

[12] Z. Lu and A. Jantsch, "Slot allocation using logical networks for TDM virtual-circuit configuration for network-on-chip." In *International Conference on Computer Aided Design (ICCAD)*, Nov. 2007.

[13] Z. Lu and A. Jantsch, "TDM virtual-circuit configuration for network-on-chip," *IEEE Transactions on Very Large Scale Integration Systems* 16(8), (August 2008).

[14] E. Nilsson and J. Öberg, "Reducing peak power and latency in 2-D mesh NoCs using globally pseudochronous locally synchronous clocking." In *Proc. of International Conference on Hardware/Software Codesign and System Synthesis*, Sep. 2004.

[15] A. Borodin, Y. Rabani, and B. Schieber, "Deterministic many-to-many hot potato routing," *IEEE Transactions on Parallel and Distributed Systems* 8(6) (1997): 587–596.

[16] R. L. Cruz, "A calculus for network delay, part I: Network elements in isolation," *IEEE Transactions on Information Theory* 37(1) (January 1991): 114–131.

[17] H. Zhang, "Service disciplines for guaranteed performance service in packet-switching networks," *Proc. IEEE*, 83 (1995): 1374–1396.

[18] D. Wiklund, "Development and performance evaluation of networks on chip," Ph.D. dissertation, Department of Electrical Engineering, Linköping University, SE-581 83 Linköping, Sweden, 2005, Linköping Studies in Science and Technology, Dissertation No. 932.

[19] E. Bolotin, I. Cidon, R. Ginosar, and A. Kolodny, "QNoC: QoS architecture and design process for network on chip," *Journal of Systems Architecture*, 50(2–3) (Feb. 2004): 105–128.

[20] W. J. Dally and B. Towles, *Principles and Practices of Interconnection Networks*. Morgan Kaufman Publishers, 2004.

3

Networks-on-Chip Protocols

Michihiro Koibuchi and Hiroki Matsutani

CONTENTS

3.1   Introduction

In this chapter, we explain NoC protocol family, that is, switching techniques, routing protocols, and flow controls. These techniques are responsible for low-latency packet transfer. They strongly affect the performance, hardware amount, and power consumption of on-chip interconnection networks. Figure 3.1 shows an example NoC that consists of 16 tiles, each of which has a processing core and a router. In these networks, source nodes (i.e., cores) generate packets that consist of a header and payload data. On-chip routers transfer these packets through connected links, whereas destination nodes decompose them. High-quality communication that never loses data within the network is required for on-chip communication, because delayed packets of inter-process communication may degrade the overall performance of the target (parallel) application.

Switching techniques, routing algorithms, and flow control have been studied for several decades for off-chip interconnection networks. General discussion about these techniques is provided by existing textbooks [1–3], and some textbooks also describe them [4,5]. We introduce them from a view point of on-chip communications, and discuss their pros and cons in terms of throughput, latency, hardware amount, and power consumption. We also survey these techniques used in various commercial and prototype NoC systems.

The rest of this chapter is organized as follows. Section 3.2 describes switching techniques and channel buffer managements, and Section 3.3 explains the routing protocols. End-to-end flow control is described in Section 3.4. Section 3.5 discusses the trends of NoC protocols, and Section 3.6 summarizes the chapter.

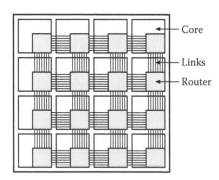

FIGURE 3.1
Network-on-Chip: routers, cores, and links.

3.2 Switch-to-Switch Flow Control

NoC can improve the performance and scalability of on-chip communication by introducing a network structure that consists of a number of packet routers and point-to-point links. However, because they perform complicated internal operations, such as routing computation and buffering, routers introduce larger packet latency at each router compared to that of a repeater buffer on a bus structure. (NoCs use routers instead of the repeater buffers on a bus structure.) These delays are caused by intra-router operation (e.g., crossbar arbitration) and inter-router operation. We focus our discussion on inter-router switching and channel buffer management techniques for low-latency communications.

3.2.1 Switching Techniques

Packets are transferred to their destination through multiple routers along the routing path in a hop-by-hop manner. Each router keeps forwarding an incoming packet to the next router until the packet reaches its final destination. Switching techniques decide when the router forwards the incoming packet to the neighboring router, therefore affecting the network performance and buffer size needed for each router.

3.2.1.1 Store-and-Forward (SAF) Switching

Every packet is split into transfer units called flits. A single flit is sent from an output port of a router at each time unit. Once a router receives a header flit, the body flits of the packet arrive every time unit. To simply avoid input-channel buffer overflow, the input buffer must be larger than the maximum packet size. The header flit is forwarded to the neighboring router after it receives the tail flit. This switching technique is called store-and-forward (SAF). The advantage of SAF switching is the simple needed control mechanism between routers due to packet-based operation [other switching techniques, such as wormhole switching that are described below use flit-based operation (Figure 3.2)]. The main drawback of SAF switching is the large needed channel buffer size that increases the hardware amount of the router. Moreover, SAF suffers from a larger latency compared with other switching techniques, because a router in every hop must wait to receive the entire packet before forwarding the header flit. Thus, SAF switching does not fit well with the requirements of NoCs.

3.2.1.2 Wormhole (WH) Switching

Taking advantage of the short link length on a chip, an inter-router hardware control mechanism that stores only fractions of a single packet [i.e., flit(s)]

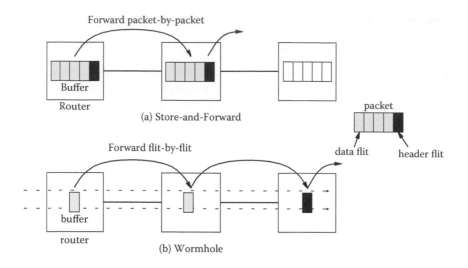

FIGURE 3.2
Store-and-forward (SAF) and wormhole (WH) switching techniques.

could be constructed with small buffers. Theoretically, the channel buffer at every router can be as small as a single flit.

In wormhole (WH) switching, a header flit can be routed and transferred to the next hop before the next flit arrives, as shown in Figure 3.2. Because each router can forward flits of a packet before receiving the entire packet, these flits are often stored in multiple routers along the routing path. Their movement looks like a worm. WH switching reduces hop latency because the header flit is processed before the arrival of the next flits.

Wormhole switching is better than SAF switching in terms of both buffer size and (unloaded) latency. The main drawback of WH switching is the performance degradation due to a chain of packet blocks. Fractions of a packet can be stored across different routers along the routing path in WH switching; so a single packet often keeps occupying buffers in multiple routers along the path, when the header of the packet cannot progress due to conflictions. Such a situation is referred to as head-of-line (HOL) blocking. Buffers occupied by the HOL blockings block other packets that want to go through the same lines, resulting in performance degradation.

3.2.1.3   *Virtual Cut-Through (VCT) Switching*

To mitigate the HOL blocking that frequently occurs in WH switching, each router should be equipped with enough channel buffers to store a whole packet. This technique is called *virtual cut-through* (VCT), and can forward the header flit before the next flit of the packet arrives. VCT switching has the advantage of both low latency and less HOL blocking.

A variation called asynchronous wormhole (AWH) switching uses channel buffers smaller than the maximum used packet size (but larger than the packet header size). When a header is blocked by another packet at a

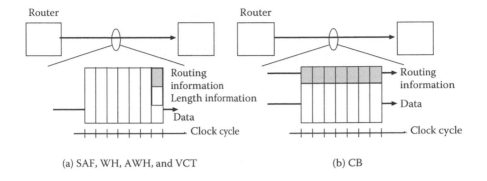

(a) SAF, WH, AWH, and VCT   (b) CB

FIGURE 3.3
Packet structure of the various switching techniques discussed in this section.

router, the router stores flits (same-sized as channel buffers). Flits of the same packet could be stored at different routers. Thus, AWH switching theoretically accepts an infinite packet length, whereas VCT switching can cope with only packets whose length is smaller than its channel buffer size.

Another variation of the VCT switching customized to NoC purposes is based on a cell structure using a fixed single flit packet [6]. This is similar to the asynchronous transfer mode (ATM) (traditional wide area network protocol). As mentioned above, the main drawback of WH routing is that the buffer is smaller in size than the maximum used packet size, which frequently causes the HOL blocking. To mitigate this problem, the cell-based (CB) switching limits the maximum packet size to a single flit with each flit having its own routing information.

To simplify the packet management procedure, the cell-based switching removes the support of variable-length packets in routers and network interfaces. Routing information is transferred on dedicated wires besides data lines in a channel with a single-flit packet structure (Figure 3.3).

The single-flit packet structure introduces a new problem; namely, the control information may decrease the ratio of raw data (payload) in each transfer unit, because it attaches control information to every transfer unit.

Table 3.1 and Figure 3.3 compare SAF, WH, VCT, AWH, and CB switching techniques.

TABLE 3.1

Comparison of the Switching Techniques Discussed in This Section

	Control	Channel Buffer Size	Throughput	(Unloaded) Latency*
SAF	Software	Maximum packet size	Low	$(h + b) \times D$
WH	Hardware	Header size	Low	$h \times D + b$
VCT	Hardware	Maximum packet size	High	$h \times D + b$
AWH	Hardware	Smaller than a packet	High	$h \times D + b$
CB	Hardware	Header size	Low	$h \times D$

*h, b, and D are the number of header flits of a packet, the number of body flits, and the diameter of the topology, respectively.

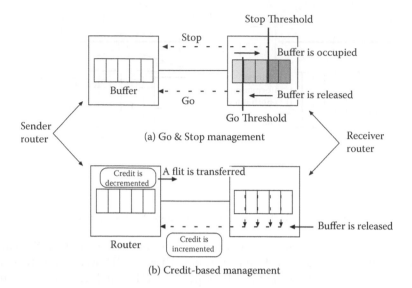

FIGURE 3.4
Channel buffer management techniques.

3.2.2 Channel Buffer Management

To implement a switching technique without buffer overflow, a channel buffer management between routers is needed.

3.2.2.1 Go & Stop Control

The simplest buffer management is the Go & Stop control, sometimes called X_{on}/X_{off} and *on/off*. As shown in Figure 3.4, the receiver router sends a *stop* signal to the sender router as soon as a certain space of its channel buffer becomes occupied to avoid channel buffer overflow. If the buffer size used by packets falls below a preset threshold, the receiver router sends a *go* signal to the sender router to resume sending.

The receiver buffer is required to store at least the number of flits that are in flight between sender and receiver routers during processing of the stop signal. Therefore, the minimum channel buffer size is calculated as follows:

$$\text{Minimum Buffer Size} = \text{Flit Size} \times (R_{\text{overhead}} + S_{\text{overhead}} + 2 \times \text{Link delay})$$

$$(3.1)$$

where R_{overhead} and S_{overhead} are, respectively, the overhead (required time units) to issue the stop signal at the receiver router and the overhead to stop sending a flit as soon as the stop signal is received.

3.2.2.2 Credit-Based Control

The Go & Stop control requires at least the buffer size calculated in Equation (3.1), and the buffer makes up most of the hardware for a lightweight router. The credit-based control makes the best use of channel buffers, and can be implemented regardless of the link length or the sender and receiver overheads.

In the case of the credit-based control, the receiver router sends a credit that allows the sender router to forward one more flit, as soon as a used buffer is released (becomes free). The sender router can send a number of flits up to the number of credits, and uses up a single credit when it sends a flit, as shown in Figure 3.4. If the credit becomes *zero*, the sender router cannot forward a flit, and must wait for a new credit from the receiver router.

The main drawback of the credit-based control is that it needs more control signals between sender and receiver routers compared to the Go & Stop control.

3.2.3 Evaluation

In this subsection, we compare the switching techniques in terms of throughput, latency, and hardware amount. The switching technique and routing protocol used are both important for determining the throughput and latency. The impact of the routing protocol used on throughput and latency is analyzed in the next section.

3.2.3.1 *Throughput and Latency*

A flit-level simulator written in C++ is used for measuring the throughput and latency, and is the same as the one used by Matsutani et al. [7,8]. Every router has three, four, or five ports for a 4×4 2-D mesh topology, and a single node connected to every router. Dimension-order routing with no virtual channels is employed (introduced in the next section). Nodes inject packets independently of each other. Packet length is set at eight flits, including one header flit. The simulation time is set to at least 200,000 cycles, and the first 1,000 cycles are ignored to avoid distortions due to the startup transient. As for the traffic patterns, we use uniform random traffic for baseline comparison.

Figure 3.5 shows the relation between the average latency and the accepted traffic of cell-based switching [CB (1)], wormhole switching with 1-flit buffer (WH.1), asynchronous wormhole switching with 2-flit buffer [AWH (2)], and virtual cut-through switching [VCT (8)]. Because the throughput and latency of VCT switching are better than those of the asynchronous and WH switchings, it can be said that the impact of the input channel buffer sizes on the throughput and latency is large. In particular, the WH switching introduces a large number of HOL blockings that degrade the throughput by 52% compared with the VCT switching. The CB switching increases latency, because each flit (single-flit packet) is routed independently, whereas only a header flit is routed in the other switching techniques.

3.2.3.2 *Amount of Hardware*

Here we compare the hardware amounts of CB, WH, with two virtual channels, and VCT switchings. We implemented these switching techniques on

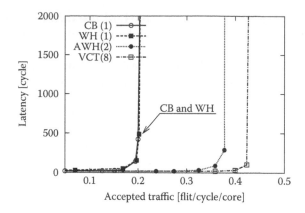

FIGURE 3.5
Throughput and latency of the switching techniques discussed in Section 3.2.3.1.

simple three-cycle on-chip routers and synthesized them with a 90 nm CMOS standard cell library. The behavior of the synthesized NoC designs was confirmed through a gate-level simulation assuming an operating frequency of 500 MHz.

Each router has five ports, which can be used for 2-D tori and meshes. One or two virtual channels are used. The flit width is set to 64 bits and the packet length is set to 8 flits, which affects the channel buffer size of VCT switching. There are several choices for the channel buffer structure: the flip-flop (FF)-based buffer, register-file (RF)-based buffer, and SRAM-based buffer. The RF- and SRAM-based ones are better if the buffer depth is deep, but they are not so efficient if the buffer depth is shallow. In our design, the FF-based one is used for the WH and CB routers that have a 4-flit FIFO buffer for each input channel. The RF-based one is used for the VCT router that has larger input buffers.

The routing decisions are stored in the header flit prior to packet injection (i.e., source routing); thus routing tables that require register files for storing routing paths are not needed in each router, resulting in a low cost router implementation. This router architecture is the same as the one used by Matsutani et al. [7,8].

Figure 3.6 shows the network logic area of 5-port routers that employ CB, WH, and VCT switching techniques, and their breakdown. As shown in the graph, the WH router uses 34% less hardware compared to the VCT router, and the CB router uses 2.6% less hardware compared to the WH router. The size of the channel buffers thus dominates the hardware amount for the router, and its control mechanism also affects the router hardware amount. Because the size of input buffer affects the implementation of channel buffers (flipflops for CB and WH switchings and 2-port register file for VCT), different implementations are employed in routers with WH and VCT switching.

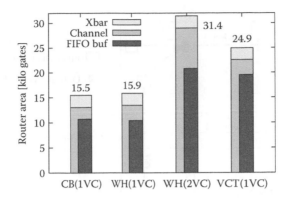

FIGURE 3.6
Hardware amount of the switching techniques discussed in Section 3.2.3.2.

Note that every virtual channel requires a buffer, and the virtual-channel mechanism makes the structure of the arbiter and crossbar more complicated, increasing the router hardware by 90%.

3.3 Packet Routing Protocols

Packet routing schemes decide routing paths between a given source and destination nodes.* The channel buffer management technique does not let packets be discarded between two neighboring routers. Similarly, a sophisticated routing protocol can prevent packets from being discarded between any pair of nodes caused by deadlocks and livelocks of packets.

3.3.1 Deadlocks and Livelocks of Packet Transfer

At the routing protocol layer of a computer network, a packet may be dropped to allow forwarding another blocked packet. Figure 3.7 shows a situation where every packet is blocked by another one, and they cannot be permanently forwarded. Such a cyclic dependency is called a *deadlock*. Once a deadlock occurs, at least a single packet within a network must be killed and resent. To avoid deadlocks, deadlock-free routing algorithms that never cause deadlocks on paths have been widely researched.

Besides the deadlock-free property, a routing protocol must have the livelock-free property to stop packets from being discarded needlessly. Packets would not arrive at their destinations if they ware to take nonminimal paths that go away from destination nodes. If it is the case, they would be permanently forwarded within NoCs. This situation is called *livelock*.

---

* We use the term "nodes" for IP cores that are connected on a chip.

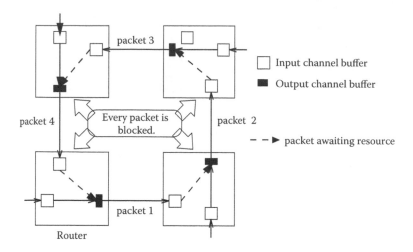

FIGURE 3.7
Deadlocks in routing protocols.

The deadlock- and livelock-free properties are not strictly required in routing algorithms in the case of traditional LANs and WANs. This is because the Ethernet usually employs a spanning tree protocol that limits the topology to that of a tree whose structure does not cause deadlocks of paths; moreover, the Internet Protocol allows packets to have time-to-live field that limits the maximum number of transfers. However, NoC routing protocols cannot simply borrow the techniques used by commodity LANs and WANs. Therefore, new research fields dedicated to NoCs have developed, similar to those in parallel computers.

3.3.2 Performance Factors of Routing Protocols

A number of performance factors are involved in designing deadlock- and livelock-free routing algorithms, in terms of throughput, amount of hardware, and energy. The implementation of routing algorithms depends on the complexity of routing algorithms, and affects the amount of hardware for the router and/or network interface. Figures 3.8 and 3.11 show the taxonomy of various routing methods.

From a view point of path hops, routing algorithms can be classified into minimal routing and nonminimal routing. A minimal routing algorithm always assigns topological minimal paths to a given source and destination pair, although a nonminimal routing algorithm could take both minimal and nonminimal paths. The performance of a routing algorithm strongly depends on two factors, average path hop and path distribution. Adaptivity allows alternative paths between the same pair of source and destination nodes (Figure 3.9). This property provides fault tolerance, because it usually enables the routing algorithm to select a path that avoids faulty network components. The different-path property bears some resemblance to adaptivity.

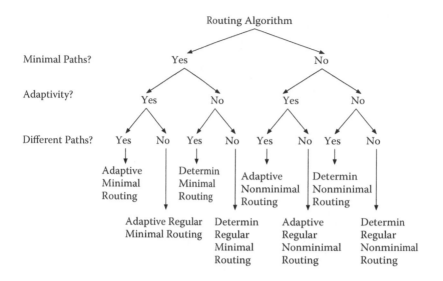

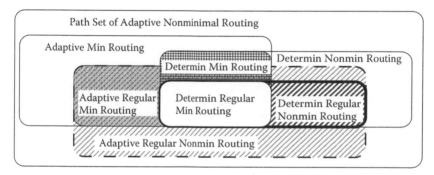

FIGURE 3.8
Taxonomy of routing algorithms.

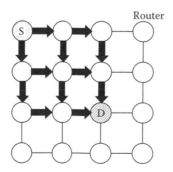

FIGURE 3.9
The adaptivity property.

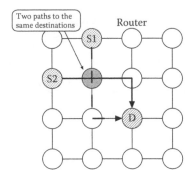

FIGURE 3.10
The different-paths property.

Unlike that of adaptivity, though, the different-path properties enable a choice to be made between different paths to a single destination node depending on the input channel, or source node (Figure 3.10). The different-paths property affects the routing table format at a router, and its entry number.

A routing algorithm that possesses the adaptivity property is called an adaptive routing algorithm, whereas the one that does not have this property is called a deterministic routing algorithm. In deterministic routing, all routing paths between any source and destination pairs are statically fixed and never changed during their flight. In adaptive routing, on the other hand, routing paths are dynamically changed during their flight in response to network conditions, such as presence of congestion or faulty links. Deterministic routing has the following advantages:

1. simple switching, without selecting an output channel dynamically from alternative channels, can be used;

2. in-order packet delivery is guaranteed, as is often required in a communication protocol.

System software including lightweight communication library and the implementation of parallel applications are sometimes optimized on the assumption that in-order packet delivery is guaranteed in the network protocol. However, adaptive routing that provides multiple paths between the same pair of nodes introduces out-of-order packet delivery. In the case of using adaptive routing, an additional sorting mechanism at the destination node is needed to guarantee in-order packet delivery at the network protocol layer.

As shown in Figure 3.8, the path set of an adaptive nonminimal routing can include that of an adaptive minimal routing, because the nonminimal path set consists of minimal paths and nonminimal paths. Similarly, the path set of an adaptive regular nonminimal routing can include that of a deterministic regular minimal routing. For example, the path set of West-First Turn Model (an adaptive regular nonminimal routing) includes that of dimension-order routing (a deterministic regular minimal routing).

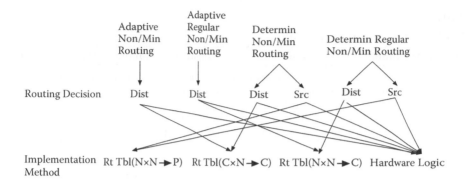

FIGURE 3.11
Taxonomy of routing implementations.

Figure 3.11 shows the taxonomy of routing algorithm implementations. Routing implementation can be classified into source routing (Src) and distributed (Dist) routing according to where their routing decisions are made. In source routing, routing decisions are made by the source node prior to the packet injection to the network. The routing information calculated by the source node is stored in the packet header, and intermediate nodes forward the packet according to the routing information stored in the header. Because a routing function is not required for each router, source routing has been used widely for NoCs to reduce the size of on-chip routers. In distributed routing, routing decisions are made by every router along the routing path.

There are three routing-table formats for distributed routing. These routing-table formats affect the amount of routing information. The simplest routing format (function) directly associates routing (destination) addresses to paths, and is based on $N(source) \times N(destination) \mapsto P$ routing relation (all-at-once) [2], where N and P are the node set and the path set, respectively. Because a routing address corresponds to a path in this relation, the routing address stored in a packet can be used to detect a source node at a destination node. The other routing functions provide information only for routing, and is based on $N \times N \mapsto C$ routing relation that only takes into account the current and destination nodes [1], or $C \times N \mapsto C$ routing relation, where C is the channel set. In the $N \times N \mapsto C$, and the $C \times N \mapsto C$ routing relations, the destination nodes cannot identify the source nodes from the routing address, and these routing relations cannot represent all complicated routing algorithms [2], unlike the $N \times N \mapsto P$ routing relation. However, their routing address can be smaller than that of the $N \times N \mapsto P$ routing relation.

3.3.3 Routing Algorithm

Table 3.2 shows typical deadlock-free routing algorithms and their features. Existing routing algorithms are usually dedicated to the target topology, such as k-ary n-cube topology. Because the set of paths on tree-based topologies

TABLE 3.2

Deadlock-Free Routing Algorithms

Routing Algorithm	Type	Topology	Minimum Number of VCs
DOR	Deterministic regular min	k-ary n-cube	1 (mesh) or 2 (torus)
Turn-Model family	Adaptive regular nonmin	k-ary n-cube	1 (mesh) or 2 (torus)
Duato's protocol	Adaptive regular min	k-ary n-cube	2 (mesh) or 3 (torus)
Up*/down*	Adaptive nonmin	Irregular top	1
VC transition	Adaptive non-/min	Irregular top	2

such as H-tree is naturally deadlock-free, we will omit discussion of routing algorithms on tree-based topologies.

3.3.3.1 k-ary n-cube Topologies

3.3.3.1.1 Dimension-Order Routing

A simple and popular deterministic routing is dimension-order routing (DOR), which transfers packets along minimal path in the visiting policy of low dimension first. For example, DOR uses y-dimension channels after using x-dimension channels in 2-D tori and meshes. DOR uniformly distributes minimal paths between all pairs of nodes. DOR is usually implemented with simple combinational logic on each router; thus, routing tables that require register files or memory cells for storing routing paths are not used.

3.3.3.1.2 Turn-Model Family

Assume that a packet moves to its destination in a 2-D mesh topology, and routing decisions are implemented as a distributed routing. The packet has two choices in each hop. That is, it decides to go straight or turn to another dimension at every hop along the routing path until it reaches its destination. However, several combinations of turns can introduce cyclic dependencies that cause deadlocks. Glass and Ni analyzed special combinations of turns that never introduce deadlocks [9]. Their model is referred to as the turn model [9].

In addition, for 2-D mesh topology, Glass and Ni proposed three deadlock-free routing algorithms by restricting the minimum sets of prohibited turns that may cause deadlocks [9]. These routing algorithms are called West-First routing, North-Last routing, and Negative-Last routing [9]. Figure 3.12 shows the minimum sets of prohibited turns in these routing algorithms. In West-First routing, for example, turns from the north or south to the west are prohibited. Glass and Ni proved that deadlock freedom is guaranteed if these two prohibited turns are not used in 2-D mesh [9].

Chiu extended the turn model into one called the odd-even turn model [10], in which nodes in odd columns and even columns prohibit different sets of turns. Figure 3.13.(a) shows a prohibited turn-set for nodes in odd columns, and Figure 3.13.(b) shows one for nodes in even columns. As reported by Chiu [10], the routing diversity of the odd-even turn model is better than those

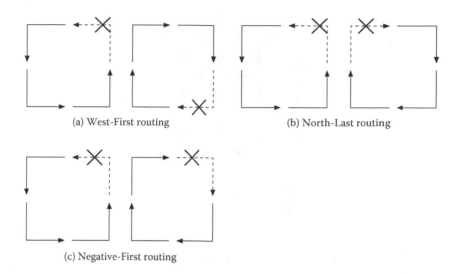

(a) West-First routing　　　　　　　　(b) North-Last routing

(c) Negative-First routing

FIGURE 3.12
Prohibited turn sets of three routing algorithms in the turn model.

of the original turn model proposed by Glass and Ni. Thus, the odd-even turn model has an advantage over the original ones, especially in networks with faulty links that require a higher path diversity to avoid them.

Turn models can guarantee deadlock freedom in 2-D mesh, but they cannot remove deadlocks in rings and tori that have wraparound channels in which cyclic dependencies can be formed. A virtual channel mechanism is typically used to cut such cyclic dependencies. That is, packets are first transferred using virtual-channel number *zero* in tori, and the virtual-channel number is then increased when the packet crosses the wraparound channels.

Moreover, turn models achieve some degree of fault tolerance. Figure 3.14 shows an example of the shortest paths avoiding the faulty link on a North-Last turn model.

3.3.3.1.3 *Duato's Protocol*

Duato gave a general theorem defining a criterion for deadlock freedom and used the theorem to develop a fully adaptive, profitable, and progressive

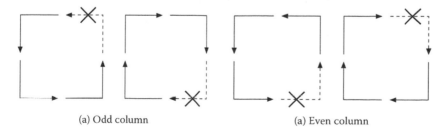

(a) Odd column　　　　　　　　(a) Even column

FIGURE 3.13
Prohibited turn set in the odd-even turn model.

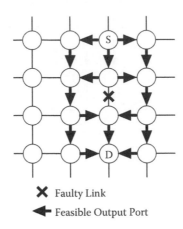

X Faulty Link

◄ Feasible Output Port

FIGURE 3.14
Paths avoiding a faulty link on a North-Last turn model.

protocol [11], called Duato's protocol or *-channel. Because the theorem states that by separating virtual channels on a link into escape and adaptive partitions, a fully adaptive routing can be performed and yet be deadlock-free. This is not restricted to a particular topology or routing algorithm. Cyclic dependencies between channels are allowed, provided that there exists a connected channel subset free of cyclic dependencies.

A simple description of Duato's protocol is as follows:

a. Provide that every packet can always find a path toward its destination whose channels are not involved in cyclic dependencies (escape path).

b. Guarantee that every packet can be sent to any destination node using an escape path and the other path on which cyclic dependency is broken by the escape path (fully adaptive path).

By selecting these two routes (escape path and fully adaptive path) adaptively, deadlocks can be prevented by minimal paths.

Three virtual channels are required on tori. Two virtual channels (we call them CA and CH) are used for DOR, and a packet that needs to use a wraparound path is allowed to use CA channel and a packet that does not need to use a wraparound path is allowed to use both CH and CA channels. Based on the above restrictions, these channels provide an escape path, whereas another virtual channel (called CF) is used for fully minimal adaptive routing.

Duato's protocol can be extended for irregular topologies by allowing more routing restrictions and nonminimal paths [12].

3.3.3.2 *Irregular Topologies*

In recent design methodologies, embedded systems and their applications are designed with system-level description languages like System-C, and

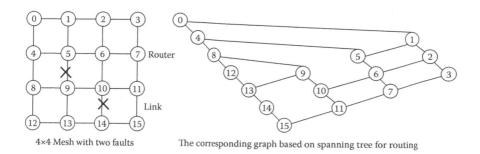

4×4 Mesh with two faults   The corresponding graph based on spanning tree for routing

FIGURE 3.15
Topology with faults.

simulated in the early stage of design. By analyzing the communication pattern between computational tasks in the target application, we can statically optimize any irregular topology [13]. The advantages of the application-specific custom topology are efficient resource usages and better energy efficiency. The irregularity of interconnection introduces difficulty in guaranteeing connectivity and deadlock-free packet transfer. As practical solutions of deadlock-free routing in irregular networks, spanning tree-based routings can be applied, as shown in Figure 3.15.

Routing algorithms for irregular topologies are required even in the case of NoCs with regular topologies, because a permanent hard failure of network resources can be caused by physical damage, which introduces irregularity, as shown in Figure 3.15.

3.3.3.2.1 Up*/Down* Routing

Up*/down* routing avoids deadlocks in irregular topologies using neither virtual channels nor buffers [14]. It is based on the assignment of direction to network channels [14]. As the basis of the assignment, a spanning tree whose nodes correspond to routers in the network is built. The "up" end of each channel is then defined as follows: (1) the end whose node is closer to the root in the spanning tree; (2) the end whose node has the lower unique identifier (UID), if both ends are on nodes at the same tree level. A legal path must traverse zero or more channels in the up direction followed by zero or more channels in the down direction, and this rule guarantees deadlock freedom while still allowing all hosts to be reached. However, an up*/down* routing algorithm tends to make imbalanced paths because it employs a 1-D directed graph.

3.3.3.2.2 Virtual-Channel Transition Routing Family

Up*/down* routing must use a number of nonminimal imbalanced paths so as not to create cycles among physical channels.

To reduce nonminimal imbalanced paths, the network is divided into layers of subnetworks with the same topology using virtual channels, and a large number of paths across multiple subnetworks are established. Enough

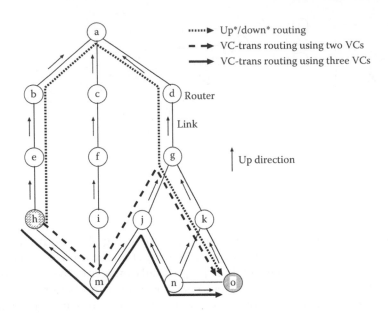

FIGURE 3.16
Example of up*/down* and VC transition routings.

restrictions on routing in each subnetwork are applied to satisfy deadlock freedom by using an existing routing algorithm, such as up*/down* routing, as long as every packet is routed inside the subnetwork. To prevent deadlocks across subnetworks, the packet transfer to a higher numbered subnetwork is prohibited. Thus, the virtual-channel transition routing takes shorter paths than up*/down* routing in most cases.

The above concept has been utilized in various virtual-channel transition routings [15,16], and a similar concept is used by Lysne et al. [17]. Figure 3.16 is an example of a virtual-channel transition routing that uses up*/down* routing within every subnetwork, from **h** to **o**. As shown in the figure, up*/down* routing requires seven hops for the packet to reach the destination ($h \rightarrow e \rightarrow b \rightarrow a \rightarrow d \rightarrow g \rightarrow k \rightarrow o$), whereas the virtual-channel transition routing with two subnetworks (virtual channels) handles the same routing in five hops ($h \rightarrow (1) \rightarrow m \rightarrow (0) \rightarrow j \rightarrow (0) \rightarrow g \rightarrow (0) \rightarrow k \rightarrow (0) \rightarrow o$). The number in parentheses indicates the subnetwork in which the packet is being transferred. Moreover with three subnetworks, the path is further reduced to four hops ($h \rightarrow (2) \rightarrow m \rightarrow (1) \rightarrow j \rightarrow (1) \rightarrow n \rightarrow (0) \rightarrow o$).

3.3.4 Subfunction of Routing Algorithms

The following subfunctions are used for routing algorithms to support various routing implementation methods.

3.3.4.1 Output Selection Function (OSF)

In adaptive routing, an output channel is dynamically selected depending on the channel conditions. If a channel is being used (i.e., it is busy), the other channel has priority over the busy channel. However, if both output channels are not being used (i.e., in the free condition), an output selection function (OSF) decides which output channel will be used. The OSF is required when an adaptive routing is implemented.

The simplest OSF is called *random selection function* [18]. It chooses an output channel from the available output channels at random. By using it, traffic will tend to be distributed to various directions randomly. "Dimension order selection function" has been proposed for k-ary n-cube [18]. It chooses an output channel, which belongs to the lowest dimension from the available output channels. For example, if there exist free output channels on the x, y dimension, this selection function chooses the one on the x direction. On the other hand, "zigzag selection function" chooses an output channel whose direction has the maximum hops to the destination [18]. These OSFs have a high probability to send a packet to a congested direction even if there exist free (legal) channels to the other directions. This comes from the understanding these OSFs take no thought of the network congestion dynamically. To address this problem, sophisticated output selection functions should use a measure that indicates the congestion of each output channel, and they dynamically decide the output only with the local congestion data inside the router. "VC-LRU selection function," which selects the least-recently-used available virtual channel [19], was proposed for networks with virtual channel mechanism.

3.3.4.2 Path Selection Algorithm

As shown in Figure 3.12, the West-First turn model includes the path set of DOR. Also, as shown in Figure 3.8, deterministic routing can be made by adaptive routing with a path selection algorithm that statically selects a single path from adaptive routing paths that include no cyclic channel dependencies.

The simplest path selection algorithm is random selection. If a path selection algorithm makes well-distributed paths, it relaxes traffic congestion around a hotspot. However, the random path selection algorithm may select a path to congestion points even if there are alternative paths that can avoid them. To alleviate this problem, a traffic balancing algorithm using a static analysis of alternative paths is proposed by Sancho et al. [20].

3.3.5 Evaluation

We use the same C++ simulator used in the previous section to compare the different routing protocols.

Figure 3.17 shows the relation between the average latency and the accepted traffic of up*/down* routing, DOR, the West-First turn model, and Duato's

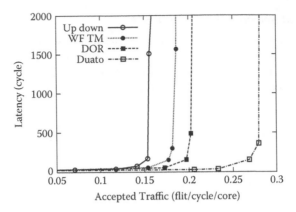

FIGURE 3.17
Throughput and latency of routing protocols discussed in Section 3.3.5.

protocol on a 4 × 4 2-D mesh. WH switching is used as a switching technique. Duato's protocol uses two virtual channels, although the other routing algorithms use no virtual channels. In the case of up*/down* routing, the selection of the root node affects its throughput on a breadth-first search (BFS) spanning tree, and we selected it so that the up*/down* routing achieves the maximum throughput.

Routing algorithms for irregular topologies usually provide lower throughput than that of regular topologies, and the up*/down* routing has the lowest throughput. An adaptive routing, the West-First turn model, achieves a lower throughput than that of the DOR on a small network with uniform traffic, though adaptive routing usually outperforms deterministic routing in large networks with high degree of traffic locality. The West-First routing algorithm can select various routing paths, and it may select a routing path set with a very poor traffic distribution that has hotspots. On the other hand, the DOR can always uniformly distribute the traffic and achieve good performance in the cases of uniform traffic, even though its path diversity is poor.

Duato's protocol that has high adaptivity provides the highest throughput, which improves by 82% compared with the up*/down* routing. It can be said that an adaptive routing with virtual channels is efficient to increase the network performance in the case of NoC domain, as well as the case of massively parallel computers.

3.4   End-to-End Flow Control

In addition to the switch-level flow control, source and destination nodes (i.e., end nodes) can manage the network congestion by adjusting injection traffic rate to a proper level.

3.4.1 Injection Limitation

In the case of interconnection networks used in microarchitecture, each sender node inserts as many packets as possible, even if the network is congested. If the offered traffic is saturated, the latency drastically increases, coming close to infinity. To make the best use of the NoC structure, the traffic load should be close to the saturation load and the amount of traffic should be managed at the node. "Injection limitation" is a technique to throttle the traffic rate at a node.

Injection limitation can be classified into static and dynamic throttling. A simple technique statically sets the threshold for allowing insertion of flits per time unit at each node. When a number of flits larger than the threshold are transferred in the time unit, the node stops inserting flits. The other technique uses dynamic network congestion information so that each node throttles its insertion of flits. Although an ideal method is to deliver information on the whole network congestion between nodes, it is difficult for each node to know whether the network is currently saturated or not. A well-known end-to-end sophisticated flow control, called windows control, manages the amount of allowed injection traffic. It is commonly used in lossy networks, such as the Internet. Injection limitation is efficient, especially when adaptive routing is employed. This is because adaptive routing drastically delays packets after the network is saturated.

3.4.2 ACK/NACK Flow Control

In addition to the throughput, hardware amount, and power consumption, the dependability is another crucial factor in the design of NoCs. Here we focus on a technique to tolerate transient soft failure that causes data to be momentarily discarded (e.g., bit error), and this loss can be recovered by a software layer. A well-known end-to-end flow control technique is based on acknowledgment (ACK) and negative-ACK (NACK) packets. As soon as the correct packets are received, the destination node sends the ACK information of the packet to the source node. If the incorrect packets (e.g., a bit error) are received, it sends the NACK information to the source node and the source node resends the packets. Also, the source node judges that the sent packet is discarded in the network, when both ACK and NACK have not been received within a certain period of time after sending it. These techniques increase dependability, and are discussed by Benini and Micheli [4].

Error control mechanisms can work for each flit or each packet. If the error rate is high, an end-to-end flow control results in a large number of retransmitted packets, which increases both power consumption and latency. To mitigate the number of retransmitted packets, an error correction mechanism is employed to every packet or flit, in addition to error detection. Instead of an end-to-end flow control, a switch-to-switch flow control can be used, which checks whether a flit or packet has an error or not at every switch. Thus, a switch-to-switch flow control achieves the smaller average latency of packets compared with that of end-to-end flow control, although each switch additionally has retransmission buffers that would increase the total power consumption. A high-performance error control mechanism, such as error correction at switch-to-switch flow control, can provide highly reliable communication with small increased latency even at the high-error rate. A sustainable low voltage usually increases the bit error rate, but it could be recovered by a high-performance error control mechanism. Thus, they could contribute to make a low-power chip, although the power/latency trade-off of the error control mechanism is sensitive. The best error recovery schemes depend on various design factors, such as flit-error rate, link delay, and packet average hops in terms of latency and power consumption [21].

3.5 Practical Issues

3.5.1 Commercial and Prototype NoC Systems

Table 3.3 summarizes the network topologies, switching techniques, virtual channels, flow controls, and routing schemes of representative NoC systems. These multiprocessor systems often have multiple networks for different purposes (Table 3.4). For example, MIT's Raw microprocessor has four networks: two for static communications (routes are scheduled at compile time) and two for dynamic communications (routes are established at run-time). Intel's Teraflops NoC employs two logical lanes (similar to virtual channels) to transfer data packets and instruction packets separately.

WH switching is the most common switching technique used in NoCs (Table 3.3). As mentioned in Section 3.2, WH switching can be implemented with small buffers so as to store at least a header flit in each hop. The buffer size of routers is an important consideration, especially in NoCs, because buffers consume a substantial area in on-chip routers and the whole chip area is shared with routers and processing cores that play a key role for applications; thus, wormhole switching is preferred for on-chip routers.

Virtual channels can be used with WH switching to mitigate frequent HOL blockings that degrade network performance. As mentioned in Section 3.2, a weak point of WH switching is the performance degradation due to the frequent HOL blockings because fractions of a packet may sometimes simultaneously occupy buffers in multiple nodes along the routing path when the header of the packet cannot progress due to the conflicts. However, adding virtual channels increases the buffer area in routers as well as adding physical channels (see Figure 3.6); thus on-chip network designers sometimes prefer to use multiple physical networks rather than virtual channels.

As for flow control techniques, both credit based and on/off (Go & Stop) schemes are popular in NoCs. As mentioned in Section 3.2, the on/off scheme is simple compared to the credit-based one, which is more sophisticated and can improve the utilization of buffers. In addition, flow control schemes for

TABLE 3.3

Switching, Flow Control, and Routing Protocols in Representative NoCs

	Ref.	Topology; Data Width	Switching; VCs	Flow Control	Routing Algorithm
MIT Raw (dynamic network)	[22] [23]	2-D mesh; 32-bit	Wormhole; no VC	Credit based	XY DOR
UPMC/LIP6 SPIN micro network	[24] [25]	Fat tree; 32-bit	Wormhole; no VC	Credit based	Up*/down* routing
QuickSilver Adaptive Computing Machine	[26]	H-tree; 32-bit	N/A;[a] no VC	Credit based	Up*/down* routing
UMass Amherst aSOC architecture	[27]	2-D mesh	PCS;[b] no VC	Timeslot based	Shortest-path
Sun UltraSparc T1 (CPX buses)	[28] [29]	Crossbar; 128-bit	N/A	Handshaking	N/A
Sony, Toshiba, IBM Cell BE EIB	[30] [31]	Ring; 128-bit	PCS;[b] no VC	Credit based	Shortest-path
UT Austin TRIPS (operand network)	[32] [33]	2-D mesh; 109-bit	N/A;[a] no VC	On/off	YX DOR
UT Austin TRIPS (on-chip network)	[33]	2-D mesh; 128-bit	Wormhole; 4 VCs	Credit based	YX DOR
Intel SCC architecture	[34] [35]	2-D torus; 32-bit	Wormhole; no VC	Stall/go	XY,YX DOR; odd-even TM[c]
Intel Teraflops NoC	[36] [37]	2-D mesh; 32-bit	Wormhole; 2 lanes[d]	On/off	Source routing (e.g., DOR)
Tilera TILE64 iMesh (dynamic networks)	[38]	2-D mesh; 32-bit	Wormhole; no VC	Credit based	XY DOR

[a] A packet contains a single flit.
[b] PCS denotes "pipelined circuit switching."
[c] TM denotes "turn model."
[d] The lanes are similar to virtual channels.

TABLE 3.4

Network Partitioning in Representative NoCs

	Ref.	Network Partitioning (Physical and Logical)
MIT Raw microprocessor	[22] [23]	Four physical networks: two for static communication and two for dynamic communications
Sony, Toshiba, IBM Cell BE EIB	[30] [31]	Four data rings: two for clockwise and two for counterclockwise
UT Austin TRIPS microprocessor	[32] [33]	Two physical networks: an on-chip network (OCN) and an operand network (OPN); OCN has four VCs.
Intel Teraflops NoC	[36] [37]	Two lanes:[a] one for data transfers and one for instruction transfers
Tilera TILE64 iMesh	[38]	Five physical networks: a user dynamic network, an I/O dynamic network, a memory dynamic network, a tile dynamic network, and a static network

[a] The lane is similar to a virtual channel.

error detection and recovery become increasingly important as process technology scales down, in order to cope with unreliable on-chip communications due to various noises, crosstalk, and process variations.

As shown in Table 3.3, DOR is widely used in NoCs that have grid-based topology. It can uniformly distribute the routing paths, and its routing function can be implemented with a small combinational logic, instead of routing tables that require registers for storing routing information. In addition to simple deterministic routing such as DOR, adaptive routing (e.g., the turn model family and up*/down* routing) is also used in NoCs. Most NoC systems employ minimal routing. Generally speaking, nonminimal paths are not efficient because they consume more network resources than the minimal paths. Moreover, power consumption is increased by the additional switching activity of routers due to nonminimal paths (the details are described in Section 3.5.2). Therefore, nonminimal paths are only used for special purposes such as avoiding network faults or congestion. This section clearly shows that the design trend of NoC protocols employed in commercial and prototype NoCs is moving towards loss-less low-latency lightweight networks.

3.5.2 Research Trend

Although commercial and prototype NoCs tend to be loss-less low-latency lightweight networks, we must not forget that the trend may change in the future, as technology continues to scale down. In the advanced research domain, various approaches have been discussed for future NoC protocols. As device technology scaling continues to be improved, the number of processing cores on a single chip will increase considerably, making reliability more important. Fault-tolerant routing algorithms, whose paths avoid hard faulty links or faulty routers, were thus proposed by Flich et al. [39] and Murali et al. [40]. These routing tables tend to be complex or large, in order to employ flexible routing paths compared with those of DOR that has the strong regularity. Thus, advanced techniques that decrease the table size have been proposed by Koibuchi et al. [6], Flich et al. [39], and Bolotin et al. [41]. Although multipaths between a same source–destination pair have the property of avoiding faulty regions, they introduce out-of-order packet delivery. Fortunately, the technique described by Murali et al. [42] and Koibuchi et al. [6] simply enforces in-order packet delivery. In addition to hard failures of NoC components, software transient error occurs. Its recovery can be done using router-to-router and end-to-end flow controls with software approach, and is discussed in terms of throughput and power consumption [43].

One of the major targets of SoCs is embedded applications, such as media processing mostly for consumer equipment. In stream processing such as Viterbi decoder, or MPEG coder, a series of processing is performed to a certain amount of data. In the processing, each task can be mapped onto each node and is performed in the pipelined manner. In this case, the communication is limited to only between pairs of the neighboring nodes. This locality leads to the possibility of extending a deadlock-free routing algorithm; namely, we can make routing algorithms that provide the deadlock-free and connectivity properties only for the set of paths used by the target application [44]. The routing algorithms explained in Section 3.3 are general techniques to establish deadlock-free paths between all pairs of nodes, and their design requirement is tighter than that of application-specific routings. Another feature of locality is the ability to determine the number of entries of routing tables and routing (address) information embedded in every packet. Because the routing address is required to identify output ports of packets generated on the application where a few pair of nodes are communicated, the routing address is assigned and optimized to the routing paths used by the target application; namely, the size of the routing (address) information can be drastically reduced [6].

Here, we focus on how the power consumption is influenced by routing algorithms, and we introduce a simple energy model. This model is useful for estimating the average energy consumption needed to transmit a single flit from a source to a destination. It can be estimated as

$$E_{\text{flit}} = w H_{\text{ave}}(E_{\text{link}} + E_{\text{sw}}), \tag{3.2}$$

where w is the flit-width, H_{ave} is the average hop count, E_{sw} is the average energy to switch the 1-bit data inside a router, and E_{link} is the 1-bit energy consumed in a link.

E_{link} can be calculated as

$$E_{\text{link}} = d V^2 C_{\text{wire}}/2, \tag{3.3}$$

where d is the 1-hop distance (in millimeters), V is the supply voltage, and C_{wire} is the wire capacitance per millimeter. These parameters can be extracted from the post place-and-route simulations of a given NoC.

Sophisticated mechanisms (e.g., virtual-channel mechanisms) and the increased number of ports make the router complex. As the switch complexity increases, E_{sw} is increased in Equation (3.2). The complex switch-to-switch flow control that uses the increased number of control signals also increases power, because of its increased channel bit width. Regarding routing protocols, Equation (3.2) shows that path hops are proportional to the energy consumption of a packet, and a nonminimal routing has the disadvantage of the energy consumption.

The energy-aware routing strategy tried to minimize the energy by improving routing algorithms [45], although other approaches make the best use of the low power network architecture. It assumes that dynamic voltage and frequency scaling (DVFS) and on/off link activation will be used in the case of NoCs [46]. The voltage and frequency scaling is a power saving technique that reduces the operating frequency and supply voltage according to the applied load. Dynamic power consumption is proportional to the square of the supply voltage; because a peak performance is not always required during the whole execution time, adjusting the frequency and supply voltage to at least achieve the required performance can reduce the dynamic power. In the paper presented by Shang et al. [47], the frequency and the voltage of network links are dynamically adjusted according to the past utilization. In an article by Stine and Carter [48], the network link voltage is scaled down by an adaptive routing to distribute the traffic load. The designer will introduce a measure of routing algorithms to assess the applicability and portability of these techniques.

Another routing approach used dynamic traffic information for improving the performance. DyAD routing forwards packets using deterministic routing at the low traffic congestion, although it forwards them using adaptive routing at the high traffic congestion [49]. The DyAD strategy achieves low latency using deterministic routing when a network is not congested, and high throughput by using adaptive routing. DyXY adaptive routing improves throughput by using the congestion information [50].

In addition to the routing protocol, advanced flow controls improve the throughput by using the traffic prediction [51] or link utilization [52] and cell-based switching optimized to the application traffic patterns [6]. Total application-specific NoC designs including deadlock-free path selection [53] or buffer spaces optimization and traffic shaping [54] have been discussed.

As device technology scaling continues to improve, new chips will have larger buffers that would lead to an increase in the number of virtual channels, and the use of VCT instead of WH switching. If power consumption continues to dominate chip design in the future, designers may try to decrease the frequency of the chip by adding router hardware (e.g, increasing the number of virtual channels or enlarging the channel buffers). In this way, the same throughput can be achieved at a lower frequency, which in turn enables the voltage to be reduced. For the same low-power purpose, adaptive routing may be used instead of deterministic routing, which has been widely used in the current NoCs. These possibilities may encourage designers to deviate greatly from the current trend of the loss-less low-latency lightweight networks. Fortunately, the fundamentals of such protocol techniques are universal and, hopefully, the readers have acquired the basic knowledge in this chapter.

3.6   Summary

This chapter presented the Networks-on-Chip (NoC) protocol family: switching techniques, routing protocols, and flow control. These techniques and protocols affect the network throughput, hardware amount, energy consumption, and reliability for on-chip communications. Discussed protocols have originally been developed for parallel computers, but now they are evolving for on-chip purposes in different ways, because the requirements for on-chip networks are different from those for off-chip systems. One of the distinctive concepts of NoCs is the loss-less, low-latency, and lightweight network architecture. Channel buffer management between neighboring routers and deadlock-free routing algorithms avoids discarding packets. Moreover, switching techniques and injection limitation enable low-latency lightweight packet transfer. This chapter also surveyed the trends in communication protocols used in current commercial and prototype NoC systems.

References

[1] J. Duato, S. Yalamanchili, and L. M. Ni, *Interconnection Networks: An Engineering Approach*. Morgan Kaufmann, 2002.

[2] W. J. Dally and B. Towles, *Principles and Practices of Interconnection Networks*. Morgan Kaufmann, 2004.

[3] J. L. Hennessy and D. A. Patterson, *Computer Architecture: A Quantitative Approach, Fourth Edition*. Morgan Kaufmann, 2007.

[4] L. Benini and G. D. Micheli, *Networks on Chips: Technology and Tools*. Morgan Kaufmann, 2006.

[5] A. Jantsch and H. Tenhunen, *Networks on Chip*. Kluwer Academic Publishers, 2003.

[6] M. Koibuchi, K. Anjo, Y. Yamada, A. Jouraku, and H. Amano, "A simple data transfer technique using local address for networks-on-chips," *IEEE Transactions on Parallel and Distributed Systems* 17 (Dec. 2006) (12): 1425–1437.

[7] H. Matsutani, M. Koibuchi, and H. Amano, "Performance, cost, and energy evaluation of Fat H-Tree: A cost-efficient tree-based on-chip network." In *Proc. of International Parallel and Distributed Processing Symposium (IPDPS'07)*, March 2007.

[8] H. Matsutani, M. Koibuchi, D. Wang, and H. Amano, "Adding slow-silent virtual channels for low-power on-chip networks." In *Proc. of International Symposium on Networks-on-Chip (NOCS'08)*, Apr. 2008, 23–32.

[9] C. J. Glass and L. M. Ni, "The turn model for adaptive routing." In *Proc. of International Symposium on Computer Architecture (ISCA'92)*, May 1992, 278–287.

[10] G.-M. Chiu, "The odd-even turn model for adaptive routing," *IEEE Transactions on Parallel and Distributed Systems* 11 (Nov. 2000) (7): 729–738.

[11] J. Duato, "A necessary and sufficient condition for deadlock-free adaptive routing in wormhole networks," *IEEE Transactions on Parallel and Distributed Systems* 6 (Jun. 1995) (10): 1055–1067.

[12] F. Silla and J. Duato, "High-performance routing in networks of workstations with irregular topology," *IEEE Transactions on Parallel and Distributed Systems* 11 (Jul. 2000) (7): 699–719.

[13] W. H. Ho and T. M. Pinkston, "A design methodology for efficient application-specific on-chip interconnects," *IEEE Transactions on Parallel and Distributed Systems* 17 (Feb. 2006) (2): 174–190.

[14] M. D. Schroeder, A. D. Birrell, M. Burrows, H. Murray, R. M. Needham, and T. L. Rodeheffer, "Autonet: A high-speed, self-configuring local area network using point-to-point links," *IEEE Journal on Selected Areas in Communications* 9 (October 1991): 1318–1335.

[15] J. C. Sancho, A. Robles, J. Flich, P. Lopez, and J. Duato, "Effective methodology for deadlock-free minimal routing in InfiniBand." In *Proc. of International Conference on Parallel Processing*, Aug. 2002, 409–418.

[16] M. Koibuchi, A. Jouraku, and H. Amano, "Descending layers routing: A deadlockfree deterministic routing using virtual channels in system area networks with irregular topologies." In *Proc. of International Conference on Parallel Processing*, Oct. 2003, 527–536.

[17] O. Lysne, T. Skeie, S.-A. Reinemo, and I. Theiss, "Layered routing in irregular networks," *IEEE Transactions on Parallel and Distributed Systems* 17 (Jan. 2006) (1): 51–65.

[18] W. J. Dally and H. Aoki, "Deadlock-free adaptive routing in multicomputer networks using virtual channels," *IEEE Transactions on Parallel and Distributed Systems* (Apr. 1993) (4): 466–475.

[19] J. C. Martinez, F. Silla, P. Lopez, and J. Duato, "On the influence of the selection function on the performance of networks of workstations." In *Proc. of International Symposium on High Performance Computing*, Oct. 2000, 292–300.

[20] J. C. Sancho, A. Robles, and J. Duato, "An effective methodology to improve the performance of the up*/down* routing algorithm," *IEEE Transactions on Parallel and Distributed Systems* 15 (Aug. 2004) (8): 740–754.

[21] S. Murali, T. Theocharides, N. Vijaykrishnan, M. J. Irwin, L. Benini, and G. D. Micheli, "Analysis of error recovery schemes for networks on chips," *IEEE Design & Test of Computers* 22 (2005) (5): 434–442.

[22] M. B. Taylor, J. S. Kim, J. E. Miller, D. Wentzlaff, F. Ghodrat, B. Greenwald, H. Hoffmann, P. Johnson, J.-W. Lee, W. Lee, A. Ma, A. Saraf, M. Seneski, N. Shnidman, V. Strumpen, M. Frank, S. P. Amarasinghe, and A. Agarwal, "The raw microprocessor: A computational fabric for software circuits and general purpose programs," *IEEE Micro* 22 (Apr. 2002) (2): 25–35.

[23] M. B. Taylor, W. Lee, J. E. Miller, D. Wentzlaff, I. Bratt, B. Greenwald, H. Hoffmann, P. Johnson, J. S. Kim, J. Psota, A. Saraf, N. Shnidman, V. Strumpen, M. Frank, S. P. Amarasinghe, and A. Agarwal, "Evaluation of the raw microprocessor: An exposed-wire-delay architecture for ILP and Streams." In *Proc. of International Symposium on Computer Architecture (ISCA'04)*, Jun. 2004, 2–13.

[24] A. Andriahantenaina, H. Charlery, A. Greiner, L. Mortiez, and C. A. Zeferino, "SPIN: A scalable, packet switched, on-chip micro-network." In *Proc. of Design Automation and Test in Europe Conference (DATE'03)*, Mar. 2003, 70–73.

[25] A. Andriahantenaina and A. Greiner, "Micro-network for SoC: Implementation of a 32-port SPIN network." In *Proc. of Design Automation and Test in Europe Conference (DATE'03)*, Mar. 2003, 1128–1129.

[26] F. Furtek, E. Hogenauer, and J. Scheuermann, "Interconnecting heterogeneous nodes in an adaptive computing machine." In *Proc. of Field-Programmable Logic and Applications (FPL'04)*, Sept. 2004, 125–134.

[27] J. Liang, A. Laffely, S. Srinivasan, and R. Tessier, "An architecture and compiler for scalable on-chip communication," *IEEE Transactions on Very Large Scale Integration Systems* 12 (July 2004) (7): 711–726.

[28] P. Kongetira, K. Aingaran, and K. Olukotun, "Niagara: A 32-way multithreaded sparc Processor," *IEEE Micro* 25 (Mar. 2005) (2): 21–29.

[29] A. S. Leon, K. W. Tam, J. L. Shin, D. Weisner, and F. Schumacher, "A power-efficient high-throughput 32-thread SPARC processor," *IEEE Journal of Solid-State Circuits* 42 (Jan. 2007) (1): 7–16.

[30] M. Kistler, M. Perrone, and F. Petrini, "Cell multiprocessor communication network: Built for speed," *IEEE Micro* 26 (May 2006) (3): 10–23.

[31] T. W. Ainsworth and T. M. Pinkston, "Characterizing the cell EIB on-chip network," *IEEE Micro*, 27 (Sept. 2007) (5): 6–14.

[32] P. Gratz, K. Sankaralingam, H. Hanson, P. Shivakumar, R. G. McDonald, S. W. Keckler, and D. Burger, "Implementation and evaluation of a dynamically routed processor operand network," In *Proc. of International Symposium on Networks-on-Chip (NOCS'07)*, May 2007, 7–17.

[33] P. Gratz, C. Kim, K. Sankaralingam, H. Hanson, P. Shivakumar, S. W. Keckler, and D. Burger, "On-chip interconnection networks of the TRIPS chip," *IEEE Micro* 27 (Sept. 2007) (5): 41–50.

[34] J. D. Hoffman, D. A. Ilitzky, A. Chun, and A. Chapyzhenka, "Architecture of the scalable communications core." In *Proc. of International Symposium on Networks-on-Chip (NOCS'07)*, May 2007, 40–52.

[35] D. A. Ilitzky, J. D. Hoffman, A. Chun, and B. P. Esparza, "Architecture of the scalable communications core's network on chip," *IEEE Micro* 27 (Sept. 2007) (5): 62–74.

[36] S. Vangal, J. Howard, G. Ruhl, S. Dighe, H. Wilson, J. Tschanz, D. Finan, P. Iyer, A. Singh, T. Jacob, S. Jain, S. Venkataraman, Y. Hoskote, and N. Borkar, "An 80-Tile 1.28TFLOPS network-on-chip in 65 nm CMOS." In *Proc. of International Solid-State Circuits Conference (ISSCC'07)*, Feb. 2007.

[37] Y. Hoskote, S. Vangal, A. Singh, N. Borkar, and S. Borkar, "A 5-GHz mesh interconnect for a teraflops processor," *IEEE Micro* 27 (Sept. 2007) (5): 51–61.

[38] D. Wentzlaff, P. Griffin, H. Hoffmann, L. Bao, B. Edwards, C. Ramey, M. Mattina, C.-C. Miao, John F. Brown III, and A. Agarwal, "On-chip interconnection architecture of the tile processor," *IEEE Micro*, 27 (Sept. 2007) (5): 15–31.

[39] J. Flich, A. Mejia, P. Lopez, and J. Duato, "Region-based routing: An efficient routing mechanism to tackle unreliable hardware in network on chips." In *Proc. of International Symposium on Networks-on-Chip (NOCS)*, May 2007, 183–194.

[40] S. Murali, D. Atienza, L. Benini, and G. D. Micheli, "A multi-path routing strategy with guaranteed in-order packet delivery and fault-tolerance for networks on chip." In *Proc. of Design Automation Conference (DAC)*, Jul. 2006, 845–848.

[41] E. Bolotin, I. Cidon, R. Ginosar, and A. Kolodny, "Routing table minimization for irregular mesh NoCs." In *Proc. of Design Automation and Test in Europe (DATE)*, Apr. 2007.

[42] M. Koibuchi, J. C. Martinez, J. Flich, A. Robles, P. Lopez, and J. Duato, "Enforcing in-order packet delivery in system area networks with adaptive routing," *Journal of Parallel and Distributed Computing (JPDC)*, 65 (Oct. 2005) (10): 1223–1236.

[43] S. Murali, T. Theocharides, N. Vijaykrishnan, M. J. Irwin, L. Benini, and G. D. Micheli, "Analysis of error recovery schemes for networks on chips," *IEEE Design & Test of Computers* 22 (Sept. 2005) (5): 434–442.

[44] H. Matsutani, M. Koibuchi, and H. Amano, "Enforcing dimension-order routing in on-chip torus networks without virtual channels." In *Proc. of International Symposium on Parallel and Distributed Processing and Applications (ISPA'06)*, Nov. 2006, 207–218.

[45] J.-C. Kao and R. Marculescu, "Energy-aware routing for e-textile applications." In *Proc. of Design Automation and Test in Europe (DATE)*, 1, 2005, 184–189.

[46] V. Soteriou and L.-S. Peh, "Exploring the design space of self-regulating power-aware on/off interconnection networks," *IEEE Transactions on Parallel and Distributed Systems* 18 (Mar. 2007) (3): 393–408.

[47] L. Shang, L.-S. Peh, and N. K. Jha, "Dynamic voltage scaling with links for power optimization of Interconnection Networks." In *Proc. of International Symposium on High-Performance Computer Architecture (HPCA'03)*, Jan. 2003, 79–90.

[48] J. M. Stine and N. P. Carter, "Comparing adaptive routing and dynamic voltage scaling for link power reduction," *IEEE Computer Architecture Letters* 3 (Jan. 2004) (1): 14–17.

[49] J. Hu and R. Marculescu, "DyAD. Smart routing for networks-on-chip." In *Proc. of Design Automation Conference (DAC'04)*, Jun. 2004, 260–263.

[50] M. Li, Q.-A. Zeng, and W.-B. Jone, "DyXY: A proximity congestion-aware deadlock-free dynamic routing method for network on chip." In *Proc. of Design Automation Conference (DAC)*, Jul. 2006, 849–852.

[51] U. Y. Ogras and R. Marculescu, "Prediction-based flow control for network-on-chip traffic." *In Proc. of Design Automation Conference (DAC)*, Jul. 2006.

[52] J. W. van den Brand, C. Ciordas, K. Goossens, and T. Basten, "Congestion-controlled best-effort communication for networks-on-chip." In *Proc. of Design Automation and Test in Europe (DATE)*, Apr. 2007.

[53] J. Hu and R. Marculescu, "Energy- and performance-aware mapping for regular NoC architectures," *IEEE Transactions on Computer-Aided Design of Integrated Circuits and Systems* 24 (Apr. 2005) (4): 551–562.

[54] S. Manolache, P. Eles, and Z. Peng, "Buffer space optimization with communication synthesis and traffic shaping for NoCs." In *Proc. of Design Automation and Test in Europe (DATE)*, 1, 2006.

4

On-Chip Processor Traffic Modeling for Network-on-Chip Design

Antoine Scherrer, Antoine Fraboulet, and Tanguy Risset

CONTENTS

95

4.1   Introduction

Next generation System-on-Chip (SoC) architectures will include many processors on a single chip, performing the entire computation that used to be
done by hardware accelerators. They are referred to as MPSoC for multiprocessor SoC. When multiprocessors were not on chip as in parallel machines 20
years ago, communication latency, synchronization, and network contention
were the most important fences for performance. This was mainly due to the
cost of communication compared to computation. For simple SoC architectures, the communication latency is kept low and the communication scheme
is simple: most of the transactions occur between the processor and the main
memory. For SoC, a Network-on-Chip (NoC), or at least a hierarchy of buses,
is needed and communication has a major influence on performance and on
power consumption of the global system. Predicting communication performance at design time is essential because it might influence physical design
parameters, such as the location of various IPs on the chip.

MPSoC are highly programmable, and can target possibly any application.
However, currently, they are mostly designed for signal processing and multimedia applications with real-time constraints, which are not as harsh as for
avionics. To meet these real-time constraints, MPSoC are composed of many
master IPs (processors) and few slave IPs (memories and peripherals).

In this chapter, we investigate on-chip processor traffic for performance
evaluation of NoC. Traffic modeling and generation of dedicated IPs (e.g.,
MPEG-2, FFT) use predictable communication schemes, such that it is possible
to generate a traffic that *looks like* the one these IPs would produce. Such a
traffic generator is usually designed together with (or even before) the IP itself
and is very different for processors. Processor traffic is much more difficult
to model for two main reasons: (1) cache behavior is difficult to predict (it
is program and data dependent) and (2) operating system interrupts lead
to nondeterministic behavior in terms of communication and contention. In
order to build an efficient tool for predicting communication performance for
a given application, it is therefore essential to have a precise modeling of the
communications induced by applications running on processors.

Predicting communication performance can be done by a precise (cycle
accurate) simulation of the complete application or by using a traffic generator instead of real IPs. Simulation is usually impossible at early stages of the
design because IPs and programs are not yet available. Note also that SoC
cycle accurate simulations are very time consuming, unless they are performed on expensive hardware emulators (based on hundreds of FPGA).
Traffic generators are preferred because they are parameterizable, faster to
simulate, and simpler to use. However, they are less precise because they do
not execute the real program.

Traffic generators can produce communications in many ways, ranging
from the replay of a previously recorded trace to the generation of sample
paths of stochastic processes, or by writing a very simple code emulating the

communications of a dedicated IP. Note that random sources can have parameters fitted to the statistical properties of the observed traffic, or parameters fixed by hand. The decisions of which communication parameter (latency, throughput, etc.) and statistical property are to be emulated are important issues that must be addressed when designing an NoC traffic modeling environment. This is the main topic of this chapter.

One of the main difficulties in modeling processor traffic is that processor activity is not stationary (its behavior is not stable). It rather corresponds to a sequence of traffic phases (corresponding to program phase [1]). In each stationary phases, data can be fitted to well-known stochastic processes with prescribed first (marginal distribution) and second (covariance) statistical orders.

The chapter is divided in two main parts: Section 4.2 gives the background of stochastic processes as well as on-chip processor traffic. In Section 4.3, we discuss in detail the various steps involved in the design of a traffic generation environment, and illustrate it with the MPTG environment [2]. Conclusion and related works are reported in Section 4.4.

4.2 Statistical Traffic Modeling

In this section, we introduce the specificities of on-chip processor traffic and show how it can be modeled with stochastic processes. We then present some statistical background, which is useful in describing and simulating NoC traffic. This theoretical background includes basic statistical modeling methods, decomposition of the traffic into stationary phases, and also an introduction to long-range dependence.

4.2.1 On-Chip Processor Traffic

On-chip processor communications are mostly issued by caches (instruction cache, data cache, or both), and not by the processor itself. The presence and the behavior of this component imply that processor traffic is the aggregation of several types of communication described hereafter. Note that a processor without a cache can be seen as a processor with a cache with one line of one single word. Transactions initiated by the cache to the NoC can be segregated into three categories:

1. **Reads.** Read transactions have the size of a cache line. The time between two reads corresponds to a time during which the processor computes only on cached data. Two flows must be distinguished, the instruction flow (binary from instruction memory) and the data flow (operands from data memory).

2. **Writes.** Write transactions can have various sizes depending on the cache writing policy: write through (one word at a time), and write

back (one line at a time). If a write buffer is present then the size is variable, as the buffer is periodically emptied.

3. **Other requests**. Requests to noncached memory parts have a size of one word, as for atomic reads/writes. If a cache coherency algorithm is implemented then additional messages are also sent among processors.

Cache performance is a very popular research field [3,4]. In the scope of embedded systems, however, low cost solutions (low area, low consumption, and low delay costs) are usually preferred. For instance, in DSP architectures, no data cache is needed because the temporal locality of data accesses is likely to be low (data flow). For the same reason, each processor has a single communication interface, meaning that the type of communication mentioned above are interleaved. In other words, the traffic generated by the processor cannot be split into data and instruction streams, these two streams are merged. Based on this assumption, we can now define more precisely how an on-chip processor communication can be modeled.

4.2.2 On-Chip Traffic Formalism

The traffic produced by a processor is modeled as a sequence of transactions composed of flits (flow transfer units) corresponding to one bus-word. The *kth* transaction is a 5-uple $T(k) = (A(k), C(k), S(k), D(k), I(k))$ meaning target address, command (read or write), size of transaction, delay, and interrequest time, respectively. This is illustrated in Figure 4.1. We also define the latency of the *kth* transaction $L(k)$ as the number of cycles between the start of a *kth* request and the arrival of the associated response. This is basically the round-trip time in the network and is used to evaluate the contention. We further define the aggregated throughput $W_\delta(i)$ as the number of transactions sent in consecutive and nonoverlapping time windows of size δ. Note that this formalism only holds for communication protocols for which each request is

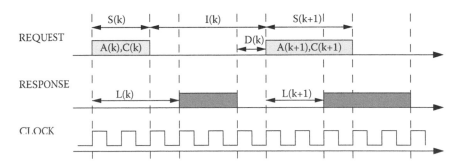

FIGURE 4.1

Traffic modeling formalism: $A(k)$ is the target address, $C(k)$ the command (read or write), $S(k)$ is the size of the transaction, $D(k)$ is the delay between the completion of one transaction and the beginning of the following one, and $I(k)$ is the interrequest time.

expecting a response (even for write requests), which is the case for most IP communication interfaces such as VCI (virtual component interface [23]).

One can distinguish two main communication schemes used by IPs: the nonsplit transactions scheme, where the IP is not able to send a request until the response to the previous one has been received, and the split transactions scheme in which new requests can be sent without waiting for the responses. The nonsplit transaction scheme is widely used by processors and caches (although, for cache, it might depend on the cache parameters), whereas the split transaction scheme is used by dedicated IPs performing computation on streams of data that are transmitted via direct memory access (DMA) modules.

4.2.3 Statistical Traffic Modeling

With the formalism introduced in the previous section, the traffic of processors consists of five time-series composing a 5-D vector sequence $(T(k))_{k \in \mathbb{N}}$. Our goal is to emulate real traffic by means of traffic generators, which will produce $T(k)$ for each k. This can be done in many ways and we present hereafter a nonexhaustive list of possibilities.

1. **Replay.** One can choose to record the complete transaction sequence $T(k)$ and to simply replay, as it is. This provides accurate simulations; however, it is limited by the length of input simulation. Furthermore, the size of the recorded simulation trace might be very large, and thus hard to load and store.

2. **Independent random vector.** One can also consider the elements of the vector as sample paths of independent stochastic processes. The statistical behavior of each element will then be described, but the correlations between them will not be considered.

3. **Random vector.** In this case, the vector is modeled by doing a statistical analysis of each element as well as correlations between each pair of elements.

4. **Hybrid approach.** From the knowledge of the processor's behavior, we can introduce some constraints on top of the stochastic modeling. For instance, if an instruction cache is present, read requests targeted to instruction memory always have the size of the cache line.

We can also distinguish between two ways of modeling, the time at which each transaction occurs, leading to different accuracy levels.

- **Delay.** Use the delay sequence $D(k)$ representing the time (in cycles) between the reception of the kth response and the start of the $(k+1)^{th}$ request.

- **Aggregated throughput.** Use the sequence of aggregated throughput of the processor $W_\delta(k)$, and transactions can be placed in various ways within the aggregation window δ.

TABLE 4.1

Some Classical Probability Distribution Functions (PDF)

PDF	Description
Gaussian	The most widely used PDF, used for aggregated throughput for instance
Exponential	Fast decay PDF, used for delay sequence for instance
Gamma	Gives intermediate PDF between exponential and Gaussian
Lognormal	Gives asymmetric PDF
Pareto	Provides *heavy-tailed* PDF (slow decay)

The statistical modeling in itself relies on signal processing tools and methods that benefit from a huge amount of references [5,6]. Let us recall that a stochastic process X is a sequence of random variables $X[i]$ (we use brackets to denote random variables). We will consider two statistical characteristics of stochastic processes: the marginal law (or probability distribution function), which represents how the values taken by the process are distributed, and the covariance function, which gives information on the correlations between the random variables of the process as a function of the time lag between them. For instance, the sequence of delays $D(k)$ can be generated as the sample path of some stochastic process $\{D[i]\}_{i \in \mathbb{N}}$, with prescribed first and second statistical orders. Typical models for probability distribution functions (PDF) and covariances are reported in Tables 4.1 and 4.2.

For each probability distribution and covariance, we use state-of-the-art parameter estimation techniques (using mainly maximum likelihood expectation) [5,6].

4.2.4 Statistical Stationarity and Traffic Phases

When we want to model one element of the transaction vector with stochastic processes we should take stationarity into account. Indeed most stochastic process models are stationary, and nonstationary models are much harder to use in terms of parameter estimation as well as model selection.

Let us first recall some background. The covariance function γ_X of a stochastic process $\{X[i]\}_{i \in \mathbb{N}}$ describes how the random variables of a process are correlated to each other as a function of the time lag between these random

TABLE 4.2

Some Classical Covariance Functions

Covariance	Description
IID (independent identically distributed)	No memory
ARMA (autoregressive move average)	Short-range dependence
FGN (fractional Gaussian noise)	Long-range dependence
FARIMA (fractional integrated ARMA)	Both short- and long-range dependence

variables. It is defined as follows ($\mathbb{E}$ is the expectation):

$$\gamma_X(i, j) = \mathbb{E}(X[i]X[j]) - \mathbb{E}(X[i])\mathbb{E}(X[j])$$

A process X is wide-sense stationary if its mean is constant ($\forall(i, j) \in \mathbb{N}^2$, $\mathbb{E}(X[i]) = \mathbb{E}(X[j]) \triangleq \mathbb{E}(X)$) and its covariance reduces to one variable function as follows:

$$\forall(i, j) \in \mathbb{N}^2, \quad \gamma_X(i, j) = \gamma_X(0, |i - j|) \triangleq \gamma_X(|i - j|)$$

So, when modeling a time series, one should carefully check that stationarity is a reasonable assumption. For on-chip processor traffic, algorithms that are executed on the processor have different phases resulting in different communication patterns, where most of the time the traffic will not be globally stationary. If signs of nonstationarity are present, one should consider building a piecewise stationary model. This implies the estimation of model parameters on several stationary phases of the data. At simulation time the generator will change the model parameters when it switches between phases.

A traffic phase is a part of the transaction sequence $T(k), i \leq k \leq j$. Because most multimedia algorithms are repetitive, it is likely that similar phases appear several times in the trace. For instance, in the MP3 decoding algorithm, each MP3 frame is decoded in a loop leading to similar treatments.

4.2.4.1 Phase Decomposition

The question is therefore to determine stationary traffic phases automatically. In general, decomposing a nonstationary process into stationary parts is very difficult. Calder et al. have developed a technique for the identification of program phases in SimPoint [7] for advanced processor architecture performance evaluation. This is a powerful technique that can dramatically accelerate simulations by simulating only one simulation point per phase and replicating that behavior during all the corresponding phases. In NoC traffic simulation, we do not pursue the same goal because we target precise traffic simulation of a given IP for NoC prototyping. Network contention needs to be precisely simulated, and as it is the result of the superposition of several communication flows, picking simulation points becomes a difficult task.

From Calder's work [7], we have developed a traffic phases discovery algorithm [8]. It uses the k-means algorithm [9], which is a classical technique to group multidimensional values in similar sets. The worst case complexity of this algorithm is exponential but in practice it is very fast. The automatic phase determination algorithm is as follows:

1. First, we select a list of M elements of the transaction sequence delay, size, command, address, etc; (see Section 4.2.2).
2. The transaction sequence is then split into nonoverlapping intervals of L transactions. Mean and variance are computed on each

interval and for each of the M selected elements. Thus we build a $2M$-dimensional representative vector used for the clustering.

3. We perform clustering in k phases using the k-means algorithm with different values of k (2 to 7 in practice). The algorithm finds k centroids in the space of representative vectors. Each interval will finally be assigned the number of its closest center (in the sense of the quadratic distance) and therefore each interval will get a phase number.

4. To evaluate different clusterings, we compute the Bayesian Information Criterion (BIC) [10]. The BIC gives a score of the clustering and a higher BIC means better clustering.

Once the phases are identified, statistical analysis is performed on each extracted phase by an automatic fitting procedure that adjusts the first and second statistical orders (for details see Scherrer et al. [11]). Examples of phases discovered by this algorithm are illustrated in Figure 4.8.

4.2.5 Long-Range Dependence

Long-range dependence (LRD) is an ubiquitous property of Internet traffic [12,13], and it has also been demonstrated on an on-chip multimedia (MPEG-2) application by Varatkar and Marculescu [14]. They have indeed found LRD in the communications between different components of an MPEG-2 hardware decoder at the macro-block level. The main interest in LRD resides in its strong impact on network performance [15]. In particular, the needed memorization in the buffers is higher when the input traffic has this property [16]. As a consequence, for macro-networks as well as for on-chip networks, LRD should be taken into account if it is found in the traffic that the network will have to handle.

Long-range dependence is a property of a stochastic process that is defined as a slow decrease of its covariance function [15]. We expect this function to be decreasing, because correlated data are more likely to be close (in time) to each other. However, if the process is long-range dependent, then the covariance decays very slowly and is not summable.

$$\sum_{k \in \mathbb{N}} \gamma_X(k) = \infty$$

Therefore, LRD reflects the ability of the process to be highly correlated with its past, because even at large lags, the covariance function is not negligible. This property is also linked to self-similarity, which is more general, and it can be shown that asymptotic second order self-similarity implies LRD [17].

A long-range dependent process is usually modeled with a power-law decay of the covariance function as follows:

$$\gamma_X(k) \underset{k \to +\infty}{\sim} ck^{-\alpha}, \quad 0 < \alpha \leq 1$$

The exponent α (also called *scaling index*) provides a parameter to tell how much a process is long-range dependent ($0 < \alpha \leq 1$). The Hurst exponent, noted H, is the classical parameter for describing self-similarity [15]. Because of the analogy between LRD and self-similarity, it can be shown that a simple relation exists between H and α: $H = (2 - \alpha)/2$. As a consequence, H ($1/2 < H < 1$) is the commonly used parameter for LRD. Note that when $H = 0.5$, there is no LRD (this is also referred to as short-range dependence).

4.2.5.1 Estimation of the Hurst Parameter

A standard wavelet-based methodology can be used for the estimation of the Hurst parameter [17]. Let $\psi_{j,k}(t) = 2^{-j/2}\psi_0(2^{-j}t - k)$ denote an orthonormal wavelet basis, derived from the mother wavelet ψ_0. The j index represents the *scale*: the larger the j, the more the wavelet is dilated. The k index is a shift in time.

For any (j, k), $d_X(j, k) = \langle \psi_{j,k}, X \rangle$ are called the *wavelet coefficients* of the stochastic process X ($\langle ., . \rangle$ is the inner product in the L^2 functional space). These wavelet coefficients enable a study of the process X at various times (values of k) and various scales (values of j). In particular, when X is a long-range dependent process with parameter H, the following limit behavior for the expectation of wavelet coefficients can be shown [17]:

$$\forall j, \quad \mathbb{E}(d_X(j,k)^2) \underset{j \to +\infty}{\sim} c2^{j(2H-1)} \tag{4.1}$$

Moreover, it can also be shown that the time averages S_j for each scale j (n_j is the number of wavelet coefficients available at scale j):

$$S_j = (1/n_j) \sum_{k=1}^{n_j} |d_X(j, k)|^2 \tag{4.2}$$

can be used as relevant, efficient, and robust estimators for $\mathbb{E}(d_X(j, k)^2)$ [17]. From Equations (4.1) and (4.2), the estimation of H is as follows: (1) plot $\log_2 S_j$ versus $\log_2 2^j = j$ and (2) perform a weighted linear regression of $\log_2 S_j$ in the coarsest scales (see for instance Figure 4.2). These plots are commonly referred to as log-scale diagrams (LD). In such diagrams, LRD is evidenced by a straight line behavior in the limit of large scales. In particular, if the line is horizontal, then $H = 0.5$ and there is no LRD.

To illustrate how we use this tool to evaluate the Hurst parameter, we provide a typical LD extracted from an Internet trace in Figure 4.2. Along the x axis are the different values of the scale j at which the process is observed. For each scale, $\log_2 S_j$ is plotted together with its confidence interval (vertical bars). The Hurst parameter can be estimated if the different points plotted are aligned on a straight line for large scales.

4.2.5.2 Synthesis of Long-Range Dependent Processes

The synthesis (generation of sample paths) of long-range dependent processes is easy if the marginal law is Gaussian [18]. The so-called Fractional Gaussian

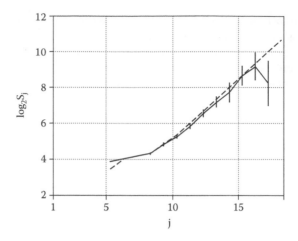

FIGURE 4.2
Example of log-scale diagram (LD), the Hurst parameter is estimated with the slope of the dashed line (here $H = 0.83$).

Noise (FGN) is commonly used for this. However, if one wants to generate a long-range dependent process whose marginal law is non-Gaussian, the problem is more complex. The inverse method [14] only guarantees an asymptotic behavior of the covariance function. We have developed, for several common laws (exponential, gamma, χ^2, etc.), an exact method of synthesis described by Scherrer et al. [11]. We can thus produce synthetic long-range dependent sample paths that can be used in traffic generation. It is important to note that most elements of the transaction sequence of on-chip processor communications have non-Gaussian distributions. For instance, delay sequences rather exhibit an exponential distribution as we expect many small delays and few big ones. With our synthesis method, we can produce a synthetic exponential process with long-range dependence. Such non-Gaussian and LRD models have been used for Internet traffic modeling as well [11].

We have introduced the major theoretical notions useful for a precise modeling of the on-chip traffic. We will now adopt a more practical vision and explain how these statistical modeling notions can be used in an SoC simulation environment.

4.3 Traffic Modeling in Practice

This section provides our practical experience of on-chip processor traffic analysis and generation. We first provide some generic guidelines that can be useful for any project leader in charge of designing an MPSoC architecture.

Then, we present a particular experimental framework called Multiphase Traffic Generator (MPTG).

4.3.1   Guidelines for Designing a Traffic Modeling Environment

Even though our experimental framework is closely related to the SystemC environment, it is possible to postulate several practical issues that any NoC prototyping environment will be confronted with. These issues should be taken into account before designing the environment itself because their absence will, at some point, break the efficiency of the traffic simulation process.

4.3.1.1   Simulation Precision

The first and important point concerns the precision of simulation. Depending on the goal of the simulation several precision levels can be targeted. There is a great emphasis today on transaction level modeling (TLM) (sometimes called *programmers view*), which basically consists of a functional simulation of the MPSoC without any time information. This level is used primarily by application developers who mainly seek a very fast simulation. It would be useless in an NoC prototyping environment: no precise network performance evaluation can be made at this level. Another possibility is to maintain a notion of global time in the simulation by stamping messages as, for instance, proposed by Chandry and Misra [19]. This level is sometimes referred to as TLM-T. Even if this level permits some performance indication, the NoC protocol is usually not precisely simulated, hence contention behavior cannot be detected.

If the simulation environment is intended to detect contention in the network, it should be *cycle accurate* at IP boundaries. In other words, a bit accurate and cycle accurate simulation of the computation itself is not necessary, but the low level protocol of the IP input/output must be emulated very precisely to model burst and cache behavior. The designer must be aware that this level of simulation implies a very slow simulation, mainly because each NoC transaction should be precisely simulated.

Also, note that the behavior of caches must be simulated very precisely, as small details of the cache protocol might have an important influence on the global on-chip traffic.

4.3.1.2   Trace Analysis

Another important issue is the power of the trace generation and trace analysis tools. All the results will be obtained via simulation traces. These traces can be huge; therefore, an efficient and parameterizable trace generation tool is needed. Traces must be compressed dynamically. Parsing and analysis is also a critical treatment. If parser generator tools are used, the grammar expressing the trace syntax must be carefully written so as to generate an efficient parser [20].

As a trace can be generated with many different parameters (size of FIFOs, latency of communication, address of communication, etc.), it should be easy to instrument the simulation platform to record any requested information. Using an existing trace format such as Value Change Dump (VCD) should be preferred.

4.3.1.3 *Platform Generation*

Describing the hardware of an MPSoC platform usually consists of connecting wires between existing IPs and deciding the memory mapping. It quickly becomes intractable to do these connections by hand, wire by wire, whereas the platform designer thinks at a coarser grain: How many processors are connected to which router. The top system simulated must be generated by some in-house script adapted to the simulation environment. Scripting should be used to generate families of platforms to prototype different system architectures or different numbers of processors. Some kind of source language should be designed for high level platform description. Enabling the connection with the Spirit [21] IP description format should be realized somewhere in the environment.

4.3.1.4 *Traffic Analysis and Synthesis Flow*

Obviously, the heart of the prototyping environment are the traffic analysis and synthesis tools, which are discussed in Section 4.2. These tools should be as generic as possible. They are basically signal processing tools that should be independent of the NoC simulation environment.

The global NoC prototyping flow should be clearly stated as soon as possible. Experiments should also be carefully classified according to a clear experimental protocol so as to be able to recover a previous result that could possibly be incoherent with a future experiment. Finally, as in any experimental framework, reproducibility is mandatory.

4.3.2 Multiphase Traffic Generation Environment

In this section, we describe the multiphase on-chip traffic generation (MPTG) environment, its integration in the SocLib simulation environment, and its key features.

We have developed our environment in the SocLib simulation environment [22]. SocLib is a library of open-source SystemC simulation models of IPs that can be interconnected through the VCI [23] interface standard [24]. VCI is a point-to-point communication protocol depicted in Figure 4.3. The simulation models available in SocLib are described at the cycle accurate level, or at the transaction level. All our experiments have been done at the cycle accurate level, as precise information was needed for NoC contention prediction. To each simulation model corresponds a synthesizable model (not necessarily open source) that can be used for designing a chip. Examples of simulation models available in SocLib are a MIPS R3000 processor (with its associated data and instruction cache), standard on-chip memories, DMA controller, and several kinds of NoC.

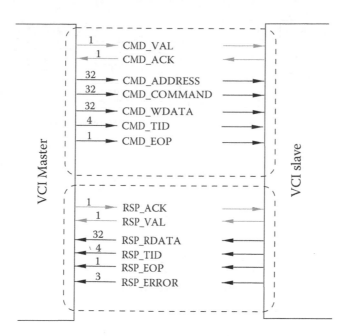

FIGURE 4.3
Example of NoC interconnect interface: Advanced VCI, defined by the OCP consortium.

Within the SocLib framework all components are connected via VCI ports to an NoC interconnection. We used the DSPIN network on chip (an evolution of SPIN [25]) which uses wormhole and credits-based contention control mechanisms. DSPIN uses a set of 4-port routers that can be interconnected in a mesh topology to provide the desired packet switched network architecture. The software running on the processors used in SocLib is compiled with the GNU GCC tool suite. A tiny open source operating system called *mutek* [26] is used when several processors run in parallel. This OS can handle multithreading on each processor.

The global MPTG flow is depicted in Figure 4.4. It is composed of three main parts, described hereafter.

- **Reference trace collection**. This is the entry point of our MPTG flow. Because we follow a trace-based approach, we perform a simulation with a fixed-latency interconnection and get a reference trace. It is important to understand that this reference trace can then be used for many platform simulations (various interconnections, IP placement, memory mapping, etc.) because with such an ideal interconnect we gather the intrinsic communication patterns of IPs. We simply make the assumption that the behavior of IPs (order of transactions, etc.) is not influenced by the latency of the network. Because our traffic generator is aware of the network latency, the reference

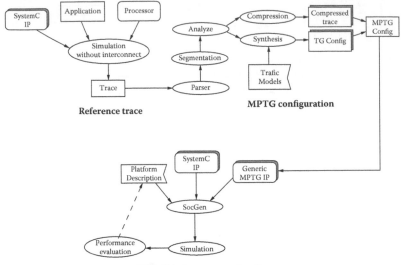

FIGURE 4.4
Multiphase traffic generation flow: An initial trace is collected by simulation with ideal interconnection. The trace is then analyzed and segmented to generate a configuration of the MPTG, then the real simulation can take place.

trace can be used to produce traffic on any interconnection system with a very small error. In the last phase (design space exploration), the real latency of the network will be simulated precisely by taking into account contention and IP placement in the SoC. This is discussed in detail in the following sections.

- **MPTG configuration.** The trace is then analyzed with a semiautomatic configuration flow, which starts by parsing the trace file to extract the transaction sequence. Then we run our phase segmentation algorithm described in Section 4.2.4. According to the designer's choice for the models, each phase is then analyzed and all parameters estimated. In the end, we obtain an MPTG configuration file such as the one reported in Figure 4.5. Note that if the designer fails to find an adequate stochastic model for the traffic, they can choose to replay the reference trace. In this case the trace is compressed to save disk space.

- **Design space exploration.** The traffic generator component with the configuration file can now be used in place of real processors to evaluate the performance of various interconnect, IP placement, memory mapping, etc. This can be done faster because the simulation of the traffic generator requires less resources than the simulation of a processor. One important point is that designers can

```
phase0{

  time:           // transaction temporal caracteristics
    mode=IA;      // selected precision is  Delay
    exponential(15);
                  // IID process following an exponential law of mean 15 flits

  content:      // transaction content caracteristics
    random("ptable",exponential(10));
                  // random generation,
                  // the ptable file contains destination addresses probability table
                  // size is modeled with an exponential law of mean 10 flits

  duration: // phase duration
    constant(10000); // 10000 transactions
  }

phase1{

  time:           // transaction temporal caracteristics
    mode=TP;      // selected precision is throughput
    deterministic("fic");
                  // trace is replayed from a previously recorded one.

  content:      // transaction content caracteristics
    cache("ptable",exponential(16));
                  // cache type generation (non-blocking writes)
                  // size is modeled with an exponential law of parameter 16

  duration:    // phase duration
    deterministic("fic"));
                  // phases duration are store in a file
  }

sequencer{      // phase switch behavior
round(10);      // round robin, repeated 10 times
}
```

FIGURE 4.5
MPTG configuration file example.

also evaluate the stability of a given architecture by investigating small changes in the configuration, for instance, the parameters of stochastic models imply small or large changes in the performance.

A generic traffic generator has been written, once for all, for the SocLib environment. This traffic generator is used as a standard IP during simulations, and provides a master VCI interface. Transactions are generated by MPTG according to a phase description file, and a sequencer is in charge of switching between phases. Each phase consists either of a replay of a recorded trace or of a stochastic model, with parameters adjusted by the fitting procedure. These traffic patterns can be described in sequence. These sequences will be used during the next runs of the simulation. Figure 4.5 illustrates such a configuration. The entry point of a configuration is the sequencer part that will schedule the different phases of the traffic. Each phase is then described in the file using its traffic shape and the associated packet size and address (destination among the IPs on the NoC).

Designer's choices made at this stage for the MPTG configuration can be categorized using the following points, also illustrated in the configuration file presented in Figure 4.5.

1. **Timing modeling.** We distinguish two types of placement of trans-actions in time, as already mentioned in Section 4.2.3. On one hand the designer can choose to model the delay $D(k)$, on the other hand they can model the aggregated throughput time series $W_\delta(i)$. This choice depends on the context and purpose of the traffic generation. Using the aggregated throughput, one loses specific information concerning the time lag between transactions but the traffic load (on a scale of time exceeding the size of the window δ) is respected. If we choose accuracy over aggregated throughput then two sub-groups will be considered to be independent: addresses, orders, and size $[A(k), C(k), S(k)]$ on one hand and aggregated throughput $[W_\delta(i)]$ on the other.

2. **Content modeling.** Once the time modeling for transactions has been decided, the designer must model the content of transactions (address, command, and size). We have defined different types of modeling to handle different situations.

   • **Random.** In this mode, each element (address, control, time, and size) is random, hence independent of the others. This can be used for generating customizable random load on the network.

   • **Cache.** In this mode, the size of the read requests is constant (equal to the size of a line cache). There is a mode for instruction on cache mixed with data cache and for data cache only.

   • **Instruction cache.** This method is specific to an instruction cache. It contains instruction specificities, meaning that accesses are only read requests of the size of a cache line.

3. **Phase duration modeling.** A phase may appear several times in a trace, therefore it is necessary to characterize the size and number of transactions for each phase.

4. **Order of phases.** This stage involves the configuration of the sequencer to choose the sequence of phases. It can basically play a given sequence of phases, or can randomly shuffle them.

On top of the traffic content, the MPTG must also define modes for memory access. Let us recall that one of the objectives of our traffic generation environment is to be able, from a reference trace collected with a simple interconnection, to generate traffic for a platform exhibiting an arbitrary interconnect. To do so, we have to prove that the communication scheme is not affected by the communication latency. From the point of view of the component, it means that communications will be the same regardless of the latency of the interconnect. In the general case of a CPU with a cache, we cannot guarantee that, because the content of transactions [$A(k)$, $C(k)$, and $S(k)$ series] may be affected by the latency of the network. This is especially due to the presence of the write buffer. The behavior of such buffer may, in some cases, cause modifications in the size of transactions sent in the network, depending on the latency, especially in the case of large sequences of consecutive writes (zero initialization of a portion of the memory for instance).

This is why we use the time $D(k)$, which relates to the receipt of the request (so that it is independent of the network latency), instead of the time between two successive transactions. However, the problem of the recovery of calculations and communications remains. We must be able to determine if the delay $D(k)$ is a time during which the component is awaiting a reply (the component is blocked waiting for the response), or if it is a time during which the component keeps running, and thus may produce new communications. This led us to define different operating modes for the traffic generator, described hereafter.

- **Blocking requests**. In this mode, regardless of the order, the traffic generator emits a burst of type $C(k)$ to address $A(k)$, and of size $S(k)$ bus-words. Once the response is received, the traffic generator waits $D(k)$ cycles before issuing the next transaction [$T(k+1)$] on the network. This characterizes a component that is blocked (pending) when making a request.

- **Nonblocking requests**. In this mode, regardless of the order, the traffic generator emits a burst of type $C(k)$ to address $A(k)$ with a size of $S(k)$ words. Once the $S(k)$ words have been sent, the traffic generator restarts after $D(k)$. Upon receipt of the answer, if $D(k)$ is in the past, then the next request is sent immediately, otherwise we wait until $D(k)$ is reached. This allows modeling of a data-flow component (e.g., hardware accelerators) that is not blocked by communications. It is likely to be used for processor traffic. We included it for sake of generality.

- **Blocking/nonblocking read and write**. We can also specify, more precisely, if read transactions and/or write transactions are blocking or not. For example, a write-through cache is not blocked by writes (the processor keeps on running). However, a processor reading block must wait before continuing its execution. The mode "nonblocking writes, blocking reads" acts as a good approximation of the behavior of a cache.

- **Full data-flow mode**. To emulate the traffic of data flow components, we have finally established a communication mode in which only the requests are considered (the arrival of the answer is not taken into account). In this mode, the traffic generator issues a request, waits $D(k)$ cycles, and makes the following request, without concern for the answer.

The definition of these operating modes allows us, without loss of generality, to be able to deal with different types of SoC platforms.

4.3.2.1 Key Features of the MPTG Environment

As a summary, we recall the key features of the MPTG environment.

- MPTG is a fully integrated, fast, and flexible NoC performance evaluation environment.
- It includes many traffic generation capacities, from trace replay to the use of advanced stochastic processes models.
- From an initial trace obtained with the simplest interconnection, a configuration file can be used extensively to evaluate the performance of any interconnection under realistic traffic patterns.

4.3.3 Experimental Analysis of NoC Traffic

In this section, we present some experimental results on processor traffic analysis and generation. We study simulation speedup, phase decomposition, and the presence of LRD.

4.3.3.1 Speedup

To evaluate the speedup of using a traffic generator instead of real IPs, we built several platforms with different number of processors and different network sizes. The results are reported in Table 4.3. We also compare our speedup factor with a traffic generation environment [27] that performs smart replay of a recorded trace.

The reference simulation time used for speedup ("S" columns) computation is the "MIPS without VCD" (processor simulation without recording the

TABLE 4.3

Speedup of the Simulation: Simulation Time in Seconds and Speedup Factor (S Columns) for Various Platforms and Traffic Generation Schemes

	Number of Processors						
	1		2		3		4	
	Mesh Size						
	0 × 0		2 × 2		3 × 3		4 × 4	
	Time	S	Time	S	Time	S	Time	S
MIPS without VCD	36.1	1	249.5	1	477.5	1	1261.3	1
MIPS with VCD	59.4	0.61	279.9	0.89	559.8	0.85	1263.8	0.95
MPTG replay	15.9	2.27	177.2	1.41	344.5	1.39	804.0	1.57
MPTG 1 phase (sto.)	19.9	1.82	177.4	1.41	337.1	1.42	790.8	1.59
MPTG 10 phase (sto.)	19.9	1.81	177.2	1.41	337.6	1.41	790.0	1.60
MPTG 1 phase (lrd)	28.8	1.25	180.2	1.39	364.1	1.31	806.5	1.56
MPTG 10 phase (lrd)	29.1	1.24	184.0	1.36	341.5	1.40	826.4	1.53
Mahadevan [27]	—	2.15	—	2.64	—	2.60	—	3.05

VCD trace file) configuration. The speedup factor for "MIPS with VCD" is less than one because recording the trace takes a fair amount of time. The simulation speedup is never greater than 2.27, which is obtained with no interconnection ("0 × 0" mesh). However, the speedup increases with the number of processors; it means, as expected, that large platforms will benefit more from traffic generators speedup than small ones. One can further note that the VCD recording impact decreases for large platforms and even becomes negligible for a "4 × 4" mesh. Generation of stochastic processes ("sto." and "lrd" lines) does not have a big impact on simulation speedup, which means that reading values from a file and generating random numbers (even LRD processes) is almost equally costly in terms of computation time. The impact of the number of phases is also very small.

Speedup factors obtained are of the same order of magnitude as the ones obtained by Mahadevan et al. [27], and are quite small. The fact is that most of the simulation time is spent in the core simulation engine and in the simulation of the interconnection system, which cannot be reduced. Note that our conclusion is opposite to Mahadevan et al. which claims a noteworthy speedup factor. On the contrary, we found that the speedup is too small to be useful for designers and we believe that the real interest of a traffic generation environment lies in its flexibility (various generation modes, easy to configure, etc.). This will be illustrated in the following paragraphs.

4.3.3.2 Simulation Setup

For the initial trace collection, we use a simple platform (Figure 4.6) in order to truly characterize the traffic of the triplet (implementation/processor/cache). If we study communications on a more complex platform, the traffic of the processor is influenced by other IPs and NoC configuration (topology, routing protocol, etc.) as already discussed in Section 4.3. This simple platform includes an MIPS r3000 processor (associated with instruction and data cache),

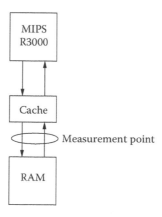

FIGURE 4.6
Simulation platforms for initial trace collection.

TABLE 4.4

Inputs Used in the Simulations

App.	Input
MPEG-2	2 images from a clip (176 × 144 color pixels)
MP3	2 frames from a sound (44.1 kHz, 128 kbps)
JPEG2000	"Lena" picture (256 × 256)

directly connected to a memory holding all necessary data. Applications with input stimuli are reported in Table 4.4

Next, to validate our environment, we run simulations on a more realistic platform shown in Figure 4.7. This platform includes five memories, a terminal type (TTY) as output peripheral, an MIPS processor executing the application, and a background traffic generator (BACK TG) used to introduce contention in the network. During design space exploration, the MIPS processor is replaced by a traffic generator producing traffic fitted to the reference trace. The simulation of the platform of Figure 4.7 (with the MIPS processor) is only used to check whether the MPTG traffic corresponds precisely to the MIPS traffic.

4.3.3.3 Multiphase

We have processed each traffic trace with the segmentation algorithm described in Section 4.2.4, using delay as the representative element and for different number of phases (k). The size of intervals is set as $L = 5000$ transactions. The choice of k is a trade-off between statistical accuracy (we need a large interval for statistical estimators to converge) and phase grain (we need many intervals to properly identify traffic phases). Figure 4.8(b), (c), and (d)

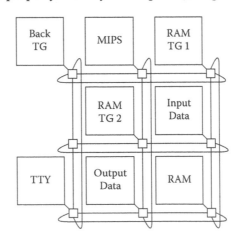

FIGURE 4.7

Simulation platforms for MPTG validation including five memories, a terminal type (TTY), a MIPS processor, and a background traffic generator (BACK TG) used to introduce contention in the network.

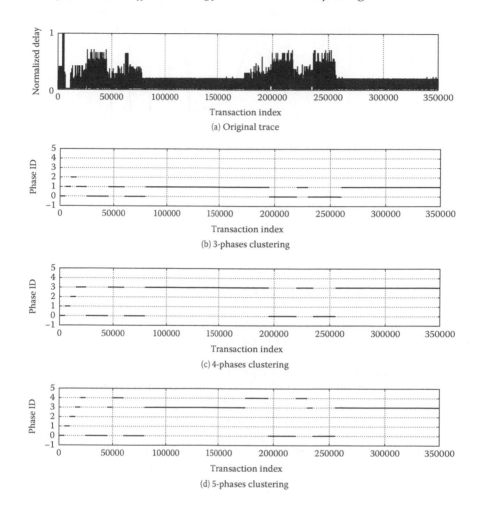

FIGURE 4.8
Phases discovered by our algorithm on the MP3 traffic trace using the delay, for different phase numbers.

show the results for various number of phases. One can see that the algorithm finds the analogy between the two frame processing, and identifies phases inside each of them. The segmentation appears to be valid and pertinent. The segmentation is done with mean and variance as representative vectors. So we expect that each identified phase is stationary, likely to be processed by a stochastic analysis.

4.3.3.4 Long-Range Dependence

We illustrate here how the presence of LRD in embedded software can be evidenced. To this end, we compute aggregated throughput time series as the number of flits sent in consecutive time-windows of size 100 cycles. This time

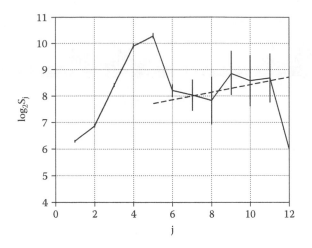

FIGURE 4.9
LD of the traffic trace corresponding to the MPEG-2 and MP3 implementations. $\hat{H} = 0.56$.

scale allows for a fine grain analysis of the traffic. For each application, we comment on the LD presented in Figures 4.9, 4.10, and 4.11.

- **MPEG-2** (Figure 4.9). The shape of the LD does not exhibit evidence for LRD. Indeed the estimated value for the Hurst parameter, $H = 0.56$, indicates that LRD is not present in the trace ($H = 0.5$ means no LRD). In this case, an IID (independent identically distributed) process would be a good approximation of the traffic. One can note a peak around scale 2^5, meaning that a recurrent operation with this periodicity is present in the algorithm, which might have an

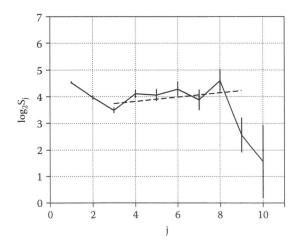

FIGURE 4.10
LD of the traffic trace corresponding to the MP3 implementation. $\hat{H} = 0.58$.

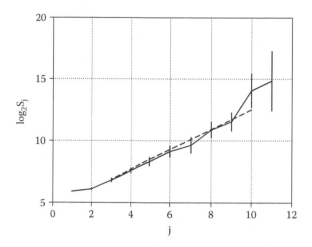

FIGURE 4.11
LD of the traffic trace corresponding to the JPEG2000 implementation. $\hat{H} = 0.89$.

impact on network contention. Such behavior could be captured by an ARMA process [6], for instance. It is interesting to note that this software implementation of MPEG-2 does not exhibit LRD whereas the hardware implementation does [14].

- **MP3** (Figure 4.10). Similar to the MPEG-2 implementation, no trace of LRD can be found in this case: the estimated Hurst parameter is close to 0.5. The other parts of the communication trace do not manifest any LRD either.

- **JPEG2000** (Figure 4.11). For this application, the traffic trace exhibits a strong nonstationarity, so that the trace must be split in rather short parts for the analysis. In some of these parts, corresponding specifically to the *Tier-1* entropic decoder of the JPEG2000 algorithm, LRD is present with an estimated Hurst parameter value between 0.85 and 0.92 (depending on the parts). In the other parts of the algorithm, no LRD could be evidenced.

We can conclude from these experiments that LRD is not an ubiquitous property of the traffic produced by a processor associated with a cache executing a multimedia application. In some parts of JPEG2000 LRD is present; however, it is combined with periodicity effects that may have an equivalent impact on the NoC performance. In this case, short-range dependent models such as ARMA [6] could be used instead of the LRD ones.

4.3.4 Traffic Modeling Accuracy

In this section, we discuss the accuracy evaluation of a traffic generator. The main idea is to indicate the difference between the traffic traces obtained from

TABLE 4.5

Accuracy of MPTG: Error (in Percent) on Various Metrics with
Respect to the Reference MIPS Simulation (NoC Platform)

Config.	Delay	Size	Cmd	Throughput	Latency
replay	1.153	0	0	0.197	0.117
random	41.278	75.242	7.709	102.316	27.825
1 phase	18.604	14.759	6.256	12.696	10.086
3 phases	17.194	8.169	3.255	6.212	0.767
5 phases	14.772	3.239	1.210	5.651	0.626

the processors' simulation and the traffic traces obtained from traffic generators. It is clear that one should not look at global metrics such as the average
delay or the average throughput. This would not highlight the interest of the
multiphase approach. As such, we define an accuracy measure by computing
the mean evolution of each transaction's element (delay, size, command, and
throughput). The mean evolution is defined as the average value of the series,
computed in consecutive time windows of size L.

To summarize the results we define the *error* as the mean of absolute values
of relative differences between two mean evolutions. Let $M_{ref}(i)$ be the mean
evolution of some element for the reference simulation. Further, let $M(i)$ be
the evolution of the same element for another simulation, and finally let n
be the number of points of both functions. The error (in percent) is: $Err =
\frac{1}{n} \sum_i |M_{ref}(i) - M(i)|/M_{ref}(i) * 100$. Note that this is a classical signal processing
technique to evaluate the distance between two signals. Furthermore, we
define the cycle error as the relative difference between numbers of simulated
cycles.

To illustrate those metrics, Table 4.5 shows accuracy results on the NoC
platform and the execution of the MP3 application.

As expected, the higher the phase number is, the more accurate the simulations are. In particular, the error on latency becomes very low when the
number of phases is greater than one. This is of major importance because
the latency of communications reflects the network state. It means that the
traffic generation from a network performance point of view is satisfactory
with multiphase traffic generation. Multiphase traffic generation provides
therefore an interesting trade-off between deterministic replay and random
traffic.

4.4   Related Work and Conclusion

SoC design companies usually have in-house NoC prototyping environments,
but it is very difficult to obtain information about these tools. Concerning
academic research, there are several works on NoC performance analysis

and design that use deterministic traffic generation (trace replay) [27–29]. For instance, the TG proposed by Mahadevan et al. [27] uses a trace compiler that can generate a program for a reduced instruction set processor that will replay the recorded transactions in a cycle accurate simulation without having to simulate the complete processor. This TG is sensitive to the network latency: changing network latency will produce a similar effect on the TG as on the original IP. This is an important point that is also taken into account in our environment.

An alternative solution for NoC performance analysis is to use stochastic traffic generators, as used in many environments [30–33]. However, none of these works proposes a fitting procedure to determine the adequate statistical parameters that should be used to simulate traffic. Recently, the work presented by Soteriou et al. [34] studies an LRD on-chip traffic model in detail with fitting procedures. To our knowledge, no NoC traffic study has introduced multiphase modeling. A complete traffic generation environment should integrate both deterministic and stochastic traffic generation techniques. From the seminal work of Varatkar and Marculescu [14], long-range dependence is used in on-chip traffic generators [35]. Marculescu et al. have isolated a long-range-dependent behavior in the communications between different parts of a hardware MPEG-2 decoder at the macro-block level.

Rapid NoC design is a major concern for next generation MPSoC design. In this field, processor traffic emulation is a real bottleneck. This chapter has investigated many issues related to the sizing of NoC. In particular it insists on the fact that a serious statistical toolbox must be used to generate realistic traffic patterns on the network.

References

[1] B. Calder, G. Hamerly, and T. Sherwood. Simpoint. Online: http://www.cse. ucsd.edu/~calder/simpoint/, April 2001.

[2] A. Scherrer. *Analyses statistiques des communications sur puces*. PhD thesis, ENS Lyon, LIP, France, Dec. 2006.

[3] J. Archibald and J. L. Baer. Cache coherence protocols: Evaluation using a multi-processor simulation model. *ACM Transactions on Computer Systems* 4. (November 1996): 273–298.

[4] R. H. Katz, S. J. Eggers, D. A. Wood, C. L. Perkins, and R. G. Sheldon. Implementing a cache consistency protocol. In *Proc. of 12th Annual International Symposium on Computer Architecture*, 276–283. Boston, MA: IEEE Computer Society Press, 1985.

[5] R. Jain. *The Art of Computer Systems Performance Analysis*. New York: John Wiley & Sons, 1991.

[6] P. J. Brockwell and R. A. Davis. *Time Series: Theory and Methods*, 2ed. Springer Series in Statistics. New York: Springer, 1991.

[7] T. Sherwood, E. Perelman, G. Hamerly, S. Sair, and B. Calder. Discovering and exploiting program phases. *IEEE Micro* 23 (2003) (6): 84–93.

[8] A. Scherrer, A. Fraboulet, and T. Risset. Automatic phase detection for stochastic on-chip traffic generation. In *CODES+ISSS*, 88–93, Seoul, South Korea, Oct. 2006.

[9] J. MacQueen. Some methods for classification and analysis of multivariate observations. In *Berkeley Symposium on Mathematical Statistics and Probability*, 281–297, Berkeley, CA, 1967.

[10] D. Pelleg and A. Moore. X-means: Extending k-means with efficient estimation of the number of clusters. In *International Conference on Machine Learning*, 727–734, San Francisco, CA, 2000.

[11] A. Scherrer, N. Larrieu, P. Borgnat, P. Owezarski, and P. Abry. Non-Gaussian and long memory statistical characterisations for Internet traffic with anomalies. *IEEE Transactions on Dependable and Secure Computing (TDSC)* 4 (2007) (1): 56–70.

[12] V. Paxon and S. Floyd. Wide-area traffic: The failure of Poisson modeling. *ACM/IEEE Transactions on Networking* 3 (June 1995) (3): 226–244.

[13] W. E. Leland, M. S. Taqqu, W. Willinger, and D. V. Wilson. On the self-similar nature of ethernet traffic (extended version). *ACM/IEEE Transactions on Networking*, 2 (Feb. 1994) (1): 1–15.

[14] G. Varatkar and R. Marculescu. On-chip traffic modeling and synthesis for MPEG-2 video applications. *IEEE Transactions on Very Large Scale Integration (VLSI) Systems*, 12 (2004) (1): 108–119.

[15] K. Park and W. Willinger, ed. *Self-Similar Network Traffic and Performance Evaluation*. New York: John Wiley & Sons, 2000.

[16] A. Erramilli, O. Narayan, and W. Willinger. Experimental queueing analysis with long-range dependent packet traffic. *ACM/IEEE Transactions on Networking*, 4 (1996) (2): 209–223.

[17] P. Abry and D. Veitch. Wavelet analysis of long-range dependent traffic. *IEEE Transaction on Information Theory*, 44 (Jan. 1998) (1): 2–15.

[18] J. M. Bardet, G. Lang, G. Oppenheim, A. Philippe, and M. S. Taqqu. *Long-range Dependence: Theory and applications*, chapter Generators of Long-range Dependent Processes: A Survey, 579–623. Birkhäuser, 2003.

[19] K. M. Chandy and J. Misra. Distributed simulation: A case study in design and verification of distributed programs. *IEEE Transaction Software Engineering*, 5 (1979) (5): 440–452.

[20] K. D. Cooper and L. Torczon. *Engineering a Compiler*. Morgan Kaufmann, 2004.

[21] The SPIRIT consortium. Enabling innovative IP re-use and design automation. Online: http://www.spiritconsortium.org/, 2008.

[22] Computer Science Laboratory of Paris IV. Soclib simulation environment. Online: http://soclib.lip6.fr/, 2006.

[23] OCP-IP. Online: http://www.ocpip.org/socket/ocpspec/, 2001.

[24] VSI Alliance. Virtual component interface standard. Online: http://www.vsi.org/library/specs/summary.html, April 2001.

[25] MEDEA+. *SPIN: A Scalable Network on Chip*, Nov. 2003.

[26] F. Pétrot and P. Gomez. Lightweight implementation of the posix threads API for an on-chip MIPs multiprocessor with VCI interconnect. In *Proc. of Design Automation and Test in Europe (DATE 03) Embedded Software Forum*, 51–56, Munchen, Germany 2003.

[27] S. Mahadevan, F. Angiolini, M. Storgaard, R. Grøndahl Olsen, Jens Sparsø, and Jan Madsen. A network traffic generator model for fast network-on-chip simulation. In *DATE 05*, 780–785, 2005.

[28] N. Genko, D. Atienza, G. De Micheli, J. M. Mendias, R. Hermida, and F. Catthoor. A complete network-on-chip emulation framework. In *DATE 05*, 246–251, Munchen, Germany 2005.

[29] M. Loghi, F. Angiolini, D. Bertozzi, L. Benini, and R. Zafalon. Analyzing on-chip communication in a MPSOC environment. In *DATE 04*, 20752, Paris, France 2004.

[30] D. Wiklund, S. Sathe, and D. Liu. Network on chip simulations for benchmarking. In *IWSOC*, 269–274, Banff, Alberts 2004.

[31] R. Thid, M. Millberg, and A. Jantsch. Evaluating NoC communication backbones with simulation. In *21th IEEE Norchip Conference*, Riga, Latvia, November 2003.

[32] K. Lahiri, A. Raghunathan, and G. Lakshminarayana. LOTTERYBUS: A new high-performance communication architecture for system-on-chip designs. In *Design Automation Conference*, 15–20, 2001.

[33] Santiago Gonzalez Pestana, Edwin Rijpkema, Andrei Rădulescu, Kees Goossens, and Om Prakash Gangwal. Cost-performance trade-offs in networks on chip: A simulation-based approach. In *DATE 04*, 20764, Paris, France 2004.

[34] V. Soteriou, H. Wang, and L. S. Peh. A statistical traffic model for on-chip interconnection networks. In *International Conference on Measurement and Simulation of Computer and Telecommunication Systems (MASCOTS '06)*, Monterey, CA, September 2006.

[35] A. Hegedus, G.M. Maggio, and L. Kocarev. A ns-2 simulator utilizing chaotic maps for network-on-chip traffic analysis. In *ISCAS*, 3375– 3378, Kobe, Japan, May 2005.

[36] A. Scherrer, A. Fraboulet, and T. Risset. Generic multi-phase on-chip traffic generator. *In ASAP*, 23–27, Steamboat Springs, CO, September 2006.

5

Security in Networks-on-Chips

Leandro Fiorin, Gianluca Palermo, Cristina Silvano,
and Mariagiovanna Sami

CONTENTS

5.1   Introduction

As computing and communications increasingly pervade our lives, security
and protection of sensitive data and systems are emerging as extremely important issues. This is especially true for embedded systems, often operating
in nonsecure environments, while at the same time being constrained by such
factors as computational capacity of microprocessor cores, memory size, and
in particular power consumption [1–3]. Due to such limitations, security solutions designed for general purpose computing are not suitable for this type
of systems.

At the same time, viruses and worms for mobile phones have been reported
recently [4], and they are foreseen to develop and spread as the targeted systems increase in offered functionalities and complexity. Known as malware,
these malicious software are currently able to spread through Bluetooth connections or MMS (Multimedia Messaging Service) messages and infect recipients' mobile phones with copies of the virus or the worm, hidden under the
appearance of common multimedia files [5,6]. As an example, the worm family *Beselo* operates on devices based on the operating system (OS) Symbian
S60 Second Edition [7]. It is able to spread via Bluetooth and MMS as Symbian
SIS installation files. The SIS file is named with MP3, JPG, or RM extensions
to trick the recipient into thinking that it is a multimedia file. If the phone user
attempts to open the file, the Symbian OS will recognize it as an installation
file and will start the application installer, thereby infecting the device.

In the context of the overall embedded System-on-Chip (SoC)/device security, security-awareness is therefore becoming a fundamental concept to be
considered at each level of the design of future systems, and to be included as
good engineering practice from the early stages of the design of software and
hardware platforms. In fact, an attacker is more likely to address its attack
to weak points of the system instead of trying to break by brute force some
complex cryptographic algorithms or secure transmission protocols in order to access/decrypt the protected information. Networks-on-Chips (NoCs)
should be considered in the secure-aware design process as well. In fact, the
advantages in term of scalability, efficiency, and reliability given by the use of
such a complex communication infrastructure may lead to new weaknesses
in the system that can be critical and should be carefully studied and evaluated. On the other hand, NoCs can contribute to the overall security of the
system, providing additional means to monitor system behavior and detect
specific attacks [8,9]. In fact, communication architectures can effectively react
to security attacks by disallowing the offending communication transactions,
or by notifying appropriate components of security violations [10].

The particular characteristics of NoC architectures make it necessary to
address the security problem in a comprehensive way, encompassing all the
aspects ranging from silicon-related to network-specific ones, both with respect to the families of attacks that should be expected and to the protective
countermeasures that must be created. To provide a guide along such lines, we

analyze and present security solutions proposed to counteract security threats at three different complementary levels of the design. We first overview typical attacks that could be carried out against an embedded system, focusing in particular on those exploiting intrinsic characteristics of the communication subsystem and NoC implementation. While the first subset of attacks is typically targeted at chip implementations, and focuses on physical characteristics, the second subset in past years has targeted networked solutions, in our case, the designer should be aware of the dangers provided by their composition. In fact, NoC architectures may allow unauthorized and possibly malicious attacks to on-chip storage due to the sharing of such storage areas among different IPs accessing it through the on-chip network. The problem of data protection is then discussed, outlining on-chip solutions to counteract attacks aiming at obtaining illegal access to protected regions of shared memories. Therefore, NoC security for reconfigurable systems is outlined, approaching the problem from the point of view of the global system. Finally, focus is on "physical" types of attacks: protection from side-channel attacks and methods to securely exchange cryptographic keys within and outside the NoC-based system are analyzed.

5.2 Attack Taxonomy

Adding specific security features to a system implies additional costs in the design stage and during the lifetime of the devices, respectively, in terms of modifications in design flow and in the need of additional hardware and software modules, as well as in performance and power consumption increase [1]. Therefore, it is mandatory to understand the requirements in terms of security of the system, that is, which security violation the system will be able to efficiently counteract. This section overviews typical attacks that could be carried out against an embedded system, providing a classification in terms of the agent used to perform the attack and its targets. It also discusses various types of security threats, namely, those exploiting software, physical and invasive techniques, and side channels techniques. After reviewing the most likely types of general attacks brought against SoCs, special attention will be given to those that may exploit the intrinsic characteristics of the communication system in an SoC based on an NoC.

5.2.1 Attacks Addressing SoCs

Figure 5.1 shows a possible classification of the attacks, in general, addressing embedded systems [11]. The given classification is based on the type of agent used to perform the attacks. One or more types of agents can be employed by a malicious entity trying to achieve its objectives on the addressed system, and can cause problems in terms of privacy of information, integrity of data and code, and availability of the system's functionalities.

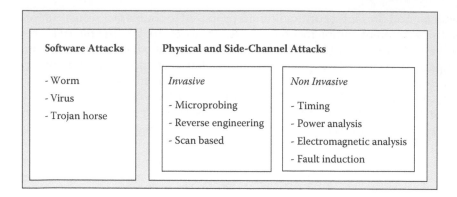

FIGURE 5.1
Attacks on embedded systems.

5.2.1.1 Software Attacks

Software attacks exploit weaknesses in system architecture, through malicious software agents such as viruses, trojan horses, and worms. These attacks address pitfalls or bugs in code, such as in the case of attacks exploiting buffer overflow or similar techniques [12]. As embedded systems software increase in complexity and functionalities offered, they are foreseen to become an ideal target for attacks exploiting software agents. Viruses for mobile phones have been reported in recent years [4], and similar attacks are likely to be extended to embedded devices in automotive electronics, domestic applications, networked sensors, and more generic pervasive applications. Due to the cheap and easy infrastructure needed by the hacker to perform a malicious task, software attacks represent the most common source of attack and the major threat to face in the challenge to secure an embedded system. Moreover, the possibility of updating functionalities and downloading new software applications, although increasing the flexibility of the system, also increases its vulnerability to external attackers and maliciously crafted application extensions. An additional challenge is also represented by the extended connectivity of embedded devices [2], which implies an increase in the number of security threats that may target the system, physical connections to access the device no longer being required.

Typical embedded system viruses will spread through the wireless communication channels offered by the device (such as Bluetooth) and install themselves in unused space in Flash ROM and EEPROM memories, immune to rebooting and reinstallation of the system software. Malicious software is in this way almost not visible to other applications on the system, and is capable of disabling selected applications, including those needed to disinfect it [6].

5.2.1.2 Physical Attacks

Physical attacks require physical intrusion into the system at some levels, in order to directly access the information stored or flowing in the device, modify it or interfere with it. These types of attacks exploit the characteristic implementation of the system or some of its properties to break the security of the device. The literature usually classifies them as *invasive* and *noninvasive* [13].

Invasive attacks require direct access to the internal components of the system. For a system implemented on a circuit board, inter-component communication can be eavesdropped by means of probes to retrieve the desired information [1]. In the case of SoC, access to the internal information of the chip implies the use of sophisticated techniques to depackage it and the use of microprobes to observe internal structure and detect values on buses, memories, and interfaces. A typical microprobing attack would employ a probing station, used in the manufacturing industry for manual testing of product line samples, and consisting of a microscope and micromanipulators for positioning microprobes on the surface of the chip. After depackaging the chip by dissolving the resin covering the silicon, the layout is reconstructed using in combination the microscope and the removal of the covering layers, inferring at various level of granularity the internal structure of the chip. Microprobes or e-beam microscopy are therefore used to observe values inside the chip. The cost of the infrastructure makes microprobing attacks difficult. However, they can be employed to gather information on some sample devices (e.g., information on the floorplan of the chip and the distribution of its main components) that can be used to perform other types of noninvasive attacks.

Noninvasive attacks exploit externally available information, unintentionally leaking from the observed system. Unlike invasive attacks, the device is not opened or damaged during the attack. There are several types of noninvasive attacks, exploiting different sources of information gained from the physical implementation of a system, such as power consumption, timing information, or electromagnetic leaks.

Timing attacks were first introduced by Kocher [14]. Figure 5.2 shows a representation of a timing attack. The attacker knows the algorithm implementation and has access to measurements of the inputs and outputs of the secure system. Its goal is to discover the secret key stored inside the secure system. The attacker exploits the observation that the execution time of computations is data-dependent, and hence secret information can be inferred from

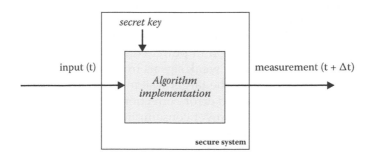

FIGURE 5.2
Representation of the timing attack.

its measurement. In those attacks, the attacker observes the time required by the device to process a set of known inputs with the goal of recovering a secret parameter (e.g., the cryptographic key inside a smart-card). The execution time for hardware blocks implementing cryptographic algorithms depends usually on the number of '1' bits in the key. Although the number of '1' bits alone is not enough to recover the key, repeated executions with the same key and different inputs can be used to perform statistical correlation analysis of timing information and therefore recover the key completely. Delaying computations to make them a multiple of the same amount of time, or adding random noise or delays, increases the number of measurements required, but does not prevent the attack. Techniques exist, however, to counteract timing attacks at the physical, technological, or algorithmic level [13].

Power analysis attacks [15] are based on the analysis of power consumption of the device while performing the encryption operation. Main contributions to power consumption are due to gate switching activity and to the parasitic capacitance of the interconnect wires. The current absorbed by the device is measured by very simple means. It is possible to distinguish between two types, of power analysis attacks: simple power analysis (SPA) and differential power analysis (DPA).

SPA involves direct interpretation of power consumption measurements collected during cryptographic operations. Observing the system's power consumption allows identifying sequences of instructions executed by the attacked microprocessor to perform a cryptographic algorithm. In those implementations of the algorithm in which the execution path depends on the data being processed, SPA can be used directly to interpret the cryptographic key employed. As an example, SPA can be used to break RSA implementations by revealing differences between multiplication and squaring operation performed during the modular exponentiation operation [15]. If the squaring operation is implemented (due to code optimization choices) differently than the multiplication, two distinct consumption patterns will be associated with the two operations, making it easier to correlate the power trace of the execution of the exponentiator to the exponent's value. Moreover, in many cases SPA attacks can help reduce the search space for brute-force attacks. Avoiding procedures that use secret intermediates or keys for conditional branching operations will help protect against this type of attack [15].

DPA attacks are harder to prevent. In addition to the large-scale power variations used in SPA, DPA exploits the correlation between the data values manipulated and the variation in power consumption. In fact, it allows adversaries to retrieve extremely weak signals from noisy sample data, often without knowing the design of the target system. To achieve this goal, these attacks use statistical analysis and error-correction statistical methods to gain information about the key. The power consumption of the target device is repeatedly and extensively sampled during the execution of the cryptographic computations. The goal of the attacker is to find the secret key used to cipher the data at the input of the device, by making guesses on a subset of the key to discover and calculating the values of the processed data in the

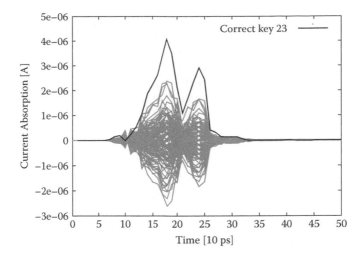

FIGURE 5.3
Power traces of a DPA attack on a Kasumi S-box. (From Regazzoni, F. et al. In *Proc. of International Symposium on Systems, Architectures, Modeling, and Simulation* (*SAMOS VII*), Somos, Greeca, July 2007).

point of the cryptographic algorithm selected for the attack. Power traces are collected and divided into two subsets, depending on the value predicted for the bit selected. The differential trace, calculated as the difference between the average trace of each subset, shows spikes in regions where the computed value is correlated to the values being processed. The correct value of the key can thus be identified from the spikes in its differential trace. As an example, Figure 5.3 shows a simulation of a DPA attack on a Kasumi S-box implemented in CMOS technology [16]. The Kasumi block cipher is a Feistel cipher with eight rounds, with a 64-bit input and a 64-bit output, and a secret key with a length of 128 bits. Kasumi is used as a standardized confidentiality algorithm in 3GPP (3rd Generation Partnership Project) [17]. In the figure it is possible to note how the differential trace of the correct key (plotted in black) presents the highest peak, being therefore clearly distinguishable from the remaining ones and showing a clear correlation to the values processed by the block cipher. For a more detailed discussion of DPA attacks, see Kocher et al. [15].

Electromagnetic analysis (EMA) attacks exploit measurements of the electromagnetic radiations emitted by a device to reveal sensitive information. This can be performed by placing coils in the neighborhood of the chip and studying the measured electromagnetic field. The information collected can therefore be analyzed with simple analysis (SEMA) and differential analysis (DEMA) or more advanced correlation attacks. Compared to power analysis attacks, EMA attacks present a much more flexible and challenging measurement phase (in some cases measurement can be carried out at a significant distance from the device—15 feet [13]), and the provided information offers a wide spectrum of potential information. A deep knowledge of the layout makes the attack much more efficient, allowing the isolation of the region around which the measurement should be performed. Moreover, depackaging the chip will avoid perturbations due to the passivation layers.

Fault induction attacks exploit some types of variations in external or environmental parameters to induce faulty behavior in the components to interrupt the normal functioning of the system or to perform privacy or precursor attacks. Faulty computations are sometimes the easiest way to discover the secret key used within the device. Results of erroneous operations and behavior can constitute the leak information related to the secret parameter to be retrieved. Faults can be induced acting on the device's environment and putting it in abnormal conditions. Typical fault induction attacks may involve variation of voltage supply, clock frequency, operating temperature, and environmental radiations and light. As an example, refer to Boneh et al. [18], where the use of the Chinese Remainder Theorem to improve performances in the execution of RSA is exploited to force a fault-based attack. Differential fault analysis (DFA) has also been introduced to attack Data Encryption Standard (DES) implementations [13].

Scan based channel attacks exploit access to scan chains to retrieve secret information stored in the device. The concept of scan design was introduced over 30 years ago by Williams and Eichelberger [19] with the basic aim of making the internal state of a finite state machine directly controllable and observable. To this end, all (D-type) flip-flops in the FSM are substituted by master-slave devices provided with a multiplexer on the data input, and when the FSM is set to test mode they are connected in a "scan path," that is, a shift register accessible from external pins. This concept has been extended for general, complex chips (and boards) through the JTAG standard (IEEE 1149.1) that allows various internal modes for the system and makes its internal operation accessible to external commands and observation—when in test mode—through the test port. JTAG compliance is by now a universal standard, given the complexity of the testing SoC. Internal scan chains are connected to the JTAG interface during the packaging of the chip, in order to provide on-chip debug capability. To prevent access after the test phase, a protection bit is set by using for instance fuses or anti-fuses, or the scan chain is left unconnected. However, both techniques can be compromised allowing the attacker to access the information stored in the scan chain [20].

5.2.2   Attacks Exploiting NoC Implementations

The attacks mentioned in Section 5.2.1 were addressed basically to any type of complex architectures. We shall now focus on attacks that exploit the specific characteristics of the NoC architecture. A security-aware design of communication architectures is becoming a necessity in the context of the overall embedded device. Although the advantages brought by the use of a communication-centric approach appear clear, an exhaustive evaluation of the possible weaknesses that, in particular, may affect an NoC-based system is still an on-going topic. The increased complexity of this type of system can provide attackers with new means of inducing security pitfalls, by exploiting the specific implementation and characteristics of the communication subsystem. In addition to the attacks discussed in the previous section, several types of attack scenarios can be identified, which exploit NoC characteristics and that derive from networking rather than from chip-based attacks [8,9,21].

5.2.2.1 Denial of Service

A denial of service attack (DoS attack) is an attempt to make the target device unavailable to its intended users. Such attacks may address the overall system or some individual component, such as the communication subsystem. The aim of the attacker is to reduce the system's performances and efficiency, up to its complete stop. This type of attack reaches particular relevance in embedded systems, where reduction in the already limited amount of available resources can constitute a not negligible problem for the device and the users. Effects of a DoS attack on an NoC-based system can appear as slowing down the network transmissions, unavailability of network and/or processing and storage cores, and disruptions in the inter-core communication. Moreover, the reduced capabilities of the communication infrastructure may compromise real-time behaviors of the system.

We consider hereafter attacks impairing bandwidth (and therefore network resources) and power availability.

Bandwidth reduction attacks aim at reducing network resources available to communicating IPs causing higher latency in on-chip transmission and consequent missing of deadlines in the system behavior. Depending on the routing strategies adopted, different attack scenarios can be identified [21]:

- **Incorrect path**. Packets with erroneous paths or invalid origin and destination information are injected into the network, with the aim of routing them to a dead end and occupying transmission channels and network resources, therefore made unavailable to other valid packets.

- **Deadlock**. Packets with routing information capable of causing deadlock with respect to the routing technique adopted are introduced into the network. These packets do not reach their destination, being blocked at some intermediate resource, which in turn, as a consequence, is not available for other transmissions. NoCs implementing wormhole switching are the most likely to suffer from this type of attacks.

- **Livelock**. Livelock, as well as deadlock, is a special case of resource starvation. Packets do not reach their destinations because they enter cyclic paths.

- **Flood (bandwidth consumption)**. Aiming at saturating the network, this type of attack is performed by injecting in the network a large number of packets or network requests, such as broadcasting or synchronization messages.

Network interfaces (NIs) provide a basic filter to requests and packets injected maliciously in the network by compromised cores. However, an illegal access to NIs' configuration registers performed by an attacker may be exploited to carry out the described types of attacks. Moreover, fault induction techniques can be applied to modify information stored in such registers and cause disruptions in inter-core communication.

Data and instructions tampering represents a serious threat for the system. Unauthorized access to data and instructions in memory can compromise the execution of programs running on the system, causing it to crash or to behave in an unpredictable way. Therefore, protection of critical data represents an essential task, in particular in multiprocessor SoC, where blocks of memory are often shared among several processing units. Tampering of data and instructions in memory can be performed when a processor writes outside the bounds of the allocated memory, for instance, in the case of an attack exploiting buffer overflow techniques [12].

Draining attacks aim at reducing the operative life of a battery-powered embedded system. In fact, the battery in mobile pervasive devices represents a point of vulnerability that must be protected. If an attacker is able to drain a device's battery, for example, by having it execute energy-hungry tasks, the device will not be of any use to the user. Literature by Martin et al. and Nash et al. [22,23] presents the following three main methods by which an attacker can drain the battery of a device:

1. **Service request power attacks**. In this scenario, repeated requests are made to the victim of the attack. In our context, the victim can be the interconnection subsystem or one or more processing or storage cores. Requests that could be made and that would address the communication infrastructure may involve the establishment of connections to valid or invalid IP cores or the range of memory addresses, as well as synchronization and broadcasting of generic messages. An example of service request power attacks to processing cores is given by the repeated sending of requests to the power manager of the core to keep it in the active state [8,24].

2. **Benign power attacks**. In this kind of attack, valid but energy-hungry tasks are forced to be executed indefinitely. Ideally invisible to the users, these tasks secretly drain the energy source. The attacker provides valid data to a program or a task to make it execute continuously and consume a considerable amount of power.

3. **Malignant power attacks**. These attacks are mainly based on viruses, worms, or trojan horses maliciously installed in the device. The attack alters the OS kernel or the application binary code in such a way that the execution consumes a higher amount of energy. Malignant power attacks can be for instance performed by a compromised core sending continuous requests to the Bluetooth module. The core will keep the module continuously executing the scan of the available devices and sending requests of connection or malicious files [6].

5.2.2.2 *Illegal Access to Sensitive Information*

This type of attack aims at reading sensitive data, critical instructions, or information kept in configuration registers on unauthorized targets. Attacks carried out by using several agents can be included under this classification. Buffer overflow can be exploited to compromise a core and use its memory access rights to access the unauthorized range addresses where sensitive data, such as cryptographic keys, are stored [10]. Moreover, side channel information leaking from the device can be detected and used to retrieve secret data or pieces of code.

5.2.2.3 *Illegal Configuration of System Resources*

In this type of attack, the aim is to alter the execution or configuration of the system to make it perform tasks set by the attacker in addition to its normal duties. Attacks can be performed as a write access in secure areas to modify the behavior or configuration of the system. The attacker takes control of one or more resources of the device, and exploits it to achieve its malicious goal. A significant example, exploiting buffer overflow to reconfigure the setting of peripheral interfaces, is described by Coburn et al. [10] for an audio CODEC adopting the IEEE 1394 interface [25]. In the application presented, the CODEC is reconfigured to send unencrypted audio samples to external unauthorized users, in order to bypass Digital Right Managements (DRM) protection.

5.2.3 Overview of Security Enhanced Embedded Architectures

Although security on NoC-based systems is a relatively new research topic, several architectures have been proposed (both by academic and industrial research groups) to enhance system security for generic SoCs. This section presents an overview of the existing approaches to improve system security in the embedded environment.

Providing OS extensions and security primitives is one of the possible ways to achieve this goal, supported by the addition of dedicated hardware to the processing element [26,27]. The work discussed in *XOM Technical Information* [26] adopts a hardware implementation of an execute-only memory (XOM) that allows instructions stored in the memory itself to be executed but not otherwise manipulated. The system supports internal compartments and does not allow a process in one compartment to read data from the other compartment. Application software loaded on the machine is protected using symmetric key cryptography, and data are protected through identification tags when they are on chip, or through encryption when stored in external memory. To prevent tampering and observation of application, even in the presence of a malicious operating system, each program is assigned a unique tag that is associated with the key used to decrypt the program's code. In this way, the OS can never read data or registers that are tagged with another program's ID. In the *AEGIS* approach [27], the processor is assumed to be trusted and protected from physical attacks, so that its internal state cannot be tampered with or observed directly by physical means. On the contrary, external memory and peripherals are assumed to be untrusted and subject to observation and tampering. Therefore, their integrity and privacy is ensured by a mechanism for integrity verification and encryption. The system is protected against untrusted OSs by a security kernel that operates with higher privileges than a regular OS, or by a hardware secure context manager that verifies the core functions of the OS.

Enhanced communication architectures have been proposed to facilitate higher security in SoCs, monitoring and detecting violations, blocking attacks, and providing diagnostic information for triggering suitable responses and recovery mechanisms [10]. This can be implemented by adding specific modules to typical communication architectures such as AMBA, to monitor access to regions on the address space, configuration of peripherals, and sequences of bus transactions.

Considering typical commercial embedded platforms, ARM's approach to enabling trusted computing within the embedded world is based on the concept of the TrustZone Platform [28]. The entire TrustZone architecture can be seen as subdivided into secure and nonsecure regions, allowing the secure code and data to run alongside an OS securely and efficiently, without being compromised or vulnerable to attack. A non-secure indicator bit (NS) determines the security operation state of the various components and can only be accessed through the "Secure Monitor" processor mode, accessible only through a limited set of entry points. This mode is allowed to switch the system between secure and nonsecure states, allowing a core in the secure state to gain higher levels of privilege. With reference to the interconnection system, the AMBA AXI Configurable Interconnect supports secure-aware transactions. Transactions requested by masters are monitored by a specific TrustZone controller, which is in charge of aborting those considered illegal. Secure-aware memory blocks are supported through the AXI TrustZone memory adapter, allowing sharing of single memory cells between secure and nonsecure storage areas. A similar solution to protect memory access is provided by *Sonics* [29] in its SMART Interconnect solutions, where an on-chip programmable security "firewall" is employed to protect the system integrity and the media content passed between on-chip processing blocks and various I/Os and the memory subsystem.

It is worth noting that the use of protected transactions is also included in the specifications defined by the Open Core Protocol International Partnership (*OCP-IP*) [30]. The standard OCP interface can be extended through a layered profile to create a secure domain across the SoC and provide protection against software and some selective hardware attacks. The secure domain might include CPU, memory, I/O, etc., which requires to be secured by using a collection of hardware and software features such as secured interrupts, secured memory, or special instructions to access the secure mode of the processor.

In multiprocessor environments, protection of preinstalled applications from native applications downloaded from untrusted sources can be assured by the adoption of security domains [31,32]. A security domain is defined as an isolated execution environment prepared for a group of applications. This technique prevents illegal access to the address spaces of other security domains, limiting the maximum amount of resources that applications on the security domain may use. Virtualization [33–35] can be used to create a virtual domain for downloaded applications. Moreover, execution in virtual environments allows the avoidance of verification of downloaded software [36], a procedure expected to become too complex for future applications. A hardware-based approach for implementing isolation of applications through security domains consists in dynamically changing the number of processors within a security domain in response to application load requirements [32]. Processors are dynamically allocated for the execution of downloaded applications, which therefore run separately from the preinstalled applications.

5.3 Data Protection for NoC-Based Systems

A typical SoC multiprocessor, and in general NoC-based SoCs, often includes blocks of memories shared among multiple IPs and accessed through the on-chip network. Once more, this represents the migration to the on-chip system of solutions previously adopted for distributed systems, and creates possible (and dangerous) security loopholes. Because memory locations store the state of a system, memory-based attacks are widely used as the basis for a relevant number of the most common types of security vulnerabilities in the last 10 years [37]. In fact, unauthorized access to information in memory can compromise the execution of programs running on the system by tampering with the information stored in a selected area or cause the extraction of critical information. Memory-based attacks discussed in Section 5.2 then become critical for NoC-based SoC, and design of ad hoc solutions protecting on-chip data storage from such attacks is mandatory.

This section discusses an architectural solution that exploits NoC characteristics to protect the system from attacks aiming at obtaining illegal access to restricted areas of memory, presenting alternative implementations and the associated overhead. No assumption is made on the specific abilities of the attacker to obtain control of processing cores to illegally access the memory, it being out of the scope of the section's topic. However, it is possible to note how, without a mindful hardware design, simple software fallacies, such as the buffer overflow, can give the attacker a way to illegally access memory with low effort.

5.3.1 The Data Protection Unit

To avoid the problem of memory attacks in NoC-based multiprocessor architectectures, a module for NoC that offers services similar to those provided by

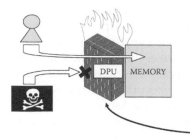

Context			Memory Address	Mask	Auth
SourceID	Role	D/I			
000	0	0	0×000A0000	0×0000FFFF	10
000	1	0	0×000A0000	0×0000FFFF	11
000	0	1	0×000A0000	0×0000FFFF	00
000	1	1	0×000A0000	0×0000FFFF	11
001	0	0	0×001A0000	0×0000FFFF	10
001	1	0	0×001A0000	0×0000FFFF	11
010	0	0	0×002A0000	0×0000FFFF	10
010	1	0	0×002A0000	0×0000FFFF	11

FIGURE 5.4
Data protection unit (DPU): basic idea.

a classical "firewall" in a data network is suggested [38]. A firewall is a dedicated module that inspects network traffic passing through it and denies or permits the passage of protocol packets, following a predefined set of rules. The module for protection of memory transactions on NoC-based multiprocessors is named data protection unit (DPU) [38]. More specifically, the DPU is a hardware module that enforces access control rules to the memory requests, specifying the way in which an IP initiating a transaction to a shared memory in the NoC can access a memory block. The partitioning of the memory into blocks allows the separation between sensitive and nonsensitive data for the different processors connected to the NoC.

Figure 5.4 shows the basic idea of the DPU. The DPU enforces the access control rules to all memory requests verifying if (authorized or not) and how (read-write, read only, write-only) an initiator in a particular context can access a memory location. As shown in Figure 5.4, the DPU uses a LUT to store the access rules. The LUT is composed of four columns: *Context*, *Memory Address*, *Mask*, and *Auth*. The *Context* column includes all the fields related to the identification of the context of the request. It includes the identification of the initiator (*SourceID*) that makes the request, its *Role* during the request (user or supervisor), and if the target of the request are Data rather than Instructions (*D/I*). In this way the DPU is able to differentiate the authorization policy on a memory block at fine grain, improving its efficiency. The *Memory Address* and the *Mask* columns are used to identify the target memory block. *Mask* is used to specify which bits of the *Memory Address* should be considered a *don't care* when identifying the memory block.

The column *Auth* represents the encoding of the request authorization. It follows the following rules:

- 00: both read and write operations are not authorized
- 01: a write operation is authorized while a read operation, not
- 10: a read operation is authorized while a write operation, not
- 11: both read and write operations are authorized

Each entry in the LUT is indexed by the concatenation of the following information derived from the memory request: the identifier of the requester

(*SourceID*), its *Role* at the time of the request, the type of the target data (data or instruction, $\overline{D}/I$)) and the target *Memory Address*.

5.3.2 DPU Microarchitectural Issues

From the architecture point of view, the choice of coupling the DPU to the memory is not a good choice. In fact, performing the rights control before each memory access implies added latency for each memory request, due to the DPU. To avoid such extra latency, the DPU has been integrated within the NI where the memory accesses are filtered through the lookup of the access rights in parallel with the protocol translations.

Figure 5.5 shows the integration of the DPU into a Multiprocessor architecture composed of 3 Initiators (μPs) and one Target (*Mem*) connected through the component of a typical NoC architecture: routers (*Rs*) and *NIs*. In this architecture, the DPU is a module embedded in the NI of the target memory (or in general of a memory-mapped peripheral) to protect, avoiding unauthorized accesses.

Figure 5.6 shows the microarchitecture details of the DPU, when it is embedded in the target NI. For this architecture, the DPU checks the header of the incoming packet to verify if the requested operation is allowed to access the target. This access control is mainly based on an LUT, where entries are indexed by the concatenation of the *SourceID*, the type of information ($\overline{D}/I$), and the starting address of the requested memory operation *MemAddr*. The number of entries in the table depends on the number of memory blocks to be protected in the system, as well as on the number of initiators. In the implementation shown in Figure 5.6, the size of the smallest memory block to be checked for the access rights is assumed to be of 4 kB. This means that all data within the same block of 4 kB have the same rights (corresponding to the 12 LSB in the memory address) and that only the 20 most significant bits of the *MemAddr* field are used for the lookup.

The LUT of the DPU is the most relevant part of the architecture and is composed of three parts.

1. A content addressable memory (CAM) [39] used for the lookup of the *SourceID* and type of data ($\overline{D}/I$).

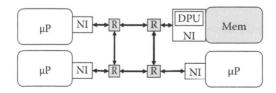

FIGURE 5.5
A simple example of a system with three initiators (μPs) and one target (*Mem*), showing the architecture using the DPU integrated at the target network interface.

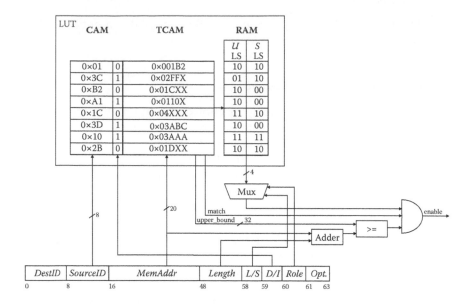

FIGURE 5.6
DPU microarchitecture integrated at the target network interface.

2. A ternary content addressable memory (TCAM) [39] used for the lookup of the *MemAddr*. With respect to the binary CAM, the TCAM is useful for grouping ranges of keys in one entry because it allows a third matching state of X (*don't care*) for one or more bits in the stored datawords, thus adding more flexibility to the search. In our context, the TCAM structure has been introduced to associate with one LUT entry memory block larger than 4 kB.

3. A simple RAM structure used to store the rights access values.

Each entry in the CAM/TCAM structure indexes a RAM line containing the access rights (*allowed*/$\overline{not\ allowed}$) for user load/store and supervisor load/store. The type of operations ($\overline{L}/S$) and its role ($\overline{U}/S$) taken from the incoming packets are the selection lines in the 4:1 multiplexer placed at the output of the RAM. Moreover, a parallel check is done to verify that the addresses involved in the data transfer are within the memory boundary of the selected entry.

If the packet header does not match any entry in the DPU, there are two possible solutions depending on the security requirements. The first one is more conservative (shown in Figure 5.6), avoiding the access to a memory block not matching any entry in the DPU LUT by using a *match* line. The second solution, less conservative, enables the access also in the case when there is no match in the DPU LUT. The latter case does not need any *match*-line and corresponds to the case when a set of memory blocks could not require any access verification.

The output-enabled line of the DPU is generated by a logic AND operation between the access right obtained by the lookup, the check on the block boundaries, and, considering the more conservative version of the DPU, the *match* on the LUT.

5.3.3 DPU Overhead Evaluation

In this section, some evaluations of the overhead introduced by the DPU architecture are presented. The synthesis and the energy estimation have been performed by using, respectively, Synopsys Design Compiler and Prime Power with 0.13 μm HCMOS9GPHS STMicroelectronics technology libraries.

Figures 5.7 shows the synthesis results in terms of delay (ns), area (mm^2), and energy (nJ) by varying the DPU entries. All the reported overhead figures are related to a working frequency for the DPU of 500 MHz, that is met for all the presented configurations, as shown in Figure 5.7(a). Figure 5.7(b) shows that the DPU area increases almost linearly with the number of entries (0.042 mm^2 for each 10 entries). This is due to the fact that the most significant area contribution is given by the CAM/TCAM included in the DPU. As expected, because the main part of the DPU is composed of a CAM/TCAM, the energy trends shown in Figure 5.7(c) by scaling the number of DPU entries are similar to those already described for the area values. Looking at Figure 5.7(b), (c) it is possible to note that, independent of the size of the DPU, the ratio energy for access-area is 1 nJ/mm^2.

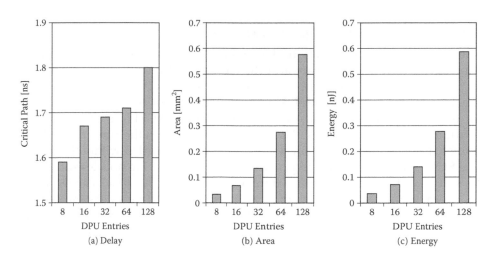

FIGURE 5.7
DPU overhead by varying the number of entries.

5.4   Security in NoC-Based Reconfigurable Architectures

The NoC paradigm offers attractive possibilities for the implementation of coarse-grained "reconfigurable architectures." Needs for reconfigurable hardware and architectures are rising in academic and industrial environments, due to several factors. A reduced time-to-market, the possibility to adapt at run-time the same platform to different applications, and, as done already for software, to fix hardware bugs after release, make the use of this technology interesting and convenient for developing new multimedia and mobile embedded devices [40]. However, it must be noticed that reconfiguration adds further possible weaknesses in terms of security, which could be exploited by an attacker to obtain control of part of the system and to have direct access to data or configuration registers. This introduces therefore a new security threat, namely, unwanted reconfiguration leading to denial of service and/or unwanted (and unpredictable, as far as the user is concerned) behavior. On the other hand, the same NoC paradigm can be employed to detect unexpected system behaviors and notify attempts of security violation. In this section, the enhancement of NoC modules for allowing security in reconfigurable systems is discussed, based on the works of Diguet et al. and Evain and Diguet [9,21].

5.4.1   System Components

The NoC solution for secure reconfigurable systems is composed of two main modules: a Security and Configuration Manager (SCM) and Secure Network Interfaces (SNIs).

5.4.1.1   Security and Configuration Manager

The Security and Configuration Manager is a dedicated core in charge of the configuration of the hardware and the communication subsystem, and collecting behaviors monitored to appropriately counteract security violations in the system. In particular, the SCM is dedicated to security control, and holds for this purpose two main tasks. In the first one, the SCM is in charge of the configuration of all the SNIs. The second task involves the run-time collection of warning messages from several monitors embedded within the SNIs. The SCM will also counteract security violations, following policies specified by software designers at design time. Depending on the attack carried out, countermeasures may involve the transmission of alert messages through a secured network communication, and/or the isolation of the compromised core by closing the corresponding SNI. For these specific and critical tasks, the SCM will represent the most sensitive resource of the system.

5.4.1.2   Secure Network Interface

Network interfaces are in charge of protocol translation between IPs and interconnection network and provide a reliable communication among the cores.

As also shown in Section 5.3, additional services can be added, in particular for security purposes. SNIs can handle several types of attacks in a distribute and dedicated way. Apart from memory and configuration registers access control, already discussed in detail in Section 5.3, SNIs can handle attack symptoms, such as the Denial of Service, and notify security alerts to the SCM. In fact, network traffic can be conveniently analyzed within the secure NI, exploiting NI proceeding delays to implement security control in parallel with data flow. Moreover, locating traffic monitoring at the NIs avoids the use of costly probes in the routers and the implementation of additional secure channels for the communication between the routers and the SCM. To separate normal data traffic between IPs from the signal used for security monitoring and configuration, distinct virtual channels and networks are used. Prioritized channels prevent normal data traffic from delaying or stopping secure service communications, in particular in the case of attempts of Denial of Service attacks.

To implement access control on the incoming traffic, the following four steps can be identified:

1. **Overflow checking.** The first packet transmitted in a transaction contains the message size, expressed in number of words. If the FIFO of the SNI initiating the transaction is not empty when the bound is reached, the FIFO is flushed and the transaction ended. Packet length is limited to the input FIFO size and the network layer control bits of the last flit are automatically set by the NI controller.

2. **Boundaries checking on local addresses.** Based on the local-based address of the transaction and on the size of the data to be transferred, a check is performed to verify that data are within the boundaries of the allowed address range for that transaction.

3. **Collection of statistics and alerts.** Data transmitted or received is monitored to discover traffic outside predefined normal behavior bounds. In case of such a detection, an alert is sent to the SCM. Available credits allocations are accumulated and compared to upper and lower bounds over a given time window.

4. **Identification of the sender.** The identification of the sender can be based on tags inserted in the packet header or the payload, which can be inserted by the NI of the processing element initiating the transaction [38]. Alternatively, paths can be used to identify request and response communications. As proposed by Evain and Diguet [21], this technique can be implemented using a routing algorithm based on the input port index of the router and from the number of turns, counter-clockwise, from the considered input port to reach the selected output port. For each connection that is set up, path information is inserted in the header. Each router crossed by the packet executes its path instruction, and then complements it with respect to its own *arity* (number of bi-directional ports of this router) through a round shift of the path information in the header.

This technique preserves routing instruction information in the packet header, allowing identification of the sender at the destination and filtering of illegal requests, as well as easy generation the backward path to the initiator. The authors provide a more detailed discussion on this subject [21].

5.4.1.3 Secure Configuration of NIs

The protocol for SNIs configuration is shown in Figure 5.8. Configuration of NIs assumes a particular relevance. In fact, if the attacker is able to modify NI's configuration registers, all security policies remain ineffective. Four phases can be identified:

1. **INIT.** At boot time, the initial hardware configuration of the system is loaded from an external ciphered memory. Both IPs and NoC cores, if implemented using FPGA technology, are configured. Ciphered configuration information is decrypted using a dedicated core. In this first phase, SoC hardware can be considered in a safe configuration.

2. **SNI.** In this phase, only read operations from the SCM to a ciphered memory containing SNI configurations, as well as prioritized communication among the SCM and the SNIs to configure NIs, are allowed. Once this phase is executed, new configurations can be dynamically performed only by the SCM.

3. **RUN.** The system is run-time monitored, based on the current hardware and software configuration. In case of security alert, SNIs can be reconfigured to counteract violations detected.

4. **DPR.** The SCM can perform run-time partial hardware reconfiguration, depending on signals received by monitors or on application requirements. Reconfiguration bitstreams are available initially or can be downloaded from a secured network connection. However, online reconfiguration management based on network download is not a trivial issue from a security perspective, and a dedicated protocol still remains to be specified.

FIGURE 5.8
Protocol used for SNIs reconfiguration.

5.4.2 Evaluation of Cost

Adopting the security solution presented in this section has a cost. Considering an FPGA implementation of a reconfigurable SoC for SetTop Box, with one SCM, seven master general purpose or dedicated processors, and 13 slave memories for data, programs, and configuration, the overhead of the secure system in term of resources occupied on the chip is around 45% [9]. Processing and storage cores are connected through a 2D mesh network composed of 4 × 3 routers with various numbers of ports. The overhead is mainly due to the additional virtual channels employed for secure transmissions, and in particular to the almost doubled dimension of the routers supporting them. However, the implementation of the secure system described slightly influences the NI's size, being the overhead associated to the implementation of additional FIFOs and registers in the NIs not significant in the overall area budget.

5.5 Protection from Side-Channel Attacks

Although SoCs offer more resistance to bus probing attacks, power/EM attacks on cores and network snooping attacks by malicious code running in compromised cores could represent a relevant problem. To avoid such a threat, architectures must be designed to protect the user's private key from attackers. Therefore, this section presents and discusses the implementation of a framework to secure the exchange of cryptographic keys within and outside the NoC-based system. Design methodologies to protect IP cores from side-channel attacks will be also discussed in this section.

5.5.1 A Framework for Cryptographic Keys Exchange in NoCs

A general SoC architecture based on NoC, required to provide support for security at several levels, is foreseen [41]. Security tasks will include the protection of sensitive data transmitted through wireless communication channels, the authentication of software applications downloaded from external sources, a secure and safe configuration of the system by service providers, and the authentication, at design time, of IP cores used within the NoC and of the NoC itself. The generic secure platform discussed in this paragraph is shown in Figure 5.9. It is composed of m secure cores ($SCore_i$ in Figure 5.9) and $n–m$ generic cores ($Core_j$). m is equal to or greater than 1, where $m = 1$ is the trivial case in which only a core is responsible for executing all the security operations of the system, that is, encryption, decryption, authentication of messages, etc. In general, a secure core is a hardware IP block implementing a dedicated encryption/decryption algorithm, such as AES or SHA-1, or in charge of executing more generic security applications.

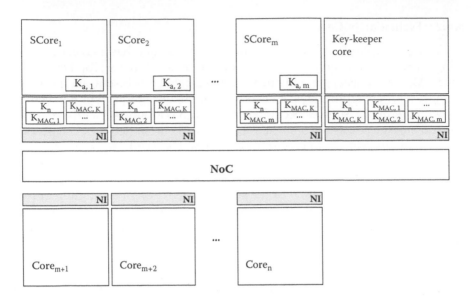

FIGURE 5.9
Framework for secure exchange of keys at the network level.

Cores are connected by an NoC. The methodology, introduced to enhance security at the network level, is based on symmetric key cryptography, adapted to an SoC architecture. It is applicable to a generic NoC, independently of the implementation of the communication infrastructure. Its goal is to protect the system from attackers aiming at extracting sensitive information from the communication channels, either by means of a direct access through external I/O pins connected to the communication network or malicious software running on a compromised core, or by the measurement of the EM radiations leaking from the system.

As shown in Figure 5.9, every secure core is provided with a security wrapper, located between the secure core and the interface to the communication infrastructure. A dedicated core, the key-keeper core, is in charge of securing the keys distribution on the NoC. Wrappers store several keys employed to perform the security operations of which it is in charge (encryption and decryption of messages, hashing, authentication). A working key K'_n is employed to encrypt/decrypt sensitive messages and it is generated at every new transmission from a master key (K_n). Both keys are stored within the wrapper in nonvolatile memory. A message authentication code (MAC) key $(K_{MAC,i})$ is stored in the wrapper, with the aim to provide identification of the sender of the messages sent/received. To easily identify messages coming from the key-keeper core, its MAC key $(K_{MAC,K})$ is also stored in the wrapper, as well as those of other security cores, if sufficient memory is available and if the running application requires a frequent exchange of information between the cores. Every secure core is identified by a unique authentication key $(K_{a,i})$, used for core and core software authentication.

The key-keeper core represents the central unit in charge of the distribution of the keys within the NoC and of updating at random time the master network key K_n. Apart from the other secure cores, it is provided with a security wrapper, and stores a certain number of encrypted keys used for individual applications, user security operations, and other general secure applications. All the MAC keys of the secure cores are equally stored for message authentication.

5.5.1.1 Secure Messages Exchange

The steps performed during the transmission and the reception of messages are shown in Figure 5.10, in the case of a message generated by secure core $SCore_i$ and directed to secure core $SCore_j$. In the figure, $h(m)$ is the x-bit hash of a message m. $E(K_e, m)$ represents the encryption of a message m using the key K_e, although $D(K_d, m)$ is the decryption of the message using the key K_d. $mac(K_{MAC}, m)$ is the operation calculating the message authentication code for the message m, using K_{MAC}. $t_{cnt,j}$ counts how many messages to secure core $SCore_j$ has been sent by secure core $SCore_i$, although $t_{cnt,i}$ represents the number of messages received by $SCore_j$ from core $SCore_i$. $c \parallel m$ represents the concatenation of message c and message m.

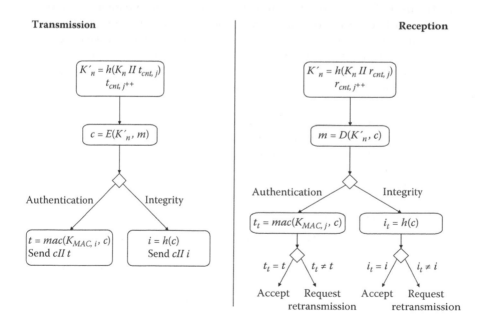

FIGURE 5.10
Protocol for secure exchange of messages within the NoC.

The following three main steps are performed before transmission of messages:

1. The working key K_n' is generated hashing the network key K_n concatenated with the number of messages sent up to the current moment. This procedure allows creating a different key for each different message.
2. The message to be sent through the network is encrypted using the current working key.
3. If the sender wants to provide authentication, the MAC of the message is created, using the $K_{MAC,i}$ stored in the wrapper. The concatenation of the encrypted message and the MAC are therefore sent to the receiving core. If proof of integrity of the message is required, the hash of the encrypted message is calculated and added to the message sent through the network.

When receiving a new message ($\{c \parallel t\}, \{c \parallel i\}$), the following steps are performed by the wrapper of the receiving secure core:

1. The working key related to the current transaction is generated.
2. The received message is decrypted, using the current working key.
3. If authentication is provided, the MAC of the message received is calculated and compared with the tag concatenate with the encrypted message, and a retransmission is requested in case they do not match. To check the integrity of the message, hash is performed and, depending on its correspondence with the tag, either the message is accepted or retransmission is requested to the sender.

The steps described aim at reducing the possibility for a compromised core of obtaining the master network key K_n. Moreover, the working keys generated at the time of each communication and the MAC keys assure a sufficient level of security, even in the case in which the master key is retrieved by the attacker.

5.5.1.2 Download of New Keys

The key-keeper core supports the download of new authenticated user private and public keys, and its distribution and use within the framework [42]. The procedure for downloading and updating a new user key is shown in Figure 5.11. K_{user}^{new} is the new user private or public key, substituting key K_{user}, stored in the system. E_{AES} is the encryption algorithm (AES) used to encrypt the new key to be downloaded. The main steps performed to update the key are the following:

1. The key-keeper core receives the new key encrypted with the old user key ($E_{AES}(K_{user}, K_{user}^{new})$), and sends it to the secure core $SCore_{AES,i}$,

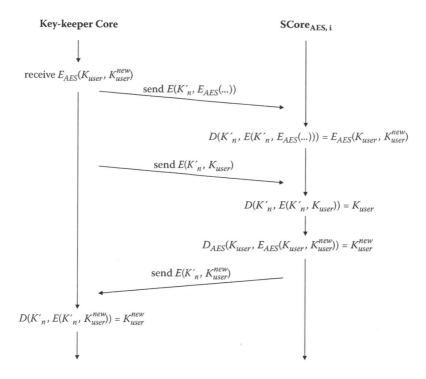

FIGURE 5.11
Protocol for user's key updating.

implementing AES encryption and decryption algorithms. As described in the previous paragraph, the message is encrypted using the working key K_n'.

2. With the same methodology, the key-keeper core sends the old user key to $SCore_{AES,i}$, in order to allow it to decrypt the message containing the new user key.

3. The wrapper of secure core $SCore_{AES,i}$ decrypts the two messages received using K_n' $[D(K_n', E_{AES}(K_{user}, K_{user}^{new}))$ and $D(K_n', K_{user})$ in the figure] and pass them to the core, which decrypts the encrypted new key applying the AES algorithm and using the old user key K_{user}.

4. The new user key is sent to the key-keeper core, encrypted by the secure wrapper using a new working key.

5. The key-keeper receives the new user key and stores it.

Authentication of the encrypted new key will also be required to verify the validity of the authority sending the message, involving in the procedure the same or other secure cores.

5.5.1.3 Other Applications

The framework described can be employed for other security purposes and applications. The secure core authentication key $K_{a,i}$ can be employed to secure the download of software upgrades into the core, as well as in securing bitstream information in case of partial reconfigurable cores based on FPGA technology. The upgrading information will be encrypted and authenticated using the key of the secure core, known only to the IP vendor, and sent to the module.

The secure core's private key can also be used to prevent an illegal use of IP cores within unauthorized design. In this case, IP vendors should include an activation key for each core, as a function of the private key of the cores in the NoC. At reset time, every secure core would receive the activation key from the key-keeper core and check if its activation is allowed, shutting down permanently in the case of an unsuccessful result.

5.5.1.4 Implementation Issues

The hardware modules within the secure wrapper that can be used to support encryption, decryption hashing, and MAC generation are mainly linear feedback shift registers (LFSRs), general shift registers, counters, XOR gates, and some nonvolatile memory to store the several keys used in the framework. Encryption and decryption operations are implemented using a stream cipher algorithm, in order to obtain a fast execution and a small area occupation. MAC generation is performed applying the encryption operation to the last n bits of the input message, while the hash applying the LFSR to the last x bits of the message [41].

A rough estimation of the area occupied by the described system can be done by taking as reference the hardware implementation of eStream phase-III stream cipher candidates [43]. Considering, for instance, an implementation of *Grain* [44], a stream cipher with a key of 128 bits, which is essentially formed around shift registers together with a combinatorial feedback and output filter functions, it is possible to note that the area occupied by the encryption/decryption block is approximately 0.028 mm^2 in 0.13 μm Standard Cell CMOS technology. Compared to common NI's implementation [45], the size of the hardware block implementing the stream cipher algorithm is around 11% of the area occupied by the NI.

5.5.2 Protection of IP Cores from Side-Channel Attacks

Protection of Individual IP cores used within the NoC should also be considered. Side-channel attacks based on DPA, timing, or EM could be easily exploited if particular care is not taken during the design phase, in particular in the case in which cores have their own individual power supply pins. It is possible to divide countermeasures for defending IP cores against DPA into two main categories [46]. Both of them aim at reducing the correlation between sensitive data and side-channel information, either trying to minimize the influence of the sensitive data (hiding) or randomizing the connection between sensitive data and the observable physical values (masking). Hardware or software implementation of the techniques can be realized. Although software implementations imply a reduction of system performance, hardware countermeasures increase the amount of area and power consumption of the system. In this section, an overview of some of these techniques to enhance IP cores security is presented.

5.5.2.1 Countermeasures to Side-Channel Attacks

The combination of parallel execution and adiabatic circuits can be used to significantly reduce power traces used in DPA attacks to retrieve secret information [41]. Adiabatic logic is based on the reuse of the capacitive charge, and therefore in the reuse of the energy temporarily stored to reduce the power consumption of the logic cells. The technique is based on a smooth switching of the nodes, achieved with no voltage applied between drain and source of transistors, and on recovering the charges stored in the circuit for later reuse [47]. Current absorbed in devices implemented adopting this technique is significantly reduced, thus increasing for the attacker the difficulty of retrieving secret values from measuring leaking information.

Randomization techniques can be used to mask inputs to a cryptographic algorithm, in order to make intermediate results of the computation uncorrelated to the inputs and useless to the attacker exploiting side-channel information [48]. This approach must guarantee that intermediate results of the computation look random to an adversary, while assuring correct results at the end of the execution of the cryptographic algorithm. Randomization can be obtained for instance with hardware implementations, by randomizing the clock signal or the power consumption, with the aim of decorrelating the external power supply from the internal power consumed by the chip. Algorithmic countermeasures include secret-sharing schemes, where each bit of the original computation is divided probabilistically into shares such that any proper subset of shares is statically independent of the bit being encoded, thus yielding no information about the bit. Methods based on the idea of masking all data and intermediate results during an encryption operation can be included [48].

Other hardware countermeasures against DPA attacks imply the use of *secure logic styles* in the implementation of encryption modules. In a secure logic style, logic gates have at all times constant power consumption, independently of signal transitions. Therefore, it is not possible anymore to distinguish between the different operations performed by the core. In Sense Amplifier Based Logic (SABL) [49], this is reached using a fixed amount of charge for every transition, including the degenerated events in which a gate does not change state. With this technique, circuits have one switching event per cycle, independent of the input value and sequence, and during the switching event, the sum of all the internal node capacitances together with one of the balanced output capacitances is discharged and charged. However, to employ SABL in the design, new standard cell libraries are necessary. Wave Dynamic

Differential Logic (WDDL) [49] aims at overcoming this drawback, obtaining the behavior of SABL gates by combining building blocks from existing standard cell libraries.

Software techniques can be employed to mask execution of basic operations in cryptographic algorithms [42]. Redundant instructions are inserted to make some basic cryptography-related operations appear equivalent from the power consumption point of view. For instance, when executing elliptic curve point multiplication, the execution of point doubling or doubling and summing is key-dependent. In order to secure the implementation of the procedure, it should not be detectable whether a point doubling or summing is being executed. Timing differences between the two executions can be removed as well as inserting redundant operations, thus making their corresponding power traces similar.

5.6   Conclusions

This chapter addressed the problem of security in embedded devices, with particular emphasis on systems adopting the NoC paradigm. We have discussed general security threats and analyzed attacks that could exploit weaknesses in the implementation of the communication infrastructure. Security should be considered at each level of the design, particularly in embedded systems, physically constrained by such factors as computational capacity of microprocessor cores, memory size, and in particular power consumption. This chapter therefore presented solutions proposed to counteract security threats at three different levels. From the point of view of the overall system design, we presented the implementation of a secure NoC-based system suitable for reconfigurable devices. Considering data transmission and secure memory transactions, we analyzed trade-offs in the implementation of on-chip data protection units. Finally, at a lower level, we addressed physical implementation of the system and physical types of attacks, discussing a framework to secure the exchange of cryptographic keys or more general sensitive messages.

Although existing work addresses specific security threats and proposes some solutions for counteracting them, security in NoC-based systems remains so far an open research topic. Lots of work still remains to be done toward the overall goal of providing a secure system at each level of the design and to address the security problem in a comprehensive way.

Future challenges in the topic of security in NoC-based SoCs are toward the direction of including security awareness at the early stages of the design of the system, in order to limit the possible fallacies that could be exploited by attackers for their malicious purposes. Moreover, modern secure systems should be able to counteract efficiently and rapidly attempts at security violations. As shown in this chapter, NoCs can represent the ideal system where malicious behaviors are monitored and detected. However, security has a cost.

Further investigation will therefore be necessary to evaluate the right trade-off between security services provided by a global system, its performance, and the overhead in terms of area, energy consumption, and cost.

5.7 Acknowledgments

Part of this work has been carried out under the MEDEA+ LoMoSA+ Project and was partially funded by KTI—The Swiss Innovation Promotion Agency—Project Nr. 7945.1 NMPP-NM. The authors would like also to acknowledge the fruitful discussions about security on embedded systems they had with Francesco Regazzoni and Slobodan Lukovic.

References

1. S. Ravi and A. Raghunathan, "Security in embedded systems: Design challenges," *ACM Transactions on Embedded Computing Systems* 3(3)(August 2004): 461–491.
2. P. Kocher, R. Lee, G. McGraw, A. Raghunathan, and S. Ravi, "Security as a new dimension in embedded system design." In *Proc. of 41st Design Automation Conference (DAC'04)*, San Diego, CA, June 2004.
3. R. Vaslin, G. Gogniat, and J. P. Diguet, "Secure architecture in embedded systems: An overview." In *Proc. of ReCoSoC'06*, Montpellier, France, July 2006.
4. "Symbos.cabir," Symantec Corporation, Technical Report, 2004.
5. J. Niemela, *Beselo—Virus Descriptions*, F-Secure, Dec. 2007. [Online]. Available: http://www.f-secure.com/v-descs/worm_symbos_beselo.shtml.
6. J. Niemela, *Skulls.D—Virus Descriptions*, F-Secure, Oct. 2005. [Online]. Available: http://www.f-secure.com/v-descs/skulls_d.shtml.
7. *Symbian OS*, Available: http://www.symbian.com.
8. L. Fiorin, C. Silvano, and M. Sami, "Security aspects in networks-on-chips: Overview and proposals for secure implementations." In *Proc. of Tenth Euromicro Conference on Digital System Design Architecture, Methods, and Tools (DSD'07)*, Lübeck, Germany, August 2007.
9. J. P. Diguet, S. Evain, R. Vaslin, G. Gogniat, and E. Juin, "NoC-centric security of reconfigurable SoC." In *Proc. of First International Symposium on Networks-on-Chips (NOCS 2007)*, Princeton, NJ, May 2007.
10. J. Coburn, S. Ravi, A. Raghunathan, and S. Chakradhar, "SECA: Security-enhanced communication architecture." In *Proc. of International Conference on Compilers, Architectures, and Synthesis for Embedded Systems*, San Francisco, CA, September 2005.
11. S. Ravi, A. Raghunathan, and S. Chakradhar, "Tamper resistance mechanism for secure embedded systems." In *Proc. of 17th International Conference on VLSI Design (VLSID'04)*, Mumbai, India, January 2004.

12. E. Chien and P. Szoe, *Blended Attacks Exploits, Vulnerabilities and Buffer Overflow Techniques in Computer Viruses*, Symantec White Paper, September 2002.

13. F. Koeune and F. X. Standaert, *Foundations of Security Analysis and Design III*. Berlin/Heidelberg: Springer, 2005.

14. P. C. Kocher, "Differential power analysis." In *Proc. of 16th International Conference on Cryptology (CRYPTO'96)*, Santa Barbara, CA, August 1996, 104–113.

15. P. C. Kocher, J. Jaffe, and B. Jun, "Differential power analysis." In *Proc. of 19th International Conference on Cryptology (CRYPTO'99)*, Santa Barbara, CA, August 1999, 388–397.

16. F. Regazzoni, S. Badel, T. Eisenbarth, J. Großschdl, A. Poschmann, Z. Toprak, M. Macchetti, et al., "A simulation-based methodology for evaluating the DPA-resistance of cryptographic functional units with application to CMOS and MCML technologies." In *Proc. of International Symposium on Systems, Architectures, Modeling and Simulation (SAMOS VII)*, Samos, Greece, July 2007.

17. *35.202 Technical Specification version 3.1.1. Kasumi S-box function specifications*, 3GPP, Technical Report, 2002, Available: http://www.3gpp.org/ftp/Specs/archive/35_series/35.202.

18. D. Boneh, R. A. DeMillo, and R. J. Lipton, "On the importance of eliminating errors in cryptographic computations," *Journal of Cryptology*, 14 (December 2001): 101–119.

19. T. W. Williams and E. B. Eichelberger, "A logic design structure for LSI testability." In *Proc. of Design Automation Conference (DAC'73)*, June 1977.

20. B. Yang, K. Wu, and R. Karri, "Scan based side channel attack on dedicated hardware implementations of Data Encryption Standard." In *Proc. of International Test Conference 2004 (ITC'04)*, Charlotte, NC, October 2004, 339–344.

21. S. Evain and J. Diguet, "From NoC security analysis to design solutions." In *Proc. of IEEE Workshop on Signal Processing Systems (SIPS'05)*, Athens, Greece, Nov. 2005, 166–171.

22. T. Martin, M. Hsiao, D. Ha, and J. Krishnaswami, "Denial-of-service attacks on battery-powered mobile computers." In *Proc. of Third International Conference on Pervasive Computing and Communications (PerCom'04)*, Orlando, FL, March 2004.

23. D. C. Nash, T. L. Martin, D. S. Ha, and M. S. Hsiao, "Towards an intrusion detection system for battery exhaustion attacks on mobile computing devices." In *Proc. of Third International Conference on Pervasive Computing and Communications (PerCom'05)*, Kauai Island, Hawaii, March 2005.

24. T. Simunic, S. P. Boyd, and P. Glynn, "Managing power consumption in network on chips," *IEEE Transactions on VLSI Systems* 12(1) (January 2004).

25. *Digital Audio over IEEE1394, White Paper*, Oxford Semiconductor, January 2003.

26. *XOM Technical Information*, Available: http://www-vlsi.stanford.edu/~lie/xom.htm.

27. G. Edward Suh, C. W. O'Donnell, I. Sachdev, and S. Devadas, "Design and implementation of the AEGIS single-chip secure processor." In *Proc. of 32nd Annual International Symposium on Computer Architecture (ISCA'05)*, Madison, WI, June 2005, 25–26.

28. T. Alves and D. Felton, *TrustZone: Integrated Hardware and Software Security*, White Paper, ARM, 2004.

29. SonicsMX SMART Interconnect Datasheet, Available: http://www.sonicsinc.com.

30. Open Core Protocol Specification 2.2, Available: http://www.ocpip.org.

31. H. Inoue, I. Akihisa, K. Masaki, S. Junji, and E. Masato, "FIDES: An advanced chip multiprocessor platform for secure next generation mobile terminals." In *Proc. of International Conference on Hardware/Software Codesign and System Synthesis (CODES+ISSS'05)*, New York, September 2005.

32. H. Inoue, I. Akihisa, A. Tsuyoshi, S. Junji, and E. Masato, "Dynamic security domain scaling on symmetric multiprocessors for future high-end embedded systems." In *Proc. of International Conference on Hardware/Software Codesign and System Synthesis (CODES+ISSS'07)*, Salzburg, Austria, October 2007.

33. J. Sugerman, G. Venkitachalam, and B.-H. Lim, "Virtualizing I/O devices on VMware workstation's hosted virtual machine monitor." In *Proc. of USENIX'01*, San Diego, CA, December 2001.

34. W. J. Armstrong, R. L. Arndt, D. C. Boutcher, R. G. Kovacs, D. Larson, K. A. Lucke, N. Nayar, and R. C. Swanberg, "Advanced virtualization capabilities of POWER5 systems," *IBM Journal of Research and Development* 49(4–5) (July 2005): 523–532.

35. P. Barham, B. Dragovic, K. Fraser, S. Hand, T. Harris, A. Ho, R. Neugebauer, I. Pratt, and A. Warfield, "Xen and the art of virtualization." In *Proc. of SOPS'03*, Bolton Landing, NY, October 2003, 164–177.

36. "BREW. The Road to Profit is Paved with Data Revenue, Internet Services White Paper," QUALCOMM, Technical Report, Jun. 2002.

37. C. Cowan, P. Wagle, C. Pu, S. Beattie, and J. Walpole, "Buffer overflows: Attacks and defenses for the vulnerability of the decade." In *Proc. of Foundations of Intrusion Tolerant Systems (OASIS'03)*, 2003, 227–237.

38. L. Fiorin, G. Palermo, S. Lukovic, and C. Silvano, "A data protection unit for NoC-based architectures." In *Proc. of International Conference on Hardware/ Software Codesign and System Synthesis (CODES+ISSS'07)*, Salzbury, Austria, October 2007.

39. K. Pagiantzis and A. Sheikholeslami, "Content-addressable memory (CAM) circuits and architectures: A tutorial and survey," *IEEE Journal of Solid-State Circuits* 41(3) (March 2006).

40. V. Rana, M. Santambrogio, and D. Sciuto, "Dynamic reconfigurability in embedded system design." In *Proc. of 34th Annual International Symposium on Computer Architecture (ISCA'07)*, San Diego, CA, June 2007.

41. C. H. Gebotys and Y. Zhang, "Security wrappers and power analysis for SoC technology." In *Proc. of International Conference on Hardware/Software Codesign and System Synthesis (CODES+ISSS'03)*, Newport Beach, CA, 2003.

42. C. H. Gebotys and R. J. Gebotys, "A framework for security on NoC technologies." In *Proc. of IEEE Computer Society Annual Symposium on VLSI (ISVLSI'03)*, Tampa, FL, June 2003, 113–117.

43. T. Good and M. Benaissa, "Hardware performance of eStream phase-III stream cipher candidates." In *Proc. of Workshop on the State of the Art of Stream Ciphers (SACS'08)*, Lausanne, Switzerland, February 2008.

44. M. Hell, T. Johansson, A. Maximov, and W. Meier, "A Stream Cipher Proposal: Grain-128." In *Proc. of 2006 IEEE International Symposium on Information Theory (ISIT'06)*, Seattle, WA, July 2006.

45. A. Radulescu, J. Dielissen, S. G. Pestana, O. Gangwal, E. Rijpkema, P. Wielage, and K. Goossens, "An efficient on-chip NI offering guaranteed services, shared-memory abstraction, and flexible network configuration," *IEEE Transactions on Computer-Aided Design of Integrated Circuits and Systems* 24(1) (January 2005): 4–17.

46. S. Tillich, C. Herbst, and S. Mangard, "Protecting AES software implementations on 32-bit processors against power analysis." In *Proc. of Fifth International Conference on Applied Cryptography and Network Security (ACNS'07)* 2 hu hai, China, June.

47. C. Piguet, ed., *Low-Power Electronics Design*, Boca Raton, FL: CRC Press, 2005.

48. J. Blömer, J. Guajardo, and V. Krummel, "Provably secure masking of AES." In *Proc. of 12th Annual Workshop on Selected Areas in Cryptography (SAC 2004)*, Waterloo, Ontario, Canada, August.

49. K. Tiri and I. Verbauwhede, "A logic level design methodology for a secure DPA resistant ASIC or FPGA implementation." In *Proc. of Design, Automation, and Test in Europe Conference (DATE'04)*, Paris, France, February 2004.

6

Formal Verification of Communications in Networks-on-Chips

Dominique Borrione, Amr Helmy, Laurence Pierre, and Julien Schmaltz

CONTENTS

Communication architectures play a central role in Systems-on-Chips (SoC) design and verification. Many initiatives are devoted to developing specific design flows and simulation techniques, while the application of formal verification methodologies to on-chip communication architectures has received attention recently. This chapter addresses the issue of the validation of communication infrastructures, especially Networks-on-Chips (NoCs), and puts the emphasis on the application of formal methods.

6.1   Introduction: Validation of NoCs

6.1.1   Main Issues in NoC Validation

Because NoCs are a relatively new paradigm, with no legacy models and no "golden" reference design, they are more subject to design errors than other components [1]. In addition to being high-risk elements in a SoC, these modules are difficult to verify using the traditional simulation techniques. Any number of nodes may be active at any given time, and may request to send a message of any length: among all possible test scenarios, a decent coverage is out of reach. This is where formal verification brings highest benefits [2]. Moreover, in platform-based design environments, where parameterized component generators are retrieved from libraries and configured for the project at hand, an NoC design should be adjustable to various sizes. Establishing its correctness *for all* sizes involves mathematical reasoning.

Like all complex design tasks, the validation of an on-chip communication infrastructure is a multi-level, multi-aspect problem [3]. The issues to be addressed for validating an NoC design range from energy efficiency and transmission reliability to the satisfaction of bandwidth and latency constraints, via the verification of correctness properties such as the absence of deadlocks and livelocks [4]. Models and tools have been developed to deal with each one of these issues at some circuit design level, or in the fields of distributed systems and computer networks (large or local area, multiprocessors). Benini and De Micheli [5] propose adapting the layered representation of micro-networks (application, transport, network, data link, and physical) to NoCs, and using the existing design methods and tools.

The major difference between NoCs and previous kinds of networks is precisely the fact that NoCs are implemented with the same technology, and layered out together with the other components of the system. This tends to blur the distinction between application-dependent characteristics and network-specific properties. It is the great merit of Ogras, Hu, and Marculescu [6] to propose distinct design dimensions for the communication infrastructure (static), the communication paradigm (dynamic), and the application mapping. Energy consumption, communication rate and volume, and the contentions owing to the interactions between concurrent communication attempts depend on the data flow between system tasks, on the mapping of tasks to computational modules, on the placement of modules, and on their intrinsic performance features. The study and optimization of power consumption, response time, effective bandwidth, and information-flow or message-dependent locks belong to the application mapping dimensions. This is where analytical techniques based on graph theory, statistics, and probabilities, previously developed for multiprocessor micro-networks and distributed systems, will best be applicable. We shall not consider them further, and will concentrate on the static and dynamic validation of the communication infrastructure.

Looking at a NoCs as an IP component, the validation of functionality should follow the conventional design specification levels: transaction level model (TLM), register transfer lever (RTL), cycle accurate bit accurate (CABA) level, logic level, and electrical level. Here again it is suggested to port the existing simulation methods and tools [4]. Going one step further, we postulate that not only dynamic but also formal methods, which have shown their efficiency and total coverage in the verification of processors, memory hierarchies, and parameterized operators, can be ported to the validation of NoCs. To achieve this goal, an appropriate isolation and modeling of the communication functionalities must be elaborated, prior to applying an adequate proof tool. This is the paradigm taken in the chapter.

6.1.2 The Generic Network-on-Chip Model

Our objective is to provide a formal foundation to the design and analysis of NoCs, from their early design phase to RTL code. We aim at developing a general model of NoCs and associated refinement methodologies. The initial abstract model supports the specification and the validation of high-level parameterized descriptions. This initial model is extended by several refinement steps until RTL code is reached. Every refined model is proven to conform with the initial specification. Such models and methodologies have not been developed yet. In this chapter, we present the initial abstract model from which we will develop the refinement methodology.

The idiosyncratic aspects of our approach are (1) to consider a generic or meta-model and (2) to provide a practical implementation. Our model is characterized by components, which are not given a definition but only characterized by properties, and how these components are interconnected. Consequently, the global correctness of this interconnection only depends on those properties, "local" to components. Our model represents all possible instances of the components, provided these instances satisfy instances of the properties. Our model is briefly introduced in Section 6.3 and a detailed presentation is given in Section 6.4.

The generic aspect of our model is its power in its implementation. The model and the proof that the local properties imply a general property of the interconnected components constituted the largest effort. The proof that particular instances also satisfy this global property reduces to the proof that they satisfy the local properties. The "implemented" model generates all these formula automatically. Moreover, our implemented model can also be executed on concrete test scenarios. The same model is used for simulation and formal validation. Section 6.5 illustrates our approach to two complete examples: the Spidergon and the HERMES NoCs.

The first section is concluded by an overview of the state-of-the-art design and analysis of NoCs and communication structures. Before presenting our approach and its applications, we introduce necessary basic notions about formal methods in Section 6.2.

6.1.3 State-of-the-Art

Intensive research efforts have been devoted to the development of performance, traffic, or behavior analyzers for NoCs. Most proposed solutions are either simulation or emulation oriented. Orion [7] is an interconnection network simulator that focuses on the analysis of power and performance characteristics. A variety of design and exploration environments have been described, such as the modeling environment for a specific NoC-based multiprocessor platform [8]. Examples of frameworks for NoC generation and simulation have been proposed: NoCGEN [9] builds different routers from parameterizable components, whereas MAIA [10] can be used for the simulation of SoC designs based on the HERMES NoC. An NoC design flow based on the Æthereal NoC [11] provides facilities for performance analysis. Genko et al. [12] describe an emulation framework implemented on an FPGA that gives an efficient way to explore NoC solutions. Two applications are reported: the emulations of a network of switches and of a full NoC.

Few approaches address the use of (semi-) formal methods, essentially toward detection of faults or debugging. A methodology based on temporal assertions [13] targets a two-level hierarchical ring structure. PSL (Property Specification Language) [14] properties are used to express interface-level requirements, and are transformed into synthesizable checkers (monitors).

In case of assertion failures, special flits are propagated to a station responsible for analyzing these failures. Goossens et al. [15] advocate communication-centric debug instead of computation-centric debug for complex SoCs, and also use a monitor-based solution. They discuss the temporal granularity at which debug can be performed, and propose a specific debug architecture with four interconnects.

As described in Section 6.2.1, formal verification is performed within a logical framework and expects the definition of the system by means of formal semantics. Widespread methods can be classified into two categories: algorithmic techniques (e.g., model checking) and deductive techniques (theorem proving). Many approaches have been proposed in the fields of protocol or network verification, in general. Most of them are based on model-checking techniques, and target very specific designs. Clarke et al. [16] use the notion of regular languages and abstraction functions to verify temporal properties of the families of systems represented by context-free network grammars, for instance the Dijkstra's token ring and a network of binary trees. Creese and Roscoe [17] exploit the inductive structure of branching tree networks, and put emphasis on data independency: data is abstracted to use the FDR model checker to prove properties of CSP specifications. Roychoudhury et al. use the SMV model checker [18] to debug an academic implementation of the AMBA AHB protocol [19]; the model is written at the RTL without any parameter. Results with theorem provers, or combinations of theorem provers and model checkers, have also been proposed, but most of them are concerned with specific architectures. The HOL theorem prover [20] is used by Curzon [21] to verify a specific network component the Fairisle ATM switching fabric. A structural description of the fabric is compared to a behavioral specification. Bharadwaj et al. [22] use the combination of the Coq theorem prover [23] and the SPIN model checker [24] to verify a broadcasting protocol in a binary tree network. Amjad [25] uses a model-checker, implemented in the HOL theorem prover, to verify the AMBA APB and AHB protocols, and their composition in a single system. Using model checking, safety properties are individually verified on each protocol, and HOL is used to verify their composition. In Gebremichael et al. [26], the Æthereal protocol of Philips is specified in the PVS logic. The main property verified is the absence of deadlock for an arbitrary number of masters and slaves.

Some research results tackle the formalization from a *generic* perspective. Moore [27] defines a formal model of asynchrony by a function in the Boyer–Moore logic [28], and shows how to use this general model to verify a biphase mark protocol. More recently, Herzberg and Broy [29] presented a formal model of stacked communication protocols, in the sense of the OSI reference model. They define operators and conditions to navigate between protocol layers, and consider all OSI layers. Thus, this work is more general than Moore's work, which is targeted at the lowest layer. In contrast, Moore provides mechanized support. Both studies focus on protocols and do not consider the underlying interconnection structure explicitly. In the context of time-triggered architectures, the seminal work of Rushby [30] proposes a general model of time-triggered implementations and their synchronous specifications. The simulation relation between these two models is proven for a large class of algorithms using an axiomatic theory. Pike recently improved the application domain of this theory [31,32]. Miner et al. [33] define a unified fault-tolerant protocol acting as a generic specification framework that can be instantiated for particular applications. These studies focus on time-triggered protocols. The framework presented in Section 6.4 aims at a more general network model, and concentrates on the actual interconnect rather than the protocols based on top of this structure. Mechanization is realized through the implementation of this model in the ACL2 theorem prover [34].

6.2 Application of Formal Methods to NoCs Verification

6.2.1 Smooth Introduction to Formal Methods

Formal verification uses mathematical techniques and provides reliable methods for the validation of functional or temporal aspects of hardware components [35–37]. A formal specification of the system under consideration and of the properties to be verified is necessary. Depending on the context, such a formal specification may require first-order logic, higher-order logic, temporal logics like LTL or CTL, etc.

Formal verification techniques are usually identified as algorithmic or deductive. Equivalence checkers and model checkers implement algorithmic methods (in particular, fixed-point computation), whereas theorem provers and proof assistants mechanize deduction (inference rules) in a given logic.

- To compare two versions of the design at different abstraction levels, typically a gate netlist compared to an RTL model, equivalence checkers are to be used (examples of commercial tools include Formality of Synopsys or FormalPro of Mentor Graphics).

- To check temporal properties expressed using a temporal logic or a specification language like PSL [14] or SVA [38], the appropriate tools are *model checkers* (e.g., RuleBase of IBM or Solidify of Averant).

- When the specification involves parameters or complex data types and operations (for instance, unbounded integers or real numbers), formal proofs require a *theorem prover* or *proof assistant* like ACL2 [34], PVS [39], or HOL [20].

Model checkers provide fully automated solutions to verify properties over finite-state systems. For infinite-state systems, or systems with billions of states such as designs with data paths, and for the validation of properties on abstract specifications, these tools can no longer be fully applied automatically: abstractions (such as data path width reduction) are necessary [40]. Theorem provers and proof assistants provide mechanized inference rules or interactive tactics, and often require user-guidance. The counterpart is their applicability to high-level or parameterized specifications.

6.2.2 Theorem Proving Features

Most up-to-date theorem provers or proof assistants are based on first-order or higher-order logic. *Propositional calculus* is a simple logic in which formulae are formed by combining atomic propositions using logical connectives: if A and B are formulae, then $A \wedge B$, $A \vee B$, $A \Rightarrow B$, $A \Leftrightarrow B$ and $\neg A$ are formulae.

First-order logic (or *predicate calculus*) additionally covers predicates and quantification. Terms in first-order logic are constants, variables, or expressions of the form $f(t_1, t_2, \ldots, t_n)$ where f is a functional symbol and $t_1, t_2, \ldots, t_n$ are terms. Then first-order logic formulae are defined as follows:

- $p(t_1, t_2, \ldots, t_n)$ is an atomic formula, where p is a predicate symbol and $t_1, t_2, \ldots, t_n$ are terms.
- If A and B are formulae, then $A \wedge B$, $A \vee B$, $A \Rightarrow B$, $A \Leftrightarrow B$, and $\neg A$ are formulae.
- If A is a formula and x is a variable, then $\forall x\ A$ and $\exists x\ A$ are formulae.

Example

The formula $\forall x\ \forall y\ (x \leq y \Leftrightarrow \exists z\ (x + z = y))$ is a first-order logic formula where x, y and z are variables, and $\leq$, $=$ and $+$ are infix representations of the corresponding functions.

In first-order logic, quantification can only be made over variables, whereas in higher-order logic it can also be made over functions. An example of higher-order formula is: $\exists g\ \forall x\ (f(x) = h(g(x)))$, where x is a variable and g is a functional symbol.

A *proof system* is formed from a set of inference rules, which can be chained together to form proofs. Predicate calculus has two inference rules: modus ponens (if $f \Rightarrow g$ has been proven and f has been proven too, then g is proven) and generalization. As soon as the proof system allows one to consider natural numbers, or other types of inductive structures such as lists or trees (this is the case for ACL2), *mathematical induction*, or more generally *structural induction* may be integrated. Mathematical induction can be stated as follows. *To prove that a statement P holds for all natural numbers n:*

- *Prove that P holds when $n = 0$ (base case).*
- *Prove that if P holds for $n = m$, then P also holds for $n = m+1$ (inductive step).*

The first mechanized proof systems date back to the 1970s. Nowadays there exists a large variety of systems, more or less automatic. They are used for proofs of mathematical theorems, verification or synthesis of programs, formal verification of hardware, etc. One of the precursors was Mike Gordon with the LCF system, with which some of the first mechanized proofs of (simple) digital circuits were performed. The descendant of LCF is HOL [20], a proof assistant for higher-order logic. One of the most famous hardware proof initiatives in the 1980s was the Viper verification project [41] that ended up with the HOL proof of the Viper microprocessor. During the same period, Robert Boyer and J. Moore designed the Boyer–Moore theorem prover Nqthm [28], the ancestor of ACL2. The automatized verification of a fully specified microprocessor, from the microinstruction level to the RT level, is reported in the technical report [42].

Coq [23] and PVS [39] are two renowned proof systems. Coq is a proof assistant based on a framework called Calculus of Inductive Constructions. It has many applications, among them the verification of hardware devices [43] and of cryptographic protocols [44]. PVS is an interactive system that supports a typed higher-order logic. One of its first applications to hardware verification was the proof of a divider circuit [45]. Many efforts have been devoted to the combined use of model checking and theorem proving techniques with PVS: a model checking decision procedure has been integrated in PVS [46], and examples of integration of PVS with model checking and abstraction techniques can be found in [47,48].

The ACL2 theorem prover [34] has been developed at the University of Texas, Austin. It supports first-order logic without quantifiers (variables in theorems are implicitly universally quantified) and is based on powerful principles, among them:

- The definition principle, which allows one to define new recursive functions that are accepted by the system only if it can be proven that they terminate (the measure of one argument or a combination of the arguments, must decrease in the recursive calls). Here is a very trivial example:

```
function TIMES(x, y) =
    if natp(x) then
        if x = 0 then return 0
        else    return y + Times(x−1, y)
        end if
    else    return 0
    end if
end function
```

This function recursively defines the multiplication over natural numbers. Provided that its first argument is actually a natural number (*natp*), it recurses on this argument. Its measure decreases in the recursive calls (x becomes $x − 1$). Hence this function is admitted by ACL2.

- The induction principle, on which the induction heuristics of the proof mechanism is based. An induction variable is automatically chosen and an induction scheme is automatically generated. For example, to prove

$$\text{natp}(x) \wedge \text{natp}(y) \Rightarrow \text{Times}(x, y+1) = x + \text{Times}(x, y) \qquad (P)$$

the induction variable is x and the induction scheme in ACL2 is as follows:
  - Prove that P holds when $x = 0$ (base case).
  - Prove that, if P holds for $x - 1$ ($x \neq 0$) then P also holds for x (inductive step).

Predefined data types are: Booleans, characters and strings, rational numbers, complex numbers, and lists. The language is not typed, that is, the types of the function parameters are not explicitly declared. Rather, typing predicates are used in the function bodies (for instance, *natp(x)* to check whether x is a natural number).

A great advantage of ACL2 with respect to the previously mentioned proof assistants is its high degree of automation (it qualifies as a theorem prover). When provided with the necessary theories and libraries of pre-proven lemmas, it may find a proof automatically: successive proof strategies are applied in a predefined order. Otherwise, the user may suggest proof hints or introduce intermediate lemmas. This automation is due to the logic it implements, gained at the cost of expressiveness, but the ACL2 logic is expressive enough for our purpose. Despite the fact that ACL2 is first-order and does not support the explicit use of quantifiers, certain kinds of originally higher-order or quantified statements can be expressed. Very powerful definition mechanisms, such as the encapsulation principle, allow one to extend the logic and reason on undefined functions that satisfy one or more constraints.

Another characteristic of interest is that the specification language is an applicative subset of the functional programming language Common Lisp. As a consequence ACL2 provides both a theorem prover and an execution engine in the same environment: theorems that express properties of the specified functions can be proven, and the same function definitions can be executed efficiently [49,50].

This prover has already been used to formally verify various complex hardware architectures, such as microprocessors [51,52], floating point operators [53], and many other structures [54]. In the next sections, we define a high-level generic formal model for NoC, and encode it into ACL2; we can then perform high-level reasoning on a large variety of structures and module generators, with possibly unbounded parameters. Our generic model is intrinsically of higher-order, for example, quantification over functions, whereas the ACL2 logic is first-order and quantifier-free. Applying a systematic and reusable mode of expression [55], our model can be entirely formalized in the ACL2 logic.

6.3   Meta-Model and Verification Methodology

The meta-model detailed in Section 6.4 represents the transmission of messages from their source to their destination, on a *generic* communication architecture, with an *arbitrary* network characterization (topology and node interfaces), routing algorithm, and switching technique. The model is composed of a collection of functions together with their *characteristic constraints*. These functions represent the main constituents of the network meta-model: interfaces, controls of network access, routing algorithms, and scheduling policies. The main function of this model, called *GeNoC*, is recursive and each recursive call represents one step of execution, where messages progress by at most one hop. Such a step defines our time unit.

A correctness theorem is associated with function *GeNoC*. It states that for all topology $\mathcal{T}$, interfaces $\mathcal{I}$, routing algorithm $\mathcal{R}$, and scheduling policy $\mathcal{S}$ that satisfy associated constraints P_1, P_2, P_3, and P_4, *GeNoC* fulfills a correctness property $\wp$.

THEOREM 6.1
$\forall \mathcal{T} \ \forall \mathcal{I} \ \forall \mathcal{R} \ \forall \mathcal{S}, \ P_1(\mathcal{T}) \wedge P_2(\mathcal{I}) \wedge P_3(\mathcal{R}) \wedge P_4(\mathcal{S}) \Rightarrow \wp(GeNoC(\mathcal{T}, \mathcal{I}, \mathcal{R}, \mathcal{S}))$
Roughly speaking, the property $\wp$ asserts that every message arrived at some node n was actually issued at some source node s and originally addressed to node n, and that it reaches its destination without modification of its content.

The constituents of the meta-model are characterized by constraints P_1, P_2, P_3, and P_4. Constraints express essential properties of the key components, for example, well-formedness of the network or termination of the routing function. The proof of Theorem 6.1 is derived from these constraints, without considering the actual definitions of the constituents. Consequently, the global correctness of the network model is preserved *for all particular definitions* satisfying the constraints. It follows that, for any instance of a network that is, for any $\mathcal{T}_0$, $\mathcal{I}_0$, $\mathcal{R}_0$, and $\mathcal{S}_0$, the property $\wp(GeNoC(\mathcal{T}_0, \mathcal{I}_0, \mathcal{R}_0, \mathcal{S}_0))$ holds provided that $P_1(\mathcal{T}_0)$, $P_2(\mathcal{I}_0)$, $P_3(\mathcal{R}_0)$, and $P_4(\mathcal{S}_0)$ are satisfied. Hence, verifying Theorem 6.1 for a given NoC is reduced to *discharging these instantiated constraints on the NoC constituents*.

The verification methodology in ACL2, for any NoC instance, proceeds by

- Giving a concrete definition to each one of the constituents, the corresponding proof obligations are automatically generated. They are the ACL2 theorems that express $P_1(\mathcal{T}_0)$, $P_2(\mathcal{I}_0)$, $P_3(\mathcal{R}_0)$, and $P_4(\mathcal{S}_0)$.
- Proving that the concrete definitions satisfy the proof obligations.

It automatically follows that the concrete network satisfies the instantiated meta-theorem Theorem 6.1.

6.4   A More Detailed View of the Model

The model formalizes, by way of proof obligations, the interactions between the three key constituents: interfaces, routing, and scheduling [56,57].

6.4.1   General Assumptions

6.4.1.1   Computations and Communications

As proposed by Rowson and Sangiovanni-Vincentelli [58], each node is divided into an *application* and an *interface*. The interface is connected to the communication architecture. Interfaces allow applications to communicate using protocols. An interface and an application communicate using *messages*, and two interfaces communicate using *frames* (messages that are sent from one application to the other are encapsulated into frames). Applications represent the computational and functional aspects of nodes. Applications are either active or passive. Typically, active applications are processors and passive applications are memories. We consider that each node contains one passive and one active application, that is, each node is capable of sending and receiving frames. As we want a general model, applications are not considered *explicitly*. Passive applications are not actually modeled, and active applications are reduced to the list of their pending communication operations. We focus on communications between distant nodes. We assume that, in every communication, the destination node is different from the source node.

6.4.1.2   Generic Node and State Models

We consider networks composed of an unbounded, but finite, number of nodes. Let *Nodes* be the set of all nodes of such a network. We assume the generic node model of Figure 6.1(a). Each node is uniquely identified by its *position* or *coordinate*. Each node has one local input port and one local output port connected to its own active and passive applications. Each node has several input and output ports connected to neighboring nodes. Tuples composed of a coordinate, a port, and its direction (i.e., tuples of the form $\langle coor, port, dir \rangle$) constitute the basis of our model. We shall refer to such an element as an *address*, abbreviated *addr*. An address is valid if (1) it belongs to a node of the network, that is, is a member of the set *Nodes*; (2) it is connected to an address of a neighboring node or it is a local port. Let *Addresses* be the set of the valid addresses of a network.

Example 1

Figure 6.1(b) shows the instantiation of the generic node model for a 2D mesh. The position is given by coordinates along the X- and Y-axis. There

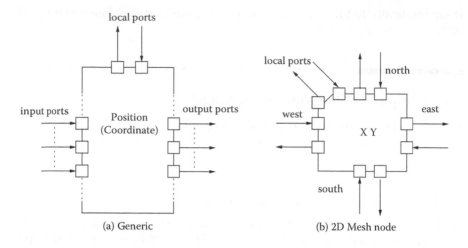

FIGURE 6.1
Generic node model and its instantiation for a simple 2D mesh.

are input and output ports for neighbors connected to all cardinal points. In a 2×2 mesh (Figure 6.2) examples of valid addresses are $\langle (0\ 0), east, o \rangle$ – which is connected to $\langle (1\ 0), west, i \rangle$ – and $\langle (0\ 1), east, i \rangle$ – which is connected to $\langle (1\ 1), west, o \rangle$. But, $\langle (0\ 0), south, o \rangle$, or $\langle (1\ 1), north, i \rangle$ are not valid addresses.

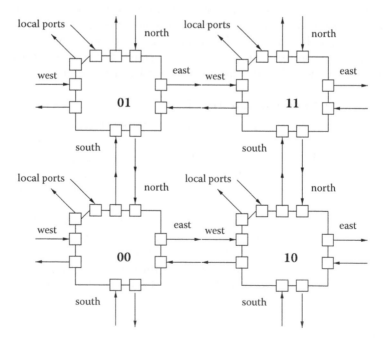

FIGURE 6.2
2D mesh.

Each address has some storage elements, noted *mem*. We make no assumption on the structure of these elements. The generic global state of a network consists of all tuples ⟨*addr, mem*⟩. Let *st* be such a state. We adopt the following notation. The *state element* of address *addr*, that is, a tuple ⟨*addr, mem*⟩, is noted *st.addr*; the *storage element* is noted *st.addr.mem*. We assume two generic functions that manipulate a global network state. Function *loadBuffer(addr, msg, st)* takes as arguments an address (*addr*), some content (*msg*), and a global state (*st*). It returns to a new state, where *msg* is added to the content of the buffer with address *addr*. Function *readBuffer (addr, st)* returns the *state element* with address *addr*, that is, *st.addr*. In Sections 6.5.1 and 6.5.2, we give instances of the generic network state.

6.4.2 Unfolding GeNoC: Data Types and Overview

Function *GeNoC* is the heart of our model. We now present its structure, which is illustrated in Figure 6.3. This structure and the different computation steps induce the main data types of our model.

6.4.2.1 Interfaces

Interfaces model the encoding and the decoding of messages and frames that are injected or received in the network. Interfaces are represented by two functions. (1) Function *send* represents the mechanisms used to encode messages into frames. (2). Function *recv* represents the mechanisms used to

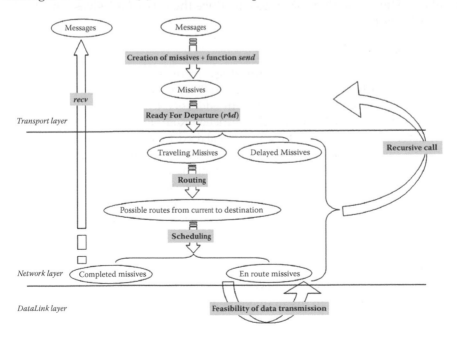

FIGURE 6.3
Unfolding function *GeNoC*.

decode frames. The main constraint associated with these functions expresses that a receiver should be able to extract the injected information, that is, the composition of functions *recv* and *send* (*recv* ∘ *send*) is the identity function.

The main input of function *GeNoC* is a list of transactions. A *transaction* is a tuple of the form ⟨*id, org, msg, dest, flit, time*⟩, where *id* is a unique identifier (e.g., a natural number), *msg* is an arbitrary message, *org* and *dest* are the *origin* and the *destination* of *msg, flit* is a natural number, which optionally denotes the number of flits in the message (*flit* is set to 1 by default), and *time* is a natural number which denotes the execution step when the message is emitted. The origin and the destination must be valid addresses, that is, be members of the set *Addresses*. The first operation of GeNoC is to encode messages into frames. It applies function *send* of the interfaces to each transaction.

A missive results from converting the message of a transaction to a frame. A missive is a transaction where the message is replaced by the frame with an additional field containing the current position of the frame. The current position must be a valid address.

6.4.2.2 *Network Access Control*

Function *r4d* (ready for departure) represents the mechanisms used to control the access to the network. Messages may be injected only at specific execution time, or under constraints on the network load (e.g., credit-based flow control). From the list of missives, function *r4d* extracts a list of missives that are authorized to enter the network. These missives constitute the *traveling* missives. The remaining missives constitute the *delayed* missives. This list is named *Delayed*.

6.4.2.3 *Routing*

The routing algorithm on a given topology is represented by function *Routing*. At each step of function *GeNoC*, function *Routing* computes for every frame all the possible routes from the current address c to the destination address d. The main constraint associated to function *Routing* is that each route from c to d actually starts in c and uses only valid addresses to end in d.

The traveling missives chosen by function *r4d* are given to function *Routing*, which computes for each frame routes from the current node to the destination. The result of this function is a list of *travels*. A travel is a tuple of the form ⟨*id, org, frm, Route, flit, time*⟩, where *Route* denotes the possible routes of the frame. The remaining fields equal the corresponding fields of the initial missive.

6.4.2.4 *Scheduling*

The switching technique is represented by function *Scheduling*. The scheduling policy participates in the management of conflicts, and computes a set of possible simultaneous communications. Formally, these communications satisfy an *invariant*. Scheduling a communication, that is, adding it to the current set of authorized communications, must preserve the invariant, at all times and in any admissible state of the network. The invariant is specific to the scheduling policy. Examples are given in Sections 6.5.1 and 6.5.2.

Function *Scheduling* represents the execution of one *network simulation step*. It takes as main arguments the list of travels produced by function *Routing*, and the current global network state. Whenever possible, function *Scheduling* moves a frame from its current address to the next address according to one of the possible routes and the current network state. It returns three main elements: the list *EnRoute* of frames that have not reached their destination, the list *Arrived* of frames that have reached their destination, and a new state *st'*.

6.4.2.5 GeNoC and GenocCore

Function *GenocCore* combines the invocations of *r4d*, *Routing*, and *Scheduling*. The travels of list *EnRoute* are converted back to missives. These missives, together with the delayed missives produced by function *r4d*, constitute the main argument of a recursive call to *GenocCore*. The frames that have reached their destination are accumulated after each recursive call. When the computation of function *GenocCore* terminates, the list *Arrived* contains all the frames that have completed their path from their source to their destination; the list *EnRoute* contains all the frames that have left their source but have not left the network; the list *Delayed* contains all the frames that are still at their origin.

6.4.2.6 Termination

To make sure that *GenocCore* terminates, we associate a *finite* number of attempts to every node. At each recursive call to *GenocCore*, every node with a pending transaction consumes one attempt. This is performed by function *ConsumeAttempts(att)*. The association list *att* stores the attempts and *att[i]* denotes the number of remaining attempts for the node *i*. Function *SumOfAtt(att)* computes the sum of the remaining attempts for all the nodes and is used as the decreasing measure of parameter *att*. Function *GenocCore* halts if all attempts have been consumed.

6.4.2.7 Final Results and Correctness

Function *GeNoC* composes the interface functions with *GenocCore*. The first output list of function *GeNoC* contains the completed transactions, that is, the messages received at some destination node. These messages are obtained by applying function *recv* of the interfaces to the frame of every travel of the list *Arrived* produced by *GenocCore*. A completed transaction is called a *completion*. A completion is a tuple ⟨*id, dest, msg*⟩ and means that address *dest* has received message *msg*. This completion corresponds to the transaction with identifier *id*. The lists *EnRoute* and *Delayed* produced by function *GenocCore* are grouped together to make the second output of function *GeNoC*.

The correctness of *GeNoC* expresses the property that every message *msg* received at a valid address *n*, was emitted at a valid address, with the same content, and destination *n* (property $\wp$ of Theorem 6.1). We give a formal and more precise statement in Section 6.4.3.

6.4.3   GeNoC and GenocCore: Formal Definition

Function *GenocCore* takes as arguments a list of missives, a list of attempts, the current execution time, and the current global network state. It also takes an accumulator to store the frames that have reached their destination, that is, the elements of list *Arrived* produced by function *Scheduling*. These accumulated arrived travels constitute the first output of function *GenocCore*. At each computation, all the frames that are still en route or delayed by function *r4d* constitute the main argument of the recursive call.* At the last computation step, these frames have not been able to enter or to leave the network. They constitute the list *Aborted*.

```
function GenocCore(Missives, att, time, st, Arrived)
    if SumOfAtt(att) = 0 then //All attempts have been consumed
            Aborted := Missives //At the end, Missives = union of en route and delayed
            return list(Arrived, Aborted)
    else
            Traveling := r4d.Traveling(Missives, time) //Extract traveling missives
            Delayed := r4d.Delayed(Missives, time) //Delayed missives
            v := Routing(Traveling) //Route and travels
            EnRoute := Scheduling.EnRoute(v, att, st)
            Arr := union(Arrived, Scheduling.Arrived(v, att, st)) //Partial result
            st' := Scheduling.st(v, att, st)
            att' := Scheduling.att(v, att, st)
            return GenocCore(union(EnRoute, Delayed), att', time + 1, st', Arr)
    end if
end function
```

The correctness of this function is expressed by Theorem 6.2 below (Property $\wp$ of Theorem 6.1). It states that for each arrived travel *atr* of list *Arrived*, there must exist exactly one missive *m* of the input argument *Missives*, such that *atr* and *m* have the same identifier, the same frame, and that the last address of the routes of *atr* equals the destination of *m*.

THEOREM 6.2

Correctness of *GenocCore*.

$$\forall atr \in Arrived, \exists! m \in Missives, \begin{cases} atr.id = m.id \quad \wedge \ atr.org = m.org \\ \wedge \ atr.frm = m.frm \wedge Last(atr.Route) = m.dest \end{cases}$$

PROOF   The proof is performed in ACL2 as follows. Function *Routing* produces valid routes (Proof Obligation 1 of Section 6.4.4), which means that for each travel there is a unique missive such that the last address of the route

---

* The union of lists *EnRoute* and *Delayed* is converted to proper missives. We do not detail this operation.

of the travel equals the destination of the missive. This is preserved by the list *Arrived* of travels produced by function *Scheduling*, because this list is a sublist of the input of *Scheduling* (Proof Obligation 3 of Section 6.4.5). For more details about a similar proof, we refer to the previous publications [56].   ■

Function *GeNoC* takes as main arguments a list of *transactions* and a global network state. It produces two lists: the list *Completed* of completions and the list *Aborted* returned by function *GenocCore*. Function *GeNoC* converts the transactions into missives by applying function *send* of the interfaces. A completion *c* is built from a travel *tr* of list *Arrived* as follows. The completion takes the identifier and the destination of the travel, which is the last address of the routes. The frame of *c* is replaced by a message using function *recv*. Finally, the completion of travel, *tr*, is the tuple $c = \langle tr.id, Last(tr.Route), recv(tr.frm) \rangle$.

Function *GeNoC* is characterized by a theorem similar to Theorem 6.2 above. The main difference is that it relates completions with transactions. Every completion must be matched by a unique transaction, which has the same identifier, the same destination, and the same message. From Theorem 6.2 and the construction of completions, we obtain the correctness of the identifier and the destination. The final message results from the application of functions *send* and *recv*. Because this composition must be the identity function, it follows that the message that is received equals the message that was sent.

We do not detail function GeNoC any further. In Section 6.5, we focus on instances of the essential functions of our model, Routing and Scheduling.

6.4.4 Routing Algorithm

We now detail the methodology to develop instances of function *Routing* and its proof obligations.

6.4.4.1 *Principle and Correctness Criteria*

Let *d* be the destination of a frame standing at node *s*. In the case of deterministic algorithms, the routing logic of a network selects a unique node as the next step in the route from *s* to *d*. This logic is represented by function $\mathcal{L}(s, d)$. The list of the visited nodes for every travel from *s* to *d* is obtained by the successive applications of function $\mathcal{L}$ until the destination is reached, that is, as long as $\mathcal{L}(s, d) \neq d$. The route from *s* to *d* is:

$$s, \mathcal{L}(s, d), \mathcal{L}(\mathcal{L}(s, d), d), \mathcal{L}(\mathcal{L}(\mathcal{L}(s, d), d), d), \ldots, d$$

A route is computed by function *routingCore*, which is the core element of function *Routing*. Function *routingCore* takes as arguments two nodes and returns a route between these two nodes. It builds a route as the recursive application of function $\mathcal{L}$.

```
function ROUTINGCORE(s, d)
    if s = d then return d //at destination
    else  return list(s, routingCore(ℒ(s, d), d)) //make one hop
    end if
end function
```

This approach can be generalized to adaptive routing algorithms as well
[56,59]. In the general case, function *routingCore* produces the list of *all* possible routes between *s* and *d*. In this chapter, we restrict our presentation to
deterministic algorithms.

Example 2

Let us consider a 3×3 mesh and the XY routing algorithm.* Function $\mathcal{L}_{xy}$
represents the routing logic of each node. It decides the next hop of a message
depending on its destination. In the following definition, s_x or d_x denotes the
coordinate along the X-axis, and s_y or d_y denotes the coordinate along the
Y-axis.

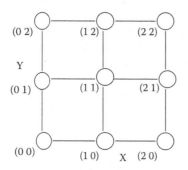

```
function ℒxy(s,d)
    if s = d then return d
    else if sx < dx then return (sx + 1, sy)
    else if sx > dx then return (sx − 1, sy)
    else if sx = dx ∧ sy < dy then return (sx, sy + 1)
    else   return (sx, sy − 1)
    end if
end function
```

6.4.4.1.1 *Routing Correctness*

A route *r* computed from a missive *m* is correct with respect to *m* if *r*
starts with the current address of *m*, ends with the destination of *m* and
every address of *r* is a valid address of the network. Every correct route
has at least two nodes. Predicate *ValidRoutep(r, m, Addresses)* checks all these
conditions.

This predicate must be satisfied by the route produced by function
routingCore. The following proof obligation has to be relieved:

Proof Obligation 1   Correctness of routes produced by *routingCore*.

$$\forall m \in Missives,\ ValidRoutep(routingCore(m.curr, m.dest), m, Addresses)\ (PO1)$$

---

* In this short example, we omit ports and directions. We refer to Section 6.5.2 for a more detailed
model of the XY routing algorithm.

6.4.4.2   Definition and Validation of Function Routing

Function *Routing* takes a missive list as argument. It returns a travel list in which a route is associated to each missive. Function *Routing* builds a travel list from the identifier, the frame, the origin, and the destination of missives.

```
function ROUTING(Missives)
    if Missives = ϵ then return ϵ //ϵ denotes the empty list
    else
            m := first(Missives) //first(l) returns the first element of list l
            t := ⟨m.id, m.org, m.frm, routingCore(m.curr, m.dest), m.flit, m.time⟩
            return list(t, Routing(tail(Missives))) //tail(l) returns l without its first element
    end if
end function
```

6.4.5   Scheduling Policy

Function *Scheduling* takes as arguments the travel list produced by function *Routing*, the list *att* of the remaining number of attempts, and the global network state. It returns a new list of the number of attempts, a new global state, and two travel lists: *EnRoute* and *Arrived*. To identify the different output arguments of function *Scheduling*, we use the following notations:

- *Scheduling.att(Travels, att, st)* returns the new list of attempts.
- *Scheduling.st(Travels, att, st)* returns the new state.
- *Scheduling.EnRoute(Travels, att, st)* returns the list *EnRoute*.
- *Scheduling.Arrived(Travels, att, st)* returns the list *Arrived*.

At each scheduling round, all travels of list *Travels* are analyzed. If several travels are associated to a single node, the node consumes one attempt for the set of its travels. At each call to *Scheduling*, an attempt is consumed at each node. If all attempts have not been consumed, the sum of the remaining attempts after the application of function *Scheduling* is strictly less than the sum of the attempts before the application of *Scheduling*. This is expressed by the following proof obligation:

Proof Obligation 2   Function *Scheduling* **consumes at least one attempt.**

Let *natt* be *Scheduling.att(Travels, att, st)*, then

$$SumOfAtt(att) \neq 0 \rightarrow SumOfAtt(natt) < SumOfAtt(att) \qquad (PO2)$$

The next two proof obligations show that there is no spontaneous generation of new travels, and that any travel of the lists *EnRoute* or *Arrived* corresponds to a unique travel of the input argument of function *Scheduling*. The first proof obligation (Proof Obligation 3) ensures that for every travel *atr* of list *Arrived*, there exists exactly one travel *v* in *Travels* such that *atr* and *v* have the same identifier, the same frame, the same origin, and that their route ends with the same destination.

Proof Obligation 3   Correctness of the arrived travels.

$\forall atr \in Scheduling.Arrived(Travels, att, st),$

$$\exists! v \in Travels, \begin{cases} atr.id = v.id \quad \land \ atr.org = v.org \\ \land \ atr.frm = v.frm \land Last(atr.Route) = Last(v.Route) \end{cases} \quad (PO3)$$

List *EnRoute* must satisfy a similar proof obligation (Proof Obligation 4):

Proof Obligation 4   Correctness of the en route travels.

$\forall etr \in Scheduling.EnRoute(Travels, att, st),$

$$\exists! v \in Travels, \begin{cases} etr.id = v.id \quad \land \ ert.org = v.org \\ \land \ etr.frm = v.frm \land Last(etr.Route) = Last(v.Route) \end{cases} \quad (PO4)$$

For clarity, we omit several proof obligations that

- state that a travel cannot be a member of both lists *EnRoute* and *Arrived*.
- constrain the behavior of function *Scheduling* when there is no attempt left: the state must be unchanged, and the list *EnRoute* must be equal to the current input list *Travels*.
- state simple type checking.

6.5 Applications

6.5.1 Spidergon Network and Its Packet-Switched Mode

6.5.1.1 *Spidergon: Architecture Overview*

The Spidergon network, designed by STMicroelectronics [60,61], is an extension of the Octagon network [62]. A basic Octagon unit consists of eight nodes and twelve bidirectional links [Figure 6.4a]. It has two main properties: the communication between any pair of nodes requires at most two hops, and it has a simple, shortest-path routing algorithm [62]. Spidergon [Figure 6.4b] extends the concept of the Octagon to an arbitrary even number of nodes. Let *NumNode* be that number. Spidergon forms a regular architecture, where all nodes are connected to three neighbors and a local IP. The maximum number of hops is $\frac{NumNode}{4}$, if *NumNode* is a multiple of four. We

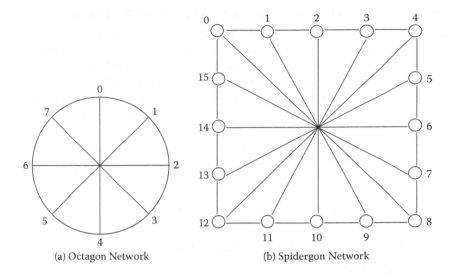

FIGURE 6.4
Octagon and Spidergon architectures.

restrict our formal model to the latter. We assume a global parameter N, such that $NumNode = 4 \cdot N$.

A *Spidergon packet* contains data that must be carried from the source node to the destination node as the result of a communication request by the source node. We consider a Spidergon network based on the packet switching technique.

The routing of a packet is accomplished as follows. Each node compares the address of a packet ($PackAd$) to its own address ($NodeAd$) to determine the next action. The node computes the relative address of a packet as

$$RelAd = (PackAd - NodeAd) \bmod (4 \cdot N)$$

At each node, the route of packets is a function of $RelAd$ as follows:

- $RelAd = 0$, process at node
- $0 < RelAd \leq N$, route clockwise
- $3 \cdot N \leq RelAd \leq 4 \cdot N$, route counterclockwise
- route across otherwise

Example 3
Consider that $N = 4$. Consider a packet *Pack* at node 2 sent to node 12. First, $12 - 2 \bmod 16 = 10$, *Pack* is routed across to 10. Then, $12 - 10 \bmod 16 = 2$, *Pack* is routed clockwise to 11, and then to node 12. Finally, $12 - 12 \bmod 16 = 0$, *Pack* has reached its final destination.

6.5.1.2   *Formal Model Preliminaries: Nodes and State Definition*

A node is divided into four ports: clockwise (*cw*), counterclockwise (*ccw*), across (*acr*), and local (*loc*). Each port is either an input (*i*) or an output (*o*) port. Finally, each node is uniquely identified by a natural number. A valid Spidergon address is a tuple ⟨*id, port, dir*⟩, where *id* is a natural number not greater than the total number of nodes *NumNode*, *port*, and *dir* are one of the valid ports and directions mentioned above. The set of all valid Spidergon addresses is noted *SpidergonAddresses*.

We assume that each node has one storage element—a *buffer*—that can store one frame. We instantiate the generic state functions with function *SpidergonLoadBuffer*(*addr, frm, st*) and function *SpidergonReadBuffer*(*addr, st*). The former updates the buffer of an address with frame *frm*, the latter reads a state. The content of an empty buffer is noted ϵ.

6.5.1.3   *Instantiating Function Routing: SpidergonRouting*

6.5.1.3.1   *Core Routing Function*

Let *s* be the current address and *d* the destination address. Each frame can move from the output port of a node to the input port of a *distant* node; or, it can move from the input port to one output port of the *same* node. The following functions define the distant or internal moves towards each one of the four ports.

> **function** CLOCKWISE(s)
>   **if** *s.dir* = *i* **then** return ⟨*s.id, cw, o*⟩
>   **else**//leave from port ⟨*s.id, cw, o*⟩
>   return ⟨(*s.id* + 1) mod (4 · *N*), *ccw, i*⟩ //enter on port ⟨*s.id* + 1, *ccw, i*⟩ of neighbor
>   **end if**
> **end function**

> **function** COUNTERCLOCKWISE(s)
>   **if** *s.dir* = *i* **then** return ⟨*s.id, ccw, o*⟩
>   **else**   return ⟨(*s.id* − 1) mod (4 · *N*), *cw, i*⟩
>   **end if**
> **end function**

> **function** ACROSS(s)
>   **if** *s.dir* = *i* **then** return ⟨*s.id, acr, o*⟩
>   **else**   return ⟨(*s.id* + 2 · *N*) mod (4 · *N*), *acr, i*⟩
>   **end if**
> **end function**

>   **function** LOCAL(s)   return ⟨*s.id, loc, o*⟩
>   **end function**

These moves are grouped into function *SpidergonLogic*, which represents the routing decision taken at each node. The relative address is *RelAd* = (*d .id* − *s.id*) mod (4 · *N*). If the current address is the destination, the packet is consumed. If the relative address is positive and less than N, the message moves clockwise. If this address is between $3N$ and $4N$, it moves counterclockwise. Otherwise, it moves across.

```
function SPIDERGONLOGIC(s,d)
    RelAd := (d.id − s.id) mod (4 . N)
    if RelAd = 0 then return Local(s) //final destination reached
    else if 0 < RelAd <= N then return Clockwise(s)//clockwise move
    else if 3 · N ≤ RelAd ≤ 4 · N then
                    return Counterclockwise(s) //counterclockwise move
    else    return Across(s) //destination in opposite half
    end if
end function
```

The core routing function *SpidergonRoutingCore* is defined as the recursive application of the unitary moves.

```
function SPIDERGONROUTINGCORE(s,d)
    if s = d then return d //at destination
    else//do one hop
            return list(s, SpidergonRoutingCore(SpidergonLogic(s,d),d))
    end if
end function
```

To show that function *SpidergonRoutingCore* constitutes a valid instance of the generic routing function, we need to prove that it produces routes that satisfy predicate *ValidRoutep* (instance of Proof Obligation 1).

THEOREM 6.3
Validity of Spidergon Routes.

$$\forall m \in Missives, ValidRoutep(SpidergonRoutingCore(m.curr, m.dest),$$
$$m, SpidergonAddresses)$$

PROOF   ACL2 performs the proof by induction on the route length.   ∎

Finally, function *SpidergonRouting* corresponds to function *Routing*:

```
function SPIDERGONROUTING(Missives)
    if Missives = ε then return ε
    else
            m := first(Missives)
            route := SpidergonRoutingCore(m.curr, m.dest)
            t := ⟨m.id, m.org, m.frm, route, m.flit, m.time⟩
            return list(t, SpidergonRouting (tail(Missives)))
    end if
end function
```

The proof obligation of Section 6.4.4 has been discharged, as well as some minor proof obligations related to type checking. Function *SpidergonRouting* is therefore a valid instance of function *Routing*.

6.5.1.4 Instantiating Function Scheduling: PacketScheduling

Our modeling of the packet switching technique proceeds in three steps: (1) to check whether the move from the current address to the neighbor address is possible, (2) to perform the move, and (3) to free the places left in step (2). We now detail the operations of each one of these steps.

We consider ports with single place buffers. A frame can make a hop if the buffer of the next address of its route is free. Let function *nxtHopFree?*(*addr, st*) return true if and only if the buffer at address *addr* is empty in state *st*. Function *GoodRoute?*(*r, st*) checks whether the next hop in route *r* is possible.

> **function** GoodRoute?(r, st)
>   return nxtHopFree(Second(r), st) //next hop is second address in route
> **end function**

To represent the effect of a frame moving to the next address, we simply remove the first address of its route, which modifies its current position. This is done by function *hop*(*v*), which updates the route of travel *v*.

> **function** hop(v) return ⟨*v.id, v.org, v.frm, tail(v.Route), v.flit, v.time*⟩
> **end function**

To empty a buffer, one simply loads it with ϵ, the empty buffer. Let *ToLeave* be the set of addresses that have been left in step (2). Function *free*(*ToLeave, st*) returns to a new state where all addresses in *ToLeave* have an empty buffer.

> **function** free(ToLeave, st)
>   if *ToLeave* = ϵ **then** return *st*
>   **else**
>   addr := first(ToLeave)
>   st' := SpidergonLoadBuffer(addr, ϵ, st)
>   return free(tail(ToLeave), st')
>   **end if**
> **end function**

Function *pktScheduler* uses functions *GoodRoute?* and *hop* to move, whenever it is possible, the frames of a list of travels. It takes as arguments a list of travels, and three accumulators: *EnRoute* to store the frames that are *en route*, *Arrived* to store the frames that have reached their destination, and *ToLeave* to store the nodes that must be emptied. It also takes as argument the current state of the network. It returns the final values of accumulators *EnRoute* and *Arrived*, and a new state.

The definition of function *pktScheduler* is given below. If a move is possible and the length of the route equals two (lines 4 to 9), the frame moves to its destination. It is added to list *Arrived*, and the two addresses of its route can be freed. If there are more than two addresses in the route, the frame makes one hop (line 10 to 14). This means that the first address of the route can be freed (line 11), the current position is modified and the frame added to list *EnRoute* (line 12), and the frame is stored in the buffer of the destination of the hop (line 13). If the frame is blocked at its current position (lines 16 to 18), the frame is simply added to list *EnRoute*.

```
 1: function PKTSCHEDULER(Travels, EnRoute, Arrived, ToLeave, st)
 2:     if Travels = ε then return list(EnRoute,Arrived, ToLeave, st)
 3:     else  v := first(Travels)
 4:         if GoodRoute?(v.Route, st) then //a move is possible
 5:             if len(v.Route) = 2 then //Next hop is final destination
 6:                 Arrived := insert(v, Arrived) //v added to list Arrived
 7:                 ToLeave := insert(v.Route, ToLeave) //2 addresses added to list ToLeave
 8:                 st' := SpidergonLoadBuffer(Second(v.Route), v.frm, st)
 9:                 return  pktScheduler(tail(Travels), EnRoute, Arrived, ToLeave, st')
10:             else//frame still en route
11:                 ToLeave := insert(first(v.Route), ToLeave) //current position has to be left
12:                 EnRoute := insert(hop(v), EnRoute) //current position = hop destination
13:                 st' := SpidergonLoadBuffer(Second(v.Route), v.frm, st)
14:                 return  pktScheduler(tail(Travels), EnRoute, Arrived, ToLeave, st')
15:             end if
16:         else//no move, no change
17:             EnRoute := insert(v, EnRoute) //v is still en route
18:             return pktScheduler(tail(Travels), EnRoute, Arrived, ToLeave, st)
19:         end if
20:     end if
21: end function
```

Finally, function *packetScheduling* uses function *free* to empty the buffers of list *ToLeave*. This function also consumes attempts.

```
function PACKETSCHEDULING(Travels, att, st)
        EnRoute := pktScheduler.EnRoute(Travels, ε, ε, ε, st)
        Arrived := pktScheduler.Arrived(Travels, ε, ε, ε, st)
        ToLeave := pktScheduler.ToLeave(Travels, ε, ε, ε, st)
        att':= ConsumeAttempts(att)
        st' := pktScheduler.st(Travels, ε, ε, ε, st)
        st" := free(ToLeave,st') //old places are freed
        return list(EnRoute, Arrived, att', st")
end function
```

The proof obligations of Section 6.4.5 have been discharged for this function.

6.5.1.5   Instantiation of the Global Function GeNocCore

We instantiate function *GenocCore* with the Spidergon and its packet switching technique. This is accomplished by replacing the functions *Routing* and *Scheduling* by their instantiated versions, functions *SpidergonRouting* and *packetScheduling*.

```
function SPIDERGONGENOCCORE(Missives, att, time, st, Arrived)
    if SumOfAtt(att) = 0 then //All attempts have been consumed
        Aborted := Missives //At the end, Missives = union of en route and delayed
        return list(Arrived, Aborted)
    else
        Traveling := r4d.Traveling(Missives, time) //Extract authorized missives
        Delayed := r4d.Delayed(Missives, time)
        v :=SpidergonRouting(Traveling) //Route and travels
        EnRoute : = packetScheduling.EnRoute(v, att, st)
        Arr:= union(Arrived, packetScheduling.Arrived(v, att, st) ) //Partial result
        st' := packetScheduling.st(v, att, st)
        att' := packetScheduling.att(v, att, st)
        return SpidergonGenocCore(union(EnRoute, Delayed), att', time + 1, st', Arr)
    end if
end function
```

Because it has been proven with ACL2 that all the instantiated functions satisfy the instantiated proof obligations, it automatically follows that function *SpidergonGenocCore* satisfies the corresponding instance of Theorem 6.2.

6.5.2   The Hermes Network and its Wormhole Switching Technique

6.5.2.1   Hermes: Architecture Overview

The Hermes network [63] is a scalable reusable interconnection fabric developed at the Universidade Catolica do Rio Grande do Sul (Brazil). Its architecture is a 2D mesh. Its basic building block, a node, is the assembly of a switch (Figure 6.5) that has five bidirectional ports (North, South, East, West, and Local) and a connected IP. The Local port establishes a connection between the switch and the local core, while each of the four other ports are connected to a neighboring switch. Each of the ports contains a parameterized input buffer for temporary storage of transient information. The control logic of the switch encodes a deterministic minimal XY routing algorithm coupled with the wormhole switching technique and a round-robin arbitration scheme for contention resolution.

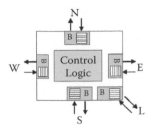

FIGURE 6.5
HERMES switch [63].

6.5.2.2 Formal Model Preliminaries: Nodes and State Definition

Each of the five bidirectional ports has two possible directions: input (i) and output (o). Each node is identified by a unique couple of natural numbers representing its coordinates. A valid address is defined as a tuple $\langle x, y, port, dir \rangle$, where x and y represent the coordinates on the X- and the Y-axis, *port* denotes one of the five possible ports, and *dir* is the direction. The set of all valid Hermes addresses is denoted *HermesAddresses*.

Hermes is described [63] with eight memory elements (buffers) for each input port, each buffer capable of storing one flit, and one buffer for each output port. Our proofs are in fact parameterized on the numbers of buffers.

Functions *2DmeshLoadBuffer(addr, frm,st)* and *2DmeshReadBuffer(addr,st)* are the instantiations of the generic functions *loadBuffer* and *readBuffer*. They play similar roles as the ones used for the Spidergon network.

6.5.2.3 Instantiating Function Routing: XYRouting

Core Routing Function. Let s be the current node and d the destination node. Four unitary moves are possible (north, south, east, west). If the flit is on an input port, it stays on the same node but moves to an output port, for example, a flit on port $\langle 0, 0, e, i \rangle$ that has to move to the north uses function *moveNorth* to move to port $\langle 0, 0, n, o \rangle$. If the flit is on an output port, for instance $\langle 0, 0, n, o \rangle$, it moves to the input port of the corresponding neighbor, for instance $\langle 0, 1, s, i \rangle$.

```
function MOVENORTH(s)
    if s.dir = i then return ⟨s.x, s.y, n, o⟩
    else    return ⟨s.x, (s.y + 1), s, i⟩
    end if
end function

function MOVESOUTH(s)
    if s.dir = i then return ⟨s.x, s.y, s, o⟩
    else    return ⟨s.x, (s.y − 1), n, i⟩
    end if
end function
```

```
function MOVEEAST(s)
    if s.dir = i then return ⟨s.x, s.y, e, o⟩
    else    return ⟨(s.x + 1), s.y, w, i⟩
    end if
end function


function MOVEWEST(s)
    if s.dir = i then return ⟨s.x, s.y, w, o⟩
    else    return ⟨(s.x − 1), s.y, e, i⟩
    end if
end function
```

These four functions are used in function *XYRoutingLogic* below. First, the X-coordinates of the two nodes are compared. If they are different, then the flit has to move to the east if the X-coordinate of the destination is higher than that of the current node, else it goes to the west. If the X-coordinates are equal then, if the Y-coordinate of the destination is higher than the current's, the flit moves to the north otherwise it moves to the south.

```
function XYROUTINGLOGIC(s,d)
    if s.x ! = d.x then //change X
        if s.x < d.x then   moveEast(s)
        else  moveWest(s)
        end if
    else//change Y
        if s.y < d.y then   moveNorth(s)
        else  moveSouth(s)
        end if
    end if
end function
```

The previous function computes the unitary movements of a flit, although the entire route is computed using function *XYRoutingCore* shown.

```
function XYROUTINGCORE(s,d)
    if s = d then return d //destination reached
    else    return list(s, XYRoutingCore(XYRoutingLogic(s,d), d)) //do one hop
    end if
end function
```

We prove that *XYRoutingCore* is a valid instance of the generic routing function, that is, it produces routes that satisfy predicate *ValidRoutep* (instance of Proof Obligation 1).

THEOREM 6.4
Validity of XY routing Routes.

$$\forall m \in Missives,\ ValidRoutep(XYRoutingCore(m.curr,\ m.dest),$$
$$m,\ HermesAddresses)$$

PROOF   ACL2 performs the proof by induction on the route length.   ■

Finally, function *XYRouting* corresponds to function *Routing*.

```
function XYROUTING(Missives)
    if Missives = ε then return ε
    else
        m := first(Missives)
        route := XYRoutingCore(m.curr, m.dest)
        t := ⟨m.id, m.org, m.frm, route, m.flit, m.time⟩
        return list(t, XYRouting(tail(Missives)))
    end if
end function
```

The same proof obligations discharged for *SpidergonRouting* are verified for *XYRouting* to prove that it is a valid instance of function *Routing*.

6.5.2.4   *Instantiating Function Scheduling: Wormhole Switching*

The modeling of the wormhole switching technique follows the same structure as the modeling of the packet switching technique. It proceeds in the same three steps: (1) to check whether the move from the current address to the neighbor address is possible, (2) to perform the move, and (3) to free the places left in step (2). Step (2) is performed by function hop defined in Section 6.5.1. We briefly discuss the difference with step (1), and then give more details about step (3).

A flit is capable of moving forward if the next hop's buffer is free. When the last flit in the buffer (last flit entered into the FIFO) is the last flit of a message as well, then a flit from another message is allowed to move into this buffer if there is a free position. Otherwise only a flit from the same message can move into the buffer. Moreover, two flits are not allowed to move into the same buffer at the same time. Function *NextHopFree?(addr, st)* encodes all these conditions.

Function *Good Route WH?* resembles the one used in the Spidergon case, and uses function *NextHopFree?*

```
function GOODROUTEWH?(r, st)
        return NextHopFree?(Second(r), st)
end function
```

In the wormhole technique, it might be the case that not all the flits of a frame can move. If the head of a worm makes one step, then all the remaining flits can also make one step. This movement is computed by function *advanceFlits(moving, st)*, which takes as arguments a list of frames, the head of which can make a step, and the current network state. It produces a new state.

In the case where the head of a frame is blocked at some address but (some of) its flits can move towards it, we use function *moveBlockedFlits(blocked, st)*, where *blocked* is a list of frames, the head of which is blocked, and *st* is the current network state. Function *moveBlockedFlits* will move flits of *blocked*, whenever it is possible. It produces a new state. These two operations define function *free*.

```
function FREE(blocked, moving, st)
        return advanceFlits (moving, (moveBlockedFlits (blocked, st)))
end function
```

Function *WormHScheduler* follows the same structure as function *pktScheduler* defined in Section 6.5.1. The main difference is that it takes two accumulators that store the frames, the head of which is blocked (list *2Bkept*), and the frames, the head of which is making one step (list *2Bmoved*). In the definition the main difference appears in the case where a head is blocked (lines 16 to 19).

```
 1: function WORMHSCHEDULER(Travels, EnRoute, Arrived, 2Bkept, 2Bmoved, st)
 2:     if Travels = ε then return list(EnRoute, Arrived, 2Bkept, 2Bmoved, st)
 3:     else  v := first(Travels)
 4:         if Good Route WH?(v.Route, st) then //a move is possible
 5:             if len(v.Route) = 2 then //Next hop is final destination
 6:                 Arrived := insert(v, Arrived)
 7:                 2Bmoved := insert(v, 2Bmoved)
 8:                 st' := 2DmeshLoadBuffer(Second(v.Route), v.frm, st)
 9:                 return  WormHScheduler(tail(Travels), EnRoute, Arrived, 2Bkept, 2Bmoved,st')
10:             else//frame still en route
11:                 2Bmoved := insert(v, 2Bmoved)
12:                 EnRoute := insert(hop(v), EnRoute) //current position = hop destination
13:                 st' := 2DmeshLoadBuffer(Second(v.Route), v.frm, st)
14:                 return  WormHScheduler(tail(Travels), EnRoute, Arrived, 2Bkept, 2Bmoved,st')
15:             end if
16:         else//no move
17:             EnRoute := insert(v, EnRoute) //v is still en route
18:             2Bkept := insert(v, 2Bkept) //flits may move towards the blocked head
19:             return WormHScheduler(tail(Travels), EnRoute,Arrived, 2Bkept, 2Bmoved,st)
20:         end if
21:     end if
22: end function
```

Finally, function *WormholeScheduling* updates the state. This function also consumes attempts.

```
function WORMHOLESCHEDULING(Travels, att, st)
        EnRoute := WormHScheduler.EnRoute(Travels, ϵ, ϵ, ϵ, ϵ, st)
        Arrived := WormHScheduler.Arrived(Travels, ϵ, ϵ, ϵ, ϵ, st)
        2Bkept := WormHScheduler.2Bkept(Travels, ϵ, ϵ, ϵ, ϵ, st)
        2Bmoved := WormHScheduler.2Bmoved(Travels, ϵ, ϵ, ϵ, ϵ, st)
        att':= ConsumeAttempts(att)
        st' := WormHScheduler.st(Travels, ϵ, ϵ, ϵ, ϵ, st)
        st" := free(2Bkept, 2Bmoved, st') //free old places and move subsequent flits
        return list(EnRoute, Arrived, att', st")
end function
```

The proof obligations of Section 6.4.5 have been discharged for this function.

6.5.2.5   *Instantiation of the Global Function GeNocCore*

Function *GenocCore* is instantiated as follows. This is done by replacing the functions *Routing* and *Scheduling* by their instantiated versions, functions *XYRouting* and *Wormhole-Scheduling*.

```
function HERMESGENOCCORE(Missives, att, time, st, Arrived)
    if SumOfAtt(att) = 0 then //All attempts have been consumed
        Aborted := Missives //At the end, Missives = union of en route and delayed
        return list(Arrived, Aborted)
    else
        Traveling := r4d.Traveling(Missives, time) //Extract authorized missives
        Delayed := r4d.Delayed(Missives, time)
        v :=XYRouting(Traveling) //Route and travels
        EnRoute := WormholeScheduling.EnRoute(v, att, st)
        Arr := union(Arrived, WormholeScheduling.Arrived(v, att, st))
        st' := WormholeScheduling.st(v, att, st)
        att' := WormholeScheduling.att(v, att, st)
        return HermesGenocCore(union(EnRoute, Delayed), att', time + 1, st', Arr)
    end if
end function
```

Because ACL2 has proved that all the instantiated functions satisfy the instantiated proof obligations, it automatically follows that function *HermesGenocCore* satisfies the corresponding instance of Theorem 6.2.

6.6   Conclusion

In this chapter, we have formalized two dimensions of the NoC design space—the communication infrastructure and the communication paradigm—as a functional model in the ACL2 logic. For each essential design decision—topology, routing algorithm, and scheduling policy—a meta-model has been given. We have identified the properties and constraints that are requested of that meta-model to guarantee the overall correctness of the message delivery over the NoC. The results thus obtained are general, and are application independent. To ensure correct message delivery on a particular NoC design, one has to instantiate the meta-model with the specific topology, routing and scheduling, and demonstrate that each one of these main instantiated function satisfies the expected properties and constraints (proof obligations). The main correctness theorem follows, because it depends on the proof obligations only, not on the detailed implementation choices. This approach has been illustrated on several NoC designs.

Although inherently higher in level, the meta-model has been implemented in the logic of ACL2, with the use of some special feature of the proof system (e.g., the "encapsulate" mechanism) that enables restricted existential quantification over functions. At the cost of added modeling effort, compared to direct higher-level logic models, the benefit is considerable.

- The effort has been done on the meta-model, but proving its instances is mechanized and largely automatic.

- Using an executable logic (we recall that the input to ACL2 is a subset of Common Lisp) allows one to visualize the advancement of messages and their interactions over the NoC on test cases, as in any conventional simulator.

Much remains to be done before this type of approach can enter a routine design flow. First, the meta-model needs to be refined and a systematic method elaborated, to progressively synthesize the very abstract view it provides into an RTL implementation. Correctness preserving transformations and possibly additional proof obligations will lead to a modeling level that can directly be translated to synthesizable HDL.

Another direction for future work concerns the proof of theorems about other application independent properties, such as absence of deadlocks and livelocks, absence of starvation, and the consideration of non-minimal adaptive routing algorithms. Again, we want to lay the ground work for the proof of properties over generic structures, and intend to proceed with a similar approach, by which a meta-model is applicable to a large class of IP generators.

References

1. J. A. Nacif, T. Silva, A. I. Tavares, A. O. Fernandes, and C. N. Coelho Jr, "Efficient allocation of verification resources using revision history information," In *Proc. of 11th IEEE Workshop on Design and Diagnostics of Electronic Circuits and Systems (DDECS'08)*. Bratislava, Slovakia: IEEE, April 2008.

2. L. Loh, "Where should I use formal functional verification," *Jasper Design Automation white paper*, July 2006. http://www.scdsource.com/download.php?id=4.
3. T. Bjerregaard and S. Mahadevan, "A survey of research and practices of network-on-chip," *ACM Computing Surveys* 38(1) (2006).
4. P. Pande, G. D. Micheli, C. Grecu, A. Ivanov, and R. Saleh, "Design, synthesis, and test of networks on chips," *Design & Test of Computers* 22 (2005) (5): 404.
5. L. Benini and G. D. Micheli, "Networks on chips: A new SoC paradigm," *Computer* 35 (2002) (1): 70.
6. U. Ogras, J. Hu, and R. Marculescu, "Key research problems in NoC design: A holistic perspective." In *Proc. of International Conference on Hardware/Software Codesign and System Synthesis (CODES+ISSS'05)*, 69. http://www.ece.cmu.edu/~sld/pubs/pagers/f175-ogras.pdf.
7. H. Wang, X. Zhu, L. Peh, and S. Malik, "Orion: A power-performance simulator for interconnection networks." In *Proc. of ACM/IEEE 35th Annual International Symposium on Microarchitecture (MICRO-35)*, 294. http://www.princeton.edu/~peh/publications/orion.pdf.
8. J. Madsen, S. Mahadevan, K. Virk, and M. Gonzalez, "Network-on-Chip modeling for system-level multiprocessor simulation." In *Proc. of the 24th IEEE Real-Time Systems Symposium (RTSS 2003)*, 2003, 265. 10.1109/REAL.2003.1253273.
9. J. Chan and S. Parameswaran, "NoCGEN: A template based reuse methodology for networks on chip architecture." In *Proc. of 17th International Conference on VLSI Design (VLSI Design 2004)*, 717. 10.1109/ICVD.2004.1261011.
10. L. Ost, A. Mello, J. Palma, F. Moraes, and N. Calazans, "MAIA—a framework for networks on chip generation and verification." In *Proc. of 2005 Conference on Asia South Pacific Design Automation (ASP-DAC 2005)*, 29. http://doi.acm.org/10.1145/1120725.1120741.
11. K. Goossens, J. Dielissen, O. P. Gangwal, S. G. Pestana, A. Radulescu, and E. Rijpkema, "A design flow for application-specific networks on chip with guaranteed performance to accelerate SoC design and verification." In *Proc. of Design, Automation, and Test in Europe Conference (DATE'05)*, 1182. http://homepages.inf.ed.ac.uk/kgoossen/2005-date.pdf.
12. N. Genko, D. Atienza, and G. D. Micheli, "NoC emulation on FPGA: HW/SW synergy for NoC features exploration." In *Proc. of International Conference on Parallel Computing (ParCo 2005)*, Malaga, Spain, September 2005.
13. J. S. Chenard, S. Bourduas, N. Azuelos, M. Boulé, and Z. Zilic, "Hardware assertion checkers in on-line detection of network-on-chip faults." In *Proc. of Workshop on Diagnostic Services in Networks-on-Chips*, Nice, France, April 2007.
14. *IEEE Std 1850-2005, IEEE Standard for Property Specification Language (PSL)*. IEEE, 2005.
15. K. Goossens, B. Vermeulen, R. van Steeden, and M. Bennebroek, "Transaction-based communication-centric debug." In *Proc. of First Annual ACM/IEEE International Symposium on Networks-on-Chip (NoCs'07)*, Princeton, NJ, May 2007, 95.
16. E. M. Clarke, O. Grumberg, and S. Jha, "Verifying parameterized networks," *ACM Transactions on Programming Languages and Systems* 19(5) (September 1997).
17. S. Creese and A. Roscoe, "Formal verification of arbitrary network topologies." In *Proc. of 1999 International Conference on Parallel and Distributed Processing Techniques and Applications (PDPTA'99)*. Las Vegas, NV: ACM/IEEE, June 1999.

18. K. L. McMillan, *Symbolic Model Checking*. Kluwer Academic Press, 1993.
19. A. Roychoudhury, T. Mitra, and S. Karri, "Using formal techniques to debug the AMBA System-on-Chip bus protocol." In *Proc. of Design, Automation, and Test in Europe Conference (DATE'03)*, Berlin, Germany: 828. http://www.comp.nus.edu.sg/~tulika/date03.pdf.
20. M. Gordon and T. Melham, eds., *Introduction to HOL: A Theorem Proving Environment for Higher Order Logic*. Cambridge, UK: Cambridge University Press, 1993.
21. P. Curzon, "Experiences formally verifying a network component." In *Proc. of Ninth Annual IEEE Conference on Computer Assurance*, 1994, 183. http://www.cl.cam.ac.uk/Research/HVG/atmproof/PAPERS/compass.ps.gz.
22. R. Bharadwaj, A. Felty, and F. Stomp, "Formalizing inductive proofs of network algorithms." In *Proc. of 1995 Asian Computing Science Conference*, Pathumthani, Thailand, December 1995, 335.
23. Y. Bertot and P. Castéran, *Interactive Theorem Proving and Program Development—Coq'Art: The Calculus of Inductive Constructions*. Berlin, Germany: Springer, 2004, see also http://coq.inria.fr.
24. G. J. Holzmann, "The model checker SPIN," *IEEE Transactions on Software Engineering* 23 (1997) (5): 279.
25. H. Amjad, "Model checking the AMBA protocol in HOL," University of Cambridge, Computer Laboratory, Technical Report, September 2004.
26. B. Gebremichael, F. W. Vaandrager, M. Zhang, K. Goossens, E. Rijpkema, and A. Radulescu, "Deadlock prevention in the Aethereal protocol." In *Proc. of 13th IFIP WG 10.5 Advanced Research Working Conference (CHARME 2005)*. http://www.ita.cs.ru.nl/publications/papers/fvaan/charme05.pdf.
27. J. S. Moore, "A formal model of asynchronous communications and its use in mechanically verifying a biphase mark protocol," *Formal Aspects of Computing* 6 (1993) (1): 60.
28. R. S. Boyer and J. S. Moore, *A Computation Logic Handbook*. London, UK: Academic Press, 1988.
29. D. Herzberg and M. Broy, "Modeling layered distributed communication systems," *Formal Aspects of Computing*, 17 (2005) (1): 1.
30. J. Rushby, "Systematic formal verification for fault-tolerant time-triggered algorithms," *IEEE Transactions on Software Engineering* 25 (1999) (5): 651.
31. L. Pike, "A note on inconsistent axioms in Rushby's systematic formal verification for fault-tolerant time-triggered algorithms," *IEEE Transactions on Software Engineering*, 32 (May 2006) (5): 347.
32. L. Pike, "Modeling time-triggered protocols and verifying their real-time schedules." In *Proc. of Formal Methods in Computer Aided Design (FMCAD'07)*, 2007. http://www.cs.indiana.edu./~lepike/pub_pages/fmcad.html.
33. P. S. Miner, A. Geser, L. Pike, and J. Maddalon, "A unified fault-tolerance protocol," In *Proc. of Formal Techniques, Modelling, and Analysis of Timed and Fault-Tolerant System* (FORMATS-FTRTFT04), LNCS 3253, 167–182, Genoble, France, September 22–24, Springer 2004, 167.
34. M. Kaufmann, P. Manolios, and J. Moore, *Computer Aided Reasoning: An Approach*. Berlin, Germany: Kluwer Academic Publishers, 2002, see also http://www.cs.utexas.edu/~moore/acl2.
35. C. Kern and M. R. Greenstreet, "Formal verification in hardware design: A survey," *ACM Transactions on Design Automation of Electronic Systems* 4(2): 123.

36. R. Dubey, "Elements of verification," *SOCcentral*, March 2005. http://www. einfochips.com./download/verification_whitepaper.pdf, Publisher: SoCcentral (www.soccentral.com).

37. L. Loh, "Formal verification: Where to use it and why," *EETimes EDA News*, July 2006. http://www.eetimes.com/news/design/showArticle. jhtml?articleID=190301228.

38. *IEEE Std 1800-2005, IEEE Standard for System Verilog: Unified Hardware Design, Specification and Verification Language.* IEEE, 2005.

39. S. Owre, J. M. Rushby, and N. Shankar, "PVS: A prototype verification system." In *Proc. of Conference on Automated Deduction (CADE 11)*, Saratoga Springs, NY, June 1992.

40. S. Merz, "Model checking: A tutorial overview." In *Modeling and Verification of Parallel Processes*, see. Lecture Notes in Computer Science, F. Cassez et al., ed., vol. 2067, 3. Berlin, Germany: Springer-Verlag, 2001.

41. A. Cohn, "The notion of proof in hardware verification," *Journal of Automated Reasoning* 5(2) (1989).

42. W. A. Hunt, "Fm8501: A verified microprocessor," Technical Report 47, Institute for Computing Science, University of Texas at Austin, February 1986.

43. C. Paulin-Mohring, "Circuits as streams in Coq: Verification of a sequential multiplier." In *Proc. of International Workshop on Types for Proofs and Programs (LNCS 1158)*, Rockport, MA, June 10–12, 1995.

44. D. Bolignano, "Towards the formal verification of electronic commerce protocols." In *Proc. of Tenth Computer Security Foundations Workshop (PCSFW)*. Washington, DC: IEEE Computer Society Press, 1997. 10.1109/CSFW.1997. 596802.

45. H. Ruess, N. Shankar, and M. Srivas, "Modular verification of SRT division." In *Proc. of CAV'96*, Aug. 1996. http://www.csl.sri.com/papers/srt-long/.

46. N. Shankar, "PVS: Combining specification, proof checking and model checking." In *Proc. of FMCAD'96*, November 1996.

47. S. Bensalem, Y. Lakhnech, and S. Owre, "InVeSt: A tool for the verification of invariance properties." In *Proc. of Eighth International Conference on Computer Aided Verification (CAV '96)*, Vancouver, BC, Canada, June 1998, 505–510. http://www.csl.sri.com/papers/cav98-tool/.

48. N. Shankar, "Combining theorem proving and model checking through symbolic analysis." In *Proc. of 11th International Conference on Concurrency Theory (CONCUR 2000)*. Lecture Notes in Computer Science, vol. 1877, New York: Springer-Verlag. 1. http://www.csl.sri.com/papers/concur2000/.

49. J. Moore, "Symbolic simulation: An ACL2 approach." In *Proc. of Formal Methods in Computer Aided Design Conference (FMCAD'98)*, Palo Alto, CA, Nov. 1998, 530. Lecture Notes in Computer Science, vol. 1522, New York: Springer-Verlag.

50. M. Wilding, D. Greve, and D. Hardin, "Efficient simulation of formal processor models," *Formal Methods in System Design* 18(3) (2001).

51. B. Brock, M. Kaufmann, and J. Moore, "ACL2 theorems about commercial microprocessors." In *Proc. of Formal Methods in Computer Aided Design Conference (FMCAD'96)*, Palo Alto, CA, November 1996, 275.

52. J. Sawada and W. A. Hunt, "Results of the verification of a complex pipelined machine model." In *Proc. of Tenth IFIP WG10.5 Advanced Research Working Conference on Correct Hardware Design and Verification Methods (CHARME'99)*, Bad Herrenalb, Germany, Sep. 1999. Lecture Notes in Computer Science, vol. 1703. New York: Springer-Verlag.

53. J. Moore, T. Lynch, and M. Kaufmann, "A mechanically checked proof of the correctness of the kernel of the AMD5K86 floating-point division algorithm," *IEEE Transactions on Computers* 47(9) (1998).

54. M. Kaufmann, P. Manolios, and J. Moore, *Computer Aided Reasoning: ACL2 Case Studies.* Berlin, Germany: Kluwer Academic Publishers, 2000.

55. J. Schmaltz and D. Borrione, "Towards a formal theory of on chip communications in the ACL2 logic." In *Proc. of Sixth International Workshop on the ACL2 Theorem Prover and its Applications ((part of FloC'06)*, Seattle, WA, August 2006.

56. J. Schmaltz and D. Borrione, "A functional formalization of on chip communications," *Formal Aspects of Computing (Springer)* 20(3): 241, May 2008.

57. D. Borrione, A. Helmy, L. Pierre, and J. Schmaltz, "A generic model for formally verifying NoC communication architectures: A case study." In *Proc. of First Annual ACM/IEEE International Symposium on Networks-on-Chip (NoCs'07)*, Princeton, NJ, May 2007, 127.

58. J. A. Rowson and A. Sangiovanni-Vincentelli, "Interface-Based Design." In *Proc. of 34th Design Automation Conference (DAC'97)*, Anaheim, CA, June 1997, 178.

59. J. Schmaltz, "Formal specification and validation of minimal routing algorithms for the 2D mesh." In *Proc. of Seventh International Workshop on the ACL2 Theorem Prover and its Applications (ACL2'07)*, Austin, TX, November 2007.

60. M. Coppola, S. Curaba, M. D. Grammatikakis, G. Maruccia, and F. Papariello, "OCCN: A network-on-chip modeling and simulation framework." In *Proc. of Design, Automation, and Test in Europe Conference (DATE'04)* 3(February 2004): 174, Paris, France.

61. M. Coppola, R. Locatelli, G. Maruccia, L. Pieralisi, and M. Grammatikakis, Spidergon: A NoC modeling paradigm. In *Model Driven Engineering for Distributed Real-Time Embedded Systems.* Paris, France: La Voisier, August 2005, ch. 13.

62. F. Karim, A. Nguyen, and S. Dey, "An interconnect architecture for networking systems on chip," *IEEE Micro* 22(5) (2002): 36–45.

63. F. Moraes, N. Calazans, A. Mello, L. Moller, and L. Ost, "HERMES: An infrastructure for low area overhead packet-switching networks on chip," *The VLSI Journal* 38 (2004) (1): 69.

7

Test and Fault Tolerance for Networks-on-Chip Infrastructures

**Partha Pratim Pande, Cristian Grecu, Amlan Ganguly,
Andre Ivanov, and Resve Saleh**

CONTENTS

7.1   Test and Fault Tolerance Issues in NoCs

Traditionally, correct fabrication of integrated circuits is verified by postmanufacturing testing using different techniques ranging from scan-based techniques to delay and current-based tests [1]. Due to their particular nature, Networks-on-Chips (NoCs) are exposed to a range of faults that can escape the classic test procedures. Such faults include crosstalk, faults in the buffers of NoC routers, and higher-level faults such as packet mis-routing and data scrambling [2]. These fault types add to the classic faults that must be tested postfabrication for all integrated circuits (stuck-at, opens, shorts, memory faults, etc.). Consequently, the test time of NoC-based systems increases considerably due to these new faults. Test time is an important component of the test cost and, implicitly, of the total fabrication cost of a chip. For large volume production, the total time that a chip requires for testing must be reduced as much as possible to keep the total cost low. The total test time of an IC is governed by the amount of test data that must be applied and the amount of controllability/observability that the design for test (DFT) techniques chosen by designers can provide. The test data increases with chip complexity and size, so the option the DFT engineers are left with is to improve the controllability/observability. Traditionally, this is achieved by increasing the number of test inputs/outputs, but this has the same effect of increasing the total cost of an IC. DFT techniques, such as scan-based tests, improve the controllability and observability of IC internal components by serializing the test input/output (I/O) data and feeding/extracting it to/from the IC through a reduced number of test pins. The trade-off is the increase in test time and test frequency, which makes at-speed test using scan-based techniques difficult. Although scan-based solutions are useful, their limitations in the particular case of NoC systems demand the development of new test data generation and transport mechanisms that reduce the total test time and at the same time do not require an increased number of test I/O pins. An effective and efficient test procedure is, however, not sufficient to guarantee the correct operation of NoC data transport infrastructures during the lifetime of the integrated circuits. Defects may appear later in the life of an IC, due to causes like electromigration, thermal effects, material aging, etc. These effects will become more important with continuous dimension downscaling of devices beyond 65 nm and moving towards the nanoscale domain. The technology projections for the next generations of nanoelectronic devices show that defect rates will be in the order of one to ten percent, and defect-tolerant techniques will have to be included in the early stages of the design flow of digital systems. Even with the defect rates indicated by the International Technology Roadmap for Semiconductors (ITRS) for upcoming CMOS processes [3], it is clear that correct fabrication is becoming more and more difficult to guarantee. An issue of concern in the case of communication-intensive platforms such as NoC is the integrity of the communication infrastructure. While addressing the reliability aspect, research must address the combination of new device-level defects or error-prone technologies within systems that must deliver high levels of reliability and dependability while satisfying other hard constraints such as low-energy consumption. By incorporating novel error correcting codes (ECC), it is possible to protect the NoC communication fabric against transient errors and at the same time lower the energy dissipation.

7.2   Test Methods and Fault Models for NoC Fabrics

The main concern for NoC/SoC test is the design of efficient test access mechanisms (TAMs) for delivering the test data to the individual cores under constraints such as test time, test power, and temperature. Among the different TAMs, TestRail [4] was one of the first to address core-based test of SoCs. Recently, a number of different research groups suggested the reuse of the communication infrastructure as a test access mechanism [5–7]. Vermeulen et al. [8] assumed the NoC fabric as fault-free, and subsequently used it to transport test data to the functional blocks; however, for large systems, this assumption can be unrealistic, considering the complexity of the design and communication protocols.

NoCs are built using a structured design approach, where a set of functional cores (processing elements, memory blocks, etc.) are interconnected through a data communication infrastructure that consists of switches and links. These cores can be organized either as regular or irregular topologies, as shown in Figure 7.1. The test strategies of NoC-based interconnect infrastructures must address three problems: (1) testing of the switch blocks; (2) testing of the interswitch wire segments; and (3) testing of the functional NoC cores. Test of both routers and links must be integrated in a streamlined fashion. First, the already-tested NoC components can be used to transport the test data toward the components under test in a recursive manner. Second, the inherent parallelism of the NoC structures allows propagating the test data simultaneously to multiple NoC elements under test. Test scheduling algorithms guarantee a minimal test time for arbitrary NoC topologies.

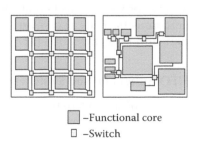

■ –Functional core
□ –Switch

FIGURE 7.1
(a) Regular (mesh architecture) NoC. (b) Irregular NoC.

7.2.1 Fault Models for NoC Infrastructure Test

When developing a test methodology for NoC fabrics, we need to start from a set of models that can realistically represent the faults specific to the nature of NoC as a data transport mechanism. As stated previously, an NoC infrastructure is built from two basic types of components: switches and interswitch links. For each type of component, we must construct test patterns that exercise its characteristic faults. Next, we describe the set of faults for the NoC switches and links.

7.2.2 Fault Models for NoC Interswitch Links

Cuviello et al. [9] proposed a novel fault model for the global interconnects of deep submicron (DSM) SoCs, which accounts for crosstalk effects between a set of aggressor lines and a victim line. This fault model is referred to as Maximum Aggressor Fault (MAF), and it occurs when the signal transition on a single interconnect line (called the victim line) is affected through crosstalk by transitions on all the other interconnect lines (called the aggressors) due to the presence of the crosstalk effect. In this model, all the aggressor lines switch in the same direction simultaneously. The MAF model is an abstract representation of the set of all defects that can lead to one of the six crosstalk errors: rising/falling delay, positive/negative glitch, and rising/falling speedup. The possible errors corresponding to the MAF fault model are presented in Figure 7.2 for a link consisting of three wires. The signals on lines Y_1 and Y_3 act as aggressors, while Y_2 is the victim line. The aggressors act collectively to produce a delay, glitch, or speedup on the victim. This abstraction covers a wide range of defects including design errors, design rules violations, process variations, and physical defects. For a link consisting of N wires, the MAF model assumes the worst-case situation with one victim line and $(N - 1)$ aggressors. For links consisting of a large number of wires, considering all such variations is prohibitive from a test coverage test time point of view [10]. The transitions needed to sensitize the MAF faults can be easily derived from Figure 7.2 based on the waveform transitions indicated. For an interswitch link consisting of N wires, a total of $6N$ faults need to be tested, requiring $6N$ 2-vector tests. These $6N$ MAF faults cover all the possible process variations and physical defects that can cause any crosstalk effect on any of the N interconnects. They also cover more traditional faults such as stuck-at, stuck open, and bridging faults [1].

7.2.3 Fault Models for FIFO Buffers in NoC Switches

NoC switches generally consist of a combinational block in charge of functions such as arbitration, routing, error control, and FIFO memory blocks that serve as communication buffers [11,12]. Figure 7.3(a) shows the generic architecture of an NoC switch. As information arrives at each of the ports, it is stored in FIFO buffers and then routed to the target destination by the routing logic block (RLB). The FIFO communication buffers for NoC fabrics

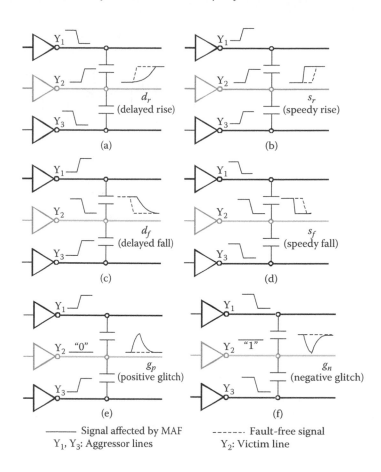

FIGURE 7.2
MAF crosstalk errors (Y_2 victim wire; Y_1, Y_3 aggressor wires).

can be implemented as register banks [13] or dedicated SRAM arrays [14]. In both cases, functional test is preferable due to its reduced time duration, good coverage, and simplicity. The block diagram of an NoC FIFO is shown in Figure 7.3(b). From a test point of view, the NoC-specific FIFOs fall under the category of restricted two-port memories. Due to the unidirectional nature of the NoC communication links, they have one write only port and one read only port, and are referred to as $(wo - ro)2P$ memories. Under these restrictions, the FIFO function can be divided in three ways: the memory cells array, the addressing mechanism, and the FIFO-specific functionality.

Memory array faults can be stuck-at, transition, data retention, or bridging faults [12]. Addressing faults on the RD/WD lines are also of importance as they may prevent cells from being read/written. In addition, functionality faults on the empty and full flags (EF and FF, respectively) are included in the set of fault models [11].

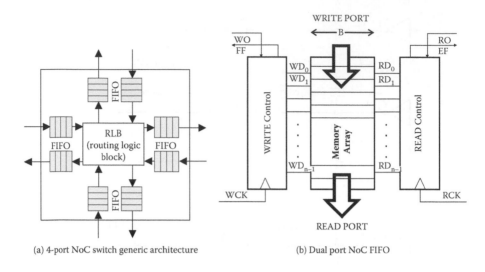

(a) 4-port NoC switch generic architecture   (b) Dual port NoC FIFO

FIGURE 7.3
(a) 4-port NoC switch-generic architecture. (b) Dual port NoC FIFO.

7.2.4 Structural Postmanufacturing Test

Once a set of fault models is selected, test data must be organized and applied to the building modules of the NoC infrastructure. In the classic SoC
test, this is accomplished by using dedicated TAMs such as TestRail. Because
NoC infrastructures are designed as specialized data transport mechanisms,
it is very efficient to reuse them as TAMs for transporting test data to functional cores [6]. The potential advantages when reusing NoC infrastructures
as TAMs are the low-cost overhead and reduced test time due to their high
degree of parallelism, which allows testing of multiple cores concurrently.
The challenges of testing the NoC infrastructure are its distributed nature
and the types of faults that must be considered. A straightforward approach
is to consider the NoC fabric as an individual core of the NoC-based system,
wrap it with an IEEE 1500 test wrapper and then use any of the core-based
test approaches. More refined methods can be used to exploit the particular
characteristics of NoC architectures. A test delivery mechanism that propagates the test data through the NoC progressively, reusing the previously
tested NoC components, was proposed by Grecu et al. [15]. The principle is
to organize test vectors as data packets and provide, for each router, a simple
BIST block that identifies the type of packets (test data) and extracts/applies
the test vectors. Test packets are organized similarly to regular data packets,
the difference being a flag in the packet header that identifies the packet as
carrying a test sequence. Test-specific control information is also embedded
into the test packets, followed by the set of test vectors. Figure 7.4 shows the
contents of a test packet.

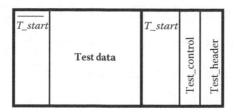

FIGURE 7.4
Test packet structure.

7.2.4.1   *Test Data Transport*

A systemwide test implementation has to satisfy the specific requirements of the NoC fabric and exploit its highly parallel and distributed nature for an efficient realization. In fact, it is advantageous to combine the testing of the NoC interswitch links with that of the other NoC components (i.e., the router blocks) to reduce the total silicon area overhead. However, special hardware may be required to implement parallel testing features.

In this section, we present the NoC modes of operation and a minimal set of features that the NoC building blocks must possess for packet-based test data transport. Each NoC switch is assigned a binary address so that the test packets can be directed to particular switches. In the case of direct-connected networks, this address is identical to the address of the IP core connected to the respective switch. In the case of indirect networks (such as BFT [13] and other hierarchical architectures) not all switches are connected to IP cores, so switches must be assigned specific addresses to be targeted by their corresponding test packets. Considering the degree of concurrency of the packets being transported through the NoC, we can distinguish two cases, described below.

7.2.4.1.1   *Unicast Mode*

The packets have a single destination. This is the most common situation and it is representative for the normal operation of an on-chip communication fabric, such as processor cores executing read/write operations from/into memory cores, or micro-engines transferring data in a pipeline [16]. As shown in Figure 7.5(a), packets arriving at a switch input port are decoded and directed to a unique output port, according to the routing information stored in the header of the packet (for simplicity, functional cores are not shown).

7.2.4.1.2   *Multicast Mode*

The packets have multiple destinations. Packets with multicast routing information are decoded at the switch input ports and then replicated identically at the switch outputs indicated by the multicast decoder. Multicast packets can reach their destinations in a more efficient and faster manner than in the

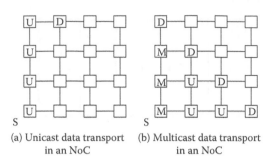

(a) Unicast data transport   (b) Multicast data transport
        in an NoC   in an NoC

FIGURE 7.5
Unicast and multicast switch modes. S and D are the source and destination nodes.

case when repeated unicast is employed to send identical data to multiple destinations. Figure 7.5(b) shows a multicast transport instance, where the data is injected at the switch source (S), replicated and retransmitted by intermediate switches in both multicast and unicast modes, and received by multiple destination switches (D). The multicast mode is especially useful for test data transport purposes, when identical blocks need to be tested as fast as possible. Several NoC platforms developed by research groups in industry and academia feature the multicast capability for functional operation [17,18]. In these cases, no modification of NoC switches hardware or addressing protocols is required to perform multicast test data transport.

If the NoC does not possess multicast capability, this can be implemented in a simplified version that only services the test packets and is transparent for the normal operation mode. As shown in Figure 7.6, the generic NoC switch structure presented in Figure 7.3(a) is modified by adding a multicast wrapper unit (MWU) whose functionality is explained below. It contains additional demultiplexers and multiplexers relative to the generic switch architecture. The MWU monitors the type of incoming packets and recognizes the packets that carry test data. An additional field in the header of the test packets identifies that they are intended for multicast distribution.

For NoCs supporting multicast for functional data transport, the routing/arbitration logic block (RLB) is responsible for identifying multicast packets, processing the multicast control information, and directing them to the corresponding output ports of the switch [13]. The multicast routing blocks can be relatively complex and hardware-intensive. For multicast test data transport only, the RLB of the switch is completely bypassed by the MWU and does not interfere with the multicast test data flow, as illustrated in Figure 7.6. The hardware implementation of the MWU is greatly simplified by the fact that the test scheduling is done off-line, that is, the path and injection time of each test packet is computed prior to performing the test operation. Therefore, for each NoC switch, the subset of input and output ports that will be involved in multicast test data transport is known *a priori*, and the implementation of this feature can be restricted to these specific subsets.

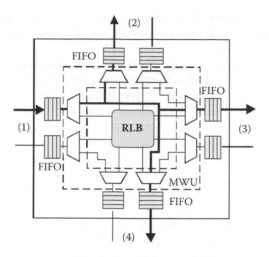

FIGURE 7.6
4-port NoC switch with multicast wrapper unit (MWU) for test data transport.

For instance, in the multicast step shown in Figure 7.5(b), only three switches must possess the multicast feature. By exploring all the necessary multicast steps to reach all destinations, we can identify the switches and ports that are involved in the multicast transport, and subsequently implement the MWU only for the required switches/ports. The header of a multidestination message must carry the destination node addresses [13]. To route a multidestination message, a switch must be equipped with a method for determining the output ports to which a multicast message must be simultaneously forwarded. The multidestination packet header encodes information that allows the switch to determine the output ports towards which the packet must be directed. When designing multicast hardware and protocols with limited purpose such as test data transport, a set of simplifying assumptions can be made to reduce the complexity of the multicast mechanism. This set of assumptions can be summarized as follows:

1. The test data traffic is fully deterministic.
2. Test traffic is scheduled off-line, prior to test application.
3. For each test packet, the multicast route can be determined exactly at all times (i.e., routing of test packets is static).
4. For each switch, the set of I/O ports involved in multicast test data transport is known and may be a subset of all I/O ports of the switch (i.e., for each switch, only a subset of I/O ports may be required to support multicast).

These assumptions help in reducing the hardware complexity of the multicast mechanism by implementing the required hardware only for those switch ports that must support multicast. For instance, in the example of Figure 7.6, if the multicast feature must be implemented exclusively from input port (1) to output ports (2), (3), and (4), then only one demultiplexer and three multiplexers are required.

Because the test data is fully deterministic and scheduled off-line, the test packets can be ordered such that the situation where two (or more) incoming packets compete for the same output port of a switch can be avoided. Therefore, no arbitration mechanism is required for multicast test packets. Also, by using this simple addressing mode, no routing tables or complex routing hardware is required.

The lack of I/O arbitration for the multicast test data has a positive impact on the transport latency of the packets. A test-only multicast implementation has lower transport latency than the functional multicast, because the only task performed by the MWU block is routing. The direct benefit is a reduced test time compared to the use of fully functional multicast, proportional to the number of processes that are eliminated. The advantages of using this simplified multicasting scheme are reduced complexity, lower silicon area required by MWU, and shorter transport latency for the test data packets.

7.2.4.1.3 *Test Time Cost-Problem Formulation*

In order to search for an optimal scheduling, we must first use the two components of the test time to determine a suitable cost function for the complete testing process. We then compute the test cost for each possible switch that can be used as a source for test packet injection. After sequencing through all the switches and evaluating their costs, we choose the one with the lowest cost as the source. We start by introducing a simple example that illustrates how the test time is computed in the two test transport modes, unicast and multicast, respectively. Consider the example in Figure 7.7, where switch S_1 and links l_1 and l_2 are already tested and fault free, and switches S_2 and S_3 are the next switches to be tested. When test data is transmitted in the unicast mode, only one NoC element goes into the test mode at a time, at any given time, as shown in Figure 7.7(a), (b).

Then, for each switch, the test time equals the sum of the transport latency and the effective test time of the switch. The latter term accounts for testing the FIFO buffers and RLB in the switches. Therefore, the total test time $T_{2,3}^u$ for testing both switches S_2 and S_3 is:

$$T_{2,3}^u = 2(T_{l,L} + T_{l,S}) + 2T_{t,S}$$

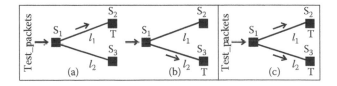

FIGURE 7.7
(a), (b) Unicast test transport. (c) Multicast test transport.

where $T_{l,L}$ is the latency of the interswitch link, $T_{l,S}$ is the switch latency [the number of cycles required for a flit (see Section 7.5.1) to traverse an NoC switch from input to output], and $T_{t,S}$ is the time required to perform the actual testing of the switch (i.e., $T_{t,S} = T_{FIFO} + T_{RLB}$). Following the same reasoning for the multicast transport case in Figure 7.7(c), the total test $T_{2,3}^m$ time for testing switches S_2 and S_3 can be written as:

$$T_{2,3}^m = (T_{l,L} + T_{l,S}) + T_{t,S}$$

From this simple example, we can infer that there are two mechanisms that can be employed for reducing the test time: reducing the transport time of test data, and reducing the effective test time of NoC components. The transport time of test patterns can be reduced in two ways: (a) by delivering the test patterns on the shortest path from the test source to the element under test; (b) by transporting multiple test patterns on nonoverlapping paths to their respective destinations.

Therefore, to reduce the test time, we would need to reevaluate the fault models or the overall test strategy (i.e., to generate test data locally for each element, with the respective incurred overhead [19]). Within the assumptions in this work (all test data is generated off-line and transported to the element under test), the only feasible way to reduce the effective test time per element is to overlap the test of more NoC components. The direct effect is the corresponding reduction of the overall test time. This can ultimately be accomplished by employing the multicast transport and applying test data simultaneously to more components. The graph representation of the NoC infrastructure used to find the minimum test transport latency is obtained by representing each NoC element as a directed graph $G = (S, L)$, where each vertex $s_i \in S$ is an NoC switch, and each edge $l_i \in L$ is an interswitch link. Each switch is tagged with a numerical pair $(T_{l,s}, T_{t,s})$ corresponding to switch latency and switch test time. Each link is similarly labeled with a pair $(T_{l,L}, T_{l,L})$ corresponding to link latency and link test time, respectively. For each edge and vertex, we define a symbolic toggle t that can take two values: $\mathbf{N}$ and $\mathbf{T}$. When $t = \mathbf{N}$, the cost (weight) associated with the edge/vertex is the latency term, which corresponds to the normal operation. When $t = \mathbf{T}$, the cost (weight) associated with the edge/vertex is the test time (of the link or switch) and corresponds to the test operation.

7.2.4.1.4   Test Output Evaluation

In classical core-based testing, test data is injected from a test source, transported and applied to the core under test, and then the test output is extracted and transported to the test sink for comparison with the expected response [4]. A more effective solution was first proposed by Grecu et al. [15], where the expected data is sent together with the input test data, and the comparison is performed locally at each component under test. A clear advantage of this approach is that, because there is no need to extract the test output and to transport it to a test sink, the total test time on the NoC infrastructure

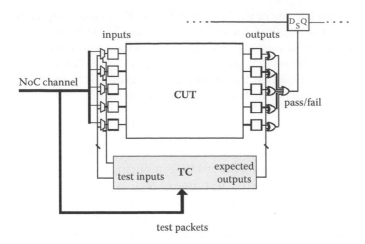

FIGURE 7.8
Test packets processing and output comparison.

can significantly be reduced. Moreover, the test protocol is also simplified, because this approach eliminates the need for a flow control of test output data (in terms of routing and addressing). The trade-off is a small increase in hardware overhead due to additional control and comparison circuitry, and increased size of the test packets (which now contain the expected output of each test vector, interleaved with test input data). As shown in Figure 7.8, the test packets are processed by test controller (TC) blocks that direct their content toward the I/Os of the component under test (CUT) and perform the synchronization of test output and expected data. This data is compared individually for each output pin and, in case of a mismatch, the component is marked as faulty by raising the pass/fail flag. The value of this flag is subsequently stored in a pass/fail flip-flop, which is a part of a shift register that connects pass/fail flops of all switches. The content of this register is serially dumped off-chip at the end of the test procedure.

7.2.5   Functional Test of NoCs

7.2.5.1   *Functional Fault Models for NoCs*

The architectural and technological complexity of NoC communication infrastructure demands the application of higher level tests for increasing the level of confidence in their correct functionality. High-level fault models for NoC infrastructures were developed taking into account the operation of NoC components (routers) and the services that an NoC must provide in terms of data delivery [19]. From a functional point of view, the services that an NoC provides can be categorized as follows:

- **Routing services.** These include forwarding data from an input port of a router to an output port according to the routing information embedded in the data packets.
- **Guaranteed/best effort services.** Data must not only be routed correctly but also performance requirements specified according to the QoS parameters in terms of throughput/latency must be satisfied for an NoC to operate correctly.
- **Network interfacing.** An NoC must be able to inject/eject data correctly at its interfaces (NIs, network interfaces), where the functional cores (processing elements, memory blocks, and other functional units) are plugged into the NoC platform. NIs also have a role in providing QoS guarantees by reserving a route from a source NI to a destination NI in the case of GT (guaranteed throughput) data.

Three high-level fault types can be defined according to the functionality described above:

1. Routing faults: Their effect is misrouting of data from an input port to an output port of the same router.
2. Router QoS faults: Their effects consist of QoS violations by a router or a set of routers.
3. NI faults: Model faults at the interaction between the functional cores and the NoC data transport fabric. These faults can be packetization/depacketization faults, or QoS faults.

NoC test based on functional fault models has several advantages compared to structural test, the most important ones being lower hardware overhead and shorter test time mainly due to a reduced set of test data that has to be applied. For a satisfactory fault coverage and yield, both structural and functional tests are required for NoC platforms.

7.3 Addressing Reliability of NoC Fabrics through Error-Control Coding

According to ITRS [3], signal integrity is expected to be an increasingly critical challenge in designing SoCs. The widespread adoption of the NoC paradigm will be facilitated if it addresses system-level signal integrity and reliability issues in addition to easing the design process and meeting all other constraints and objectives. With shrinking feature size, one of the major factors affecting signal integrity is transient errors, arising due to temporary conditions of the SoC and environmental factors. Among the transient failure mechanisms are crosstalk, electromagnetic interference, alpha particle hits, cosmic radiation, etc. [20,21]. These failures can alter the behavior of the NoC fabrics and degrade the signal integrity. Providing resilience against such failures is critical for the operation of NoC-based chips. There are many ways to achieve signal integrity. Among different practical methods, the use of new materials for device and interconnect, and tight control of device layouts may be adopted in the NoC domain. Here, we propose to tackle this problem at the design stage. Instead of depending on postdesign methods, we propose to incorporate corrective intelligence in the NoC design flow. This will help reduce the number of postdesign iterations. The corrective intelligence can be incorporated into the NoC data stream by adding error control codes to decrease vulnerability to transient errors. The basic operations of NoC infrastructures are governed by on-chip packet-switched networks. As NoCs are built on packet-switching, it is easy to modify the data packets by adding extra bits of coded information in space and time to protect against transient malfunctions.

In the face of increased gate counts, designers are compelled to reduce the power supply voltage to keep energy dissipation to a tolerable limit, thus reducing noise margins [20]. The interconnects become more closely packed and this increases mutual crosstalk effects. Faster switching can also cause ground bounce. The switching current can cause the already low-supply voltage to instantaneously go even lower, thus causing timing violations. All these factors can cause transient errors in the ultra deep submicron (UDSM) ICs [20]. Crosstalk is a prominent source of transient malfunction in NoC interconnects. Crosstalk avoidance coding (CAC) schemes are effective ways of reducing the worst-case switching capacitance of a wire by ensuring that a transition from one codeword to another does not cause adjacent wires to switch in opposite directions. Though CACs are effective in reducing mutual interwire coupling capacitance, they do not protect against any other transient errors. To make the system robust, in addition to CAC we need to incorporate forward error correction coding (FEC) into the NoC data stream. Among different FECs, single error correction codes (SECs) are the simplest to implement. There are various joint CAC/SEC codes proposed by different research groups. But aggressive supply-voltage scaling and increase in DSM noise in future-generation NoCs will prevent these joint CAC/SEC codes from satisfying reliability requirements. Hence, low-complexity joint crosstalk avoidance and multiple error correction codes (CAC/MEC) suitable for applying to NoC fabrics need to be designed. Below we elaborate characteristics of different CAC, joint CAC/SEC and CAC/MEC codes.

7.3.1 Crosstalk Avoidance Coding

Crosstalk is one of the prime causes of the transient random errors in the interswitch wire segments causing timing violations. Crosstalk occurs when adjacent wires transition (0 to 1 or 1 to 0) in opposite directions or even when adjacent wires have different slew rates although they are transitioning in the same direction. These two situations are shown in Figure 7.9(a), (b).

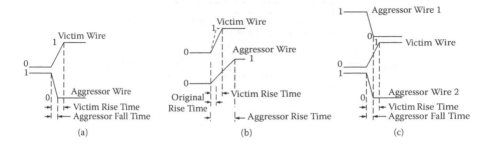

FIGURE 7.9
Different types of transitions causing crosstalk between adjacent wires.

The worst-case crosstalk occurs when two aggressors on either side of the victim wire transition in the opposite direction to the victim, as shown in Figure 7.9(c). Such a pattern of opposite transitions always increases the delay by increasing the mutual switching capacitance between the wires. In addition, it also causes extra energy dissipation due to the increase in switching capacitance. One of the common crosstalk avoidance techniques is to increase the spacing between adjacent wires. However, this doubles the wire layout area [22]. For global wires in the higher metal layers that do not scale as fast as the device geometries, this doubling of area is hard to justify. Another simple technique can be shielding the individual wires with a grounded wire in between them. Although this is effective in reducing crosstalk to the same extent as increased spacing, it also necessitates the same overhead in terms of wire routing requirements. By incorporating coding mechanisms, the same reduction in crosstalk can be achieved at a lower overhead of routing area [23]. These coding schemes, broadly termed as the class of crosstalk avoidance coding (CAC), prevent worst-case crosstalk between adjacent wires by preventing opposite transitions in the neighbors. Thus CACs enhance system reliability by reducing the probabilities of crosstalk–induced soft errors and also reduce the energy dissipation in UDSM buses and global wires by reducing the coupling capacitance between adjacent wires.

Different crosstalk avoidance codes [24] are proposed in the literature. Here, characteristics of three representative CACs that achieve different degrees of reduction in coupling capacitance are described.

7.3.2 Forbidden Overlap Condition (FOC) Codes

A wire has a worst-case switching capacitance of $(1 + 4\lambda)C_L$ [25] when it executes a rising (falling) transition and its neighbors execute falling (rising) transitions, where λ is the ratio of the coupling capacitance to the bulk capacitance and C_L is the load capacitance, including the self-capacitance of the wire. If these worst-case transitions are avoided, the maximum coupling can be reduced to $(1 + 3\lambda)C_L$ [25]. This condition can be satisfied if and only if a

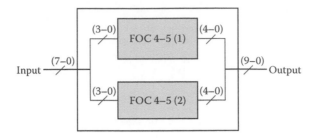

FIGURE 7.10
Block diagram of combining adjacent subchannels in FOC coding.

codeword having the bit pattern 010 does not make a transition to a codeword having the pattern 101 at the same bit positions. The codes that satisfy the above condition are referred to as forbidden overlap condition (FOC) codes. The simplest method of satisfying the forbidden overlap condition is half-shielding, in which a grounded wire is inserted after every two signal wires. Though simple, this method has the disadvantage of requiring a significant number of extra wires. Another solution is to encode the data links such that the codewords satisfy the forbidden overlap (FO) condition. However, encoding all the bits at once is not feasible for wide links due to prohibitive size and complexity of the coder-decoder (codec) hardware. In practice, partial coding is adopted, in which the links are divided into subchannels that are encoded using FOC. The subchannels are then combined in such a way as to avoid forbidden patterns at their boundaries. In this case, two subchannels can be placed next to each other without any shielding, as well as not violating the FO condition as shown in Figure 7.10. The Boolean expressions relating to the original input (d_3 to d_0) and coded bits (c_4 to c_0) for the FOC scheme are expressed as follows:

$$c_0 = d_1 + d_2\overline{d}_3$$
$$c_1 = d_2\overline{d}_3$$
$$c_2 = d_0$$
$$c_3 = d_2d_3$$
$$c_4 = d_1d_2 + d_3$$

7.3.3 Forbidden Transition Condition (FTC) Codes

The maximum capacitive coupling and, hence, the maximum delay, can be reduced even further by extending the list of nonpermissible transitions. By ensuring that the transitions between two successive codes do not cause adjacent wires to switch in opposite directions (i.e., if a codeword has a 01 bit pattern, the subsequent codeword cannot have a 10 pattern at the same bit position, and vice versa), then the coupling capacitance can be reduced to $(1 + 2\lambda)C_L$ [25]. This condition is referred to as forbidden transition

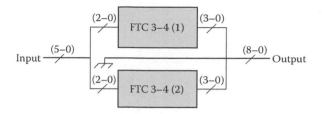

FIGURE 7.11
Block diagram of combining adjacent subchannels in FTC coding.

condition (FTC), and the CACs satisfying it are known as FTC codes. For wider communication links, the message words can be subdivided into multiple subchannels, each having a three-bit width, and then each coded subwords recombined following the scheme shown in Figure 7.11. This scheme of recombination simply places a shielded wire between each subchannel. This ensures no forbidden transitions even at the boundaries of the subchannels.

The Boolean expressions relating to the original input and coded bits for the FTC scheme are expressed as follows:

$$c_0 = d_1 + d_2\overline{d}_0$$
$$c_1 = d_0d_1d_2 + \overline{d}_0\overline{d}_1d_2$$
$$c_2 = d_0 + d_2$$
$$c_3 = d_0d_2 + d_1d_2$$

7.3.4 Forbidden Pattern Condition Codes

The same reduction of the coupling factor as for FTCs can be achieved by avoiding 010 and 101 bit patterns for each of the code words. This condition is referred to as forbidden pattern condition (FPC), and the corresponding CACs are known as FPC codes. As before, while combining the subchannels we need to confirm that there is no forbidden pattern at the boundaries.

Figure 7.12 depicts the scheme of avoiding forbidden patterns at the boundaries, considering four-bit subchannels. The MSB of a subchannel is fed to the LSB of the adjacent one. This method is more efficient than simply placing shielding wires between the encoded subchannels, and consequently results in lower overhead.

The Boolean expressions relating to the original input (d_3 to d_0) and coded bits (c_4 to c_0) for the FPC scheme are expressed as follows:

$$c_0 = d_0$$
$$c_1 = d_1d_1 + d_2d_1 + d_1\overline{d}_3 + d_0d_2\overline{d}_3$$
$$c_2 = d_2\overline{d}_3 + d_1d_2 + \overline{d}_0d_2 + d_1\overline{d}_0\overline{d}_3$$
$$c_3 = d_2d_3 + \overline{d}_0d_2 + d_2d_1 + d_1d_3\overline{d}_0$$
$$c_4 = d_3$$

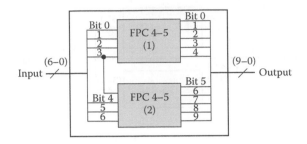

FIGURE 7.12
Block diagram of combining adjacent subchannels after FPC coding.

7.4 Joint Crosstalk Avoidance and Error-Control Coding

Besides crosstalk, there are several other sources of transient errors, as discussed earlier, like electromagnetic interference, alpha particle hits and cosmic radiation which can alter the behavior of NoC fabrics and degrade signal intégrity. Providing resilience against such failures is critical for the operation of NoC-based chips. Once again, these transient errors can be addressed by incorporating error-control coding to provide higher levels of reliability in the NoC communication fabric [26,27]. FEC or error detection (ED) followed by retransmission-based mechanisms or a hybrid combination of both can be used to protect against transient errors. The SEC codes are the simplest to implement among the FECs. These can be implemented using Hamming codes for single error correction. Parity check codes and cyclic redundancy codes also provide error resilience by forward error correction. Error detection codes can be used to detect any uncorrectable error pattern and send an automatic repeat request (ARQ) for retransmission of the data, thus reducing the possibilities of dropped information packets. Murali et al. [28] addressed error resiliency in NoC fabrics and the trade-offs involved in various error recovery schemes. In this work, the authors investigated simple error detection codes like parity or cyclic redundancy check (CRC) codes and single error correcting, double error detecting Hamming codes. One specific problem pertaining to coding in NoCs is highlighted by Bertozzi et al. [29]. They concluded that error detection followed by retransmission is more energy efficient than forward error correction. But this work was done in a much older technology generation (0.25 μm technology) than the UDSM regime, where the problems arising out of transient noise will be the most severe. As mentioned in the concluding remarks by Bertozzi et al. [29], in the UDSM domain communication energy is going to overcome computation energy. Retransmission will give rise to multiple communications over the same link and hence ultimately it will not be energy efficient. In systems dominated by retransmission, additional error correction mechanisms for the control signals also need to be incorporated. One class of codes that have achieved considerable attention in the recent past is the joint coding schemes that attempt to minimize crosstalk while also performing forward error correction. These are called joint crosstalk avoidance and single error correction codes (CAC/SEC) [30]. A few of these joint codes have been proposed in the literature for on-chip buses. These codes can be adopted in the NoC domain too. These include duplicate add parity (DAP) [31], boundary shift code (BSC) [32], or modified duplicate add parity (MDR) [33]. These are joint crosstalk-avoiding, single error correcting codes. These coding schemes achieve the dual function of reducing crosstalk and also increase the resilience against multiple sources of transient errors.

Most of the above work depended on simple SEC codes. But with technology scaling, SECs are not sufficient to protect NoCs from varied sources of transient noise. This was acknowledged for the first time by Sridhara and Shanbhag [30] in the context of traditional bus-based systems. It was pointed out that with aggressive supply scaling and increase in DSM noise, more powerful error correction schemes than the simple joint CAC/SEC codes will be needed to satisfy reliability requirements. But aggressive supply-voltage scaling and increase in deep submicron noise in future-generation NoCs will prevent joint CAC/SEC codes from satisfying reliability requirements. Hence, further investigations into the performance of joint CAC/MEC codes in NoC fabrics need to be made. A particular example of a joint crosstalk avoidance and double error correction Code (CADEC) is discussed in details. Below, the characteristics of the joint crosstalk avoidance and error correction coding schemes and their implementation principles are discussed in details.

7.4.1 Duplicate Add Parity and Modified Dual Rail Code

The *duplicate add parity (DAP) code* is a joint coding scheme that uses duplication to reduce crosstalk [31]. Duplication results in reducing the crosstalk-induced coupling capacitance of a wire segment from $(1 + 4\lambda)C_L$ to $(1 + 2\lambda)C_L$ [30]. Also, by duplication, we can achieve a Hamming distance of two and with the addition of a single parity bit, the Hamming distance [31] increases to three. Consequently, DAP has single error correction capability. The DAP encoder and decoder are shown in Figure 7.13(a), (b), respectively. Encoding involves calculating the parity and duplicating the bits of the incoming word. Similarly, in decoding, the parity bit is recreated from a set of the data flit. As shown in Figure 7.13(b), bit y_8 is the previously calculated parity, and the other signal entering the exclusive-or gate is the newly calculated parity of the more significant set (bits y_1, y_3, y_5, and y_8). The new parity is compared with the original parity calculated in the encoder, and the error-free set is chosen. For example, in case of an error in the more significant set, the parities will differ, and the less significant set will be chosen as the decoded flit. On the other hand, if the error occurs in the less significant set, the more significant set will be chosen. Thus, considering a link of k information bits, $m = k + 1$ check bits are added, leading to a code word length of $n = k + m = 2k + 1$.

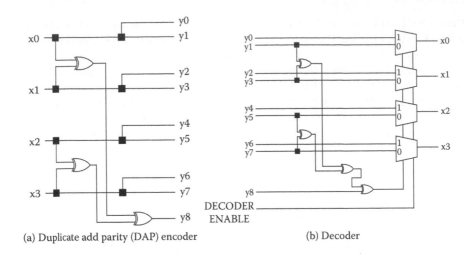

(a) Duplicate add parity (DAP) encoder   (b) Decoder

FIGURE 7.13
Duplicate add parity encoder and decoder.

We define the $k + 1$ check bits with the following equations:

$$c_i = d_i, \text{ for } i = 0, \ k - 1$$
$$c_k = d_0 \oplus d_1 \oplus \cdots \oplus d_{k-1}$$

The *modified dual rail (MDR)* code is very similar to the DAP [33]. In the MDR code, two copies of parity bit C_k are placed adjacent to the other codeword bits to reduce crosstalk.

7.4.2 Boundary Shift Code

The *boundary shift code (BSC)* is a coding scheme that attempts to reduce crosstalk-induced delay by avoiding a dependent boundary between successive codewords. The dependent boundary in a word is defined as a place where two adjacent bits differ, and denoted by the position of the leftmost bit of the boundary. As shown by Patel and Markov [32], this technique achieves a reduction in the worst-case crosstalk-induced switching capacitance from $(1 + 4\lambda)C_L$ to $(1 + 2\lambda)C_L$. It is very similar to DAP in that it uses duplication and one parity bit to achieve crosstalk avoidance and single error correction. However, the fundamental difference is that at each clock cycle, the parity bit is placed on the opposite side of the encoded flit. Encoding is achieved by duplicating bits and completing a parity calculation as in DAP. However, every second clock cycle will result in a one-bit shift. Similarly, the decoding structure is equivalent to that of DAP with the addition of a one-bit shift every other clock cycle before the parity check. Figure 7.14(a), (b) depicts the encoder and decoder, respectively.

One of the principal differences between the CAC schemes and the joint codes is that for the joint codes we do not have to divide the whole link

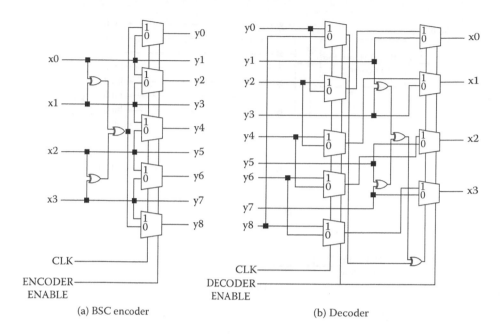

(a) BSC encoder

(b) Decoder

FIGURE 7.14
BSC encoder and decoder.

into different subchannels and then perform partial coding. We can perform DAP/BSC/MDR coding/decoding on the link as a whole.

7.4.3 Joint Crosstalk Avoidance and Double Error Correction Code

The CADEC is a joint coding scheme that performs crosstalk avoidance and double error correction simultaneously. It achieves crosstalk avoidance by duplication of the bits [34]. The same technique also increases the minimum hamming distances between codewords enabling a higher error correction capability.

CADEC encoder. The encoder is a simple combination of Hamming coding followed by DAP or BSC encoding to provide protection against crosstalk. As shown in Figure 7.15(a), the incoming 32-bit flit is first encoded using a standard (38, 32) shortened Hamming code, and then each bit of the 38-bit Hamming codeword is duplicated and appended with a parity. The standard Hamming code has a Hamming distance of 3 between adjacent code words. On duplication, this becomes 6 and, after adding the extra parity bit, this distance becomes 7. A Hamming distance of 7 enables triple error correction, but at a somewhat higher complexity cost than the double error correcting schemes considered here. Consequently, as a first step, we considered only the double error correction capability. The extra parity bit, which is a part of

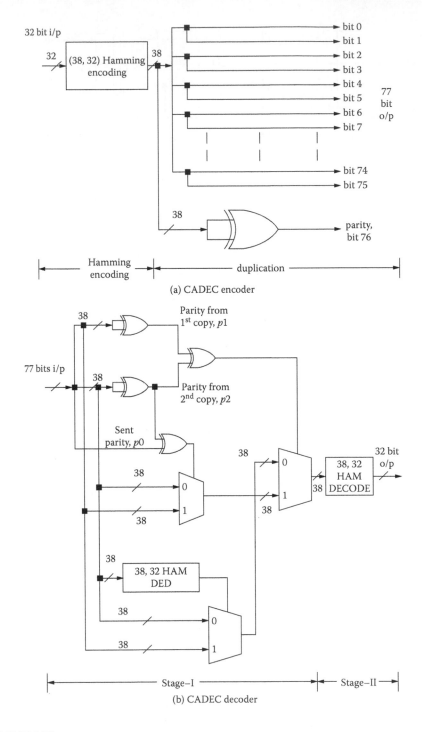

(a) CADEC encoder

(b) CADEC decoder

FIGURE 7.15
CADEC encoder and decoder.

DAP or BSC schemes, is added to make the decoding process very energy efficient as explained below.

CADEC decoder. The decoding procedure for the CADEC encoded flit can be explained with the help of the flow diagram shown in Figure 7.16. The decoding algorithm consists of the following simple steps:

1. The parity bits of the individual Hamming copies are calculated and compared with the sent parity.
2. If these two parities obtained in step (1) differ, then the copy whose parity matches with the transmitted parity is selected as the output copy of the first stage.
3. If the two parities are equal, then any one copy is sent forward for double error detection (DED) by the Hamming Syndrome detection block (38, 32).
4. If the syndrome from the DED block obtained for this copy is zero, then this copy is selected as the output of the first stage. Otherwise, the alternate copy is selected.
5. The output of the first stage is sent for single error correcting Hamming decoding (38, 32), finally producing the decoded CADEC output.

The circuit implementing the decoder is shown schematically in Figure 7.15(b). The use of the DAP or BSC parity bit actually makes the decoder more energy efficient, compared to a scheme without the parity bit, which always requires a syndrome to be computed on both copies. When the parity bits generated from individual Hamming copies fail to match, the DED-syndrome block need not be used at all, thus on average making the overall decoding process more energy efficient. This situation arises when there is single error in either one of the two Hamming copies, which generally will be the most probable case.

7.5 Performance Metrics

In order to quantify the effectiveness of the coding schemes described in earlier sections, their performance in different NoC architectures needs to be studied. The principal metrics of interest are the energy dissipation profile and latency characteristics of the NoCs, in presence of coding.

7.5.1 Energy Savings Profile of NoCs in Presence of Coding

In NoCs, the interconnect structure dissipates a large percentage of energy. In certain applications [35], this percentage has been shown to approach

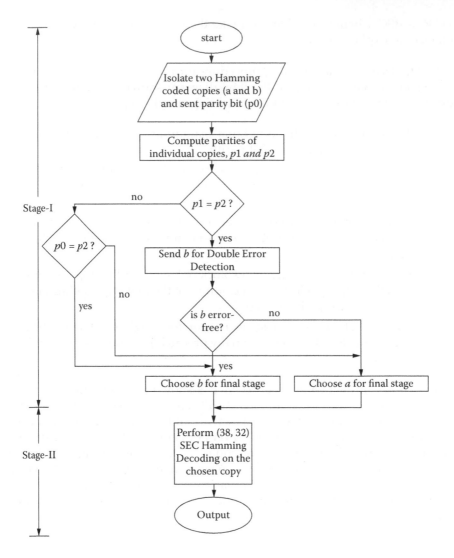

FIGURE 7.16
CADEC decoding algorithm.

50 percent. Consequently, the most important performance metric to be considered in presence of coding in an NoC is the communication energy. The data communication between the embedded cores in an NoC takes place in the form of packets routed through a wormhole switching mechanism [14]. The packets are broken down into fixed length flow control units or flits. The header flits carry the relevant routing information. Consequently header decoding enables the establishment of a path that the subsequent payload flits simply follow in a pipelined fashion. The transmitted flits are encoded to guard against possible transient errors. When flits travel on the interconnection

network, both the interswitch wires and the logic gates in the switches toggle, resulting in energy dissipation. The flits from the source nodes need to traverse multiple hops consisting of switches and wires to reach destinations.

The motivation behind incorporating CAC in the NoC fabric is to reduce switching capacitance of the interswitch wires and hence make communication among different blocks more energy efficient. But this reduction in energy dissipation is linear with the switching capacitance of the wires. By incorporating the joint coding schemes in an NoC data stream, the reliability of the system is enhanced. Consequently, the supply voltage can be reduced without compromising system reliability. As energy dissipation depends quadratically on the supply voltage, a significantly higher amount of savings is possible to achieve by incorporating the joint codes. To quantify this possible reduction in supply voltage, a Gaussian distributed noise voltage of magnitude V_N and variance or power of σ_N^2 is considered that represents the cumulative effect of all the different sources of UDSM noise. This gives the probability of bit error, ϵ, also called the bit error rate (BER) as

$$\epsilon = Q\left(\frac{V_{dd}}{2\sigma_N}\right) \tag{7.1}$$

where, the Q-function is given by

$$Q(x) = \frac{1}{\sqrt{2\pi}} \int_x^\infty e^{\frac{y^2}{2}} dy \tag{7.2}$$

The word error probability is a function of the channel BER, ϵ. If $P_{unc}(\epsilon)$ is the probability of word error in the uncoded case and $P_{ecc}(\epsilon)$ is the residual probability of word error with error control coding, then it is desirable that $P_{ecc}(\epsilon) \leq P_{unc}(\epsilon)$. Using Equation (7.1), we can reduce the supply voltage in presence of coding to $\hat{V}_{dd}$, given by

$$\hat{V}_{dd} = V_{dd}\frac{Q^{-1}(\hat{\epsilon})}{Q^{-1}(\epsilon)} \tag{7.3}$$

In Equation (7.3), V_{dd} is the nominal supply voltage in the absence of any coding such that $P_{ecc}(\hat{\epsilon}) = P_{unc}(\epsilon)$. Therefore, to compute the $\hat{V}_{dd}$ for the joint CAC and SEC, the residual word error probability of these schemes has to be computed. The various residual word error probabilities in terms of BER, ϵ, are listed in Table 7.1.

Figure 7.17 shows the plot of possible voltage swing reduction for different joint codes discussed here with increasing word error rates. As CADEC has the highest error correction capability, it allows maximum voltage swing reduction.

So, the metric of interest is the average savings in energy per flit with coding compared to the uncoded case. All the schemes have different number of bits in the encoded flit. A fair comparison in terms of energy savings demands that the redundant wires be also taken into account while comparing the energy dissipation profiles. The relevant metric used for comparison should

TABLE 7.1

Residual Word Error Probabilities of Different Coding
Schemes

Coding Scheme	Probability of Residual Word Error
Sole error detection (ED)	$P_{ED}(\epsilon) = (n-k)\epsilon^2$
DAP/BSC	$P_{DAP}(\epsilon) = \frac{3k(k+1)}{2}\epsilon^2$
CADEC	$P_{CADEC}(\epsilon)^2(n-4)\epsilon^3$

Note: The codes are assumed to be (n, k) codes with corresponding
values of n and k for individual schemes as mentioned under code
descriptions.

take into account the savings in energy due to the reduced crosstalk, reduced
voltage swing on the interconnects, and additional energy dissipated in the
extra redundant wires and the codecs. The savings in energy per flit per hop
is given by

$$E_{savings,j} = E_{link,uncoded} - (E_{link,coded} + E_{codec}) \tag{7.4}$$

where $E_{link,uncoded}$ and $E_{link,coded}$ are the energy dissipated by the uncoded flit
and the coded flit in each interswitch link, respectively. E_{codec} is the energy
dissipated by each codec. The energy savings in transporting a single flit, the
i^{th} flit, through h_i hops can be calculated as

$$E_{savings,i} = \sum_{j=1}^{h_i} E_{savings,j} \tag{7.5}$$

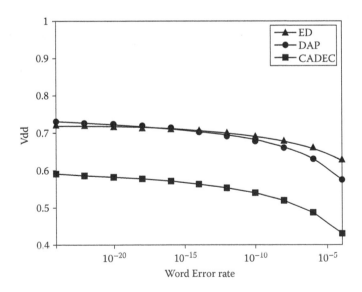

FIGURE 7.17
Variation of achievable voltage swing with bit error rate for different coding schemes.

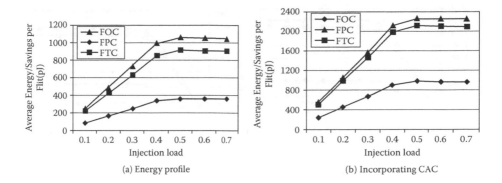

(a) Energy profile

(b) Incorporating CAC

FIGURE 7.18
Energy savings profile for a mesh-based NoC by incorporating CAC at (a) $\lambda = 1$; (b) $\lambda = 6$.

The average energy savings per flit in transporting a packet consisting of P such flits through h_i hops for each flit will be given as,

$$\overline{E}_{savings} = \frac{\sum_{i=1}^{P} \sum_{j=1}^{h_i} E_{savings,j}}{P} \tag{7.6}$$

The metric $\overline{E}_{savings}$ is independent of the specific switch implementation, which may vary based on the design. To quantify the energy savings profile for an NoC interconnect architecture, the energy dissipated in each codec, E_{codec}, can be determined by using Synopsys® Prime Power on the gate-level netlist of the codec blocks. To determine the interswitch link energy in presence and absence of coding, that is, $E_{link,coded}$ and $E_{link,uncoded}$, respectively, the capacitance of each interconnect stage, $C_{interconnect}$, can be calculated taking into account the specific layout of each topology [36]. In the presence of CACs or the Joint CAC/SEC or CAC/MEC schemes, $C_{interconnect}$ will be reduced according to the adopted coding scheme [26].

The energy savings profile of a 64-IP mesh-based NoC at the 90-nm technology node in presence of CACs, already discussed, is shown in Figure 7.18. The energy dissipation, and hence savings in energy of each interswitch wire segment, is a function of λ, the ratio of the coupling capacitance to the bulk capacitance. For a given interconnect geometry, the value of λ depends on the metal coverage in the upper and lower metal layers [30]. In 90-nm node, λ varies between 1 and 6. Consequently, the energy dissipation profile is shown for the two extreme values of λ. The average energy dissipation profile for any NoC follows a saturating trend with injection load [13]. As a result, the profile of energy savings will maintain the same trend. It is evident that among all the CACs, FOC provides us with the maximum savings.

In presence of the joint codes, in addition to reducing the capacitance of the interconnect, the voltage swing is also reduced. This reduction in voltage swing contributes significantly to the energy savings after implementation of the coding schemes. Figure 7.19 shows the performance of ED, DAP, and CADEC schemes in terms of the energy savings with injection load in the same

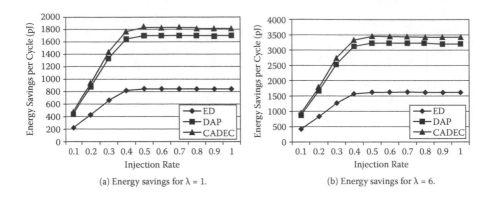

(a) Energy savings for λ = 1.   (b) Energy savings for λ = 6.

FIGURE 7.19
Energy savings profile for a mesh-based NoC by incorporating the joint codes at (a) $\lambda = 1$;
(b) $\lambda = 6$.

NoC system as used before. Because the performances of BSC and MDR are
very similar to DAP, they are omitted from the plot for clarity.

As shown in the figure, the CADEC scheme achieves more energy savings
than the other joint codes. This happens due to the fact that the residual word
error probability of CADEC is much less as it can correct up to 2-bit errors in
the flits, and hence can tolerate a much lower voltage swing.

7.5.2 Timing Constraints in NoC Interconnection Fabrics in Presence of Coding

Incorporation of coding might add timing overhead in NoC communication fabrics. The exchange of data among the constituent blocks in an SoC is
becoming an increasingly difficult task because of growing system size and
nonscalable global wire delay. To cope with these issues, designers must divide the end-to-end communication medium into multiple pipelined stages,
with the delay in each stage comparable to the clock-cycle budget. In NoC
architectures, the interswitch wire segments, along with the switch blocks,
constitute a highly pipelined communication medium characterized by link
pipelining, deeply pipelined switches, and latency insensitive component
design [37–39].

In any NoC between a source and destination pair, there is a path consisting of multiple switch blocks involving several interswitch and intraswitch
stages. The number of intraswitch stages can vary with the design style and
the features incorporated within the switch blocks. It may consist of a single
stage for a low-latency switch design or may be deeply pipelined [37,39]. In
the best case, we need at least one intra- and one interswitch stage [39]. In
accordance with ITRS, a generally accepted rule-of-thumb is that the clock
cycle of high-performance SoCs will saturate at a value in the range of 10–15
FO4 (Fanout of 4) delay units. The codec blocks might be considered as
additional pipelined stages within a switch. If the delay of the codec blocks can

be constrained within the one clock cycle limit, then the pipelined nature of the communication will be maintained. However, there is an increasing drive in the NoC research community for design of low-latency NoCs adopting numerous techniques both at the routing as well as NI level [38,40]. Therefore, it is not sufficient to just fit the codecs into separate pipelined stages as this will increase message latency. To further enhance the performance, if the delay of the codecs can be constrained so much that they can be merged with existing stages of the NoC switch, then there will be no latency penalty at all. Due to the crosstalk avoidance characteristic of the codes, the crosstalk induced bus delay (CIBD) of the interswitch wire segments will decrease. Hence, alternatively by constraining the delay of the codec blocks, if they can be merged into the interswitch link traversal stages irrespective of the switch architecture, then also there will be no additional latency penalty.

7.6 Summary

NoC is emerging as a revolutionary methodology to integrate numerous blocks in a single chip. It is the digital communications backbone that interconnects the components on a multicore SoC. It is well known that with shrinking geometries, NoCs will be increasingly exposed to permanent and transient sources of error that could degrade manufacturability, signal integrity, and system reliability. The challenges of NoC testing lie in achieving sufficient fault coverage under a set of fault models relevant to NoC characteristics, under constraints such as test time, test power dissipation, low-area overhead and test complexity. A fine balance must be achieved between test quality and test resources.

To accomplish these goals, NoCs are augmented with design-for-test features that allow efficient test data transport, built-in test data generation and comparison, and postmanufacturing yield tuning. One of the effective ways to protect the future nanoscale systems from transient errors is to apply coding techniques similar to the domain of communication engineering. By incorporating joint crosstalk avoidance and multiple error correction codes, it is possible to protect the NoC fabrics against varied sources of transient noise and yet lower the overall energy dissipation.

References

[1] M. L. Bushnell and V. D. Agrawal, *Essentials of Electronic Testing for Digital, Memory, and Mixed-Signal VLSI Circuits.* New York: Springer, 2000.

[2] A. Alaghi, N. Karimi, M. Sedghi, and Z. Navabi, "Online NoC switch fault detection and diagnosis using a high level fault model." In *Proc. of 22nd IEEE*

International Symposium on Defect and Fault-Tolerance in VLSI Systems, Rome, Italy, September 26–28, 2007, 21–29.

[3] International technology roadmap for semiconductors 2006 edition. Technical Document. (2006, Dec.) [Online]. Available: http://www.itrs.net/Links/2006Update/2006UpdateFinal.htm.

[4] E. J. Marinissen, R. Arendsen, G. Bos, H. Dingemanse, M. Lousberg, and C. Wouters, "A structured and scalable mechanism for test access to embedded reusable cores." In *Proc. of 1998 International Test Conference (ITC'98)*, Washington, DC, October 19–21, 1998, 284–293.

[5] E. Cota, L. Caro, F. Wagner, and M. Lubaszewski, "Power aware NoC reuse on the testing of core-based systems." In *Proc. of 2003 International Test Conference (ITC'03)*, Charlotte, NC, October 2003, 612–621.

[6] C. Liu, V. Iyengar, J. Shi, and E. Cota, "Power-aware test scheduling in NoC using variable-rate on-chip clocking." In *Proc. of 23rd IEEE VLSI Test Symposium (IEEE VTS'05)*, Palm Springs, CA, May 1–5, 2005, 349–354.

[7] C. Liu, Z. Link, and D. K. Pradhan, "Reuse-based test access and integrated test scheduling for network-on-chip." In *Proc. of Design, Automation and Test in Europe Conference (DATE'06)*, Munich, Germany, March 6–10, 2006, 303–308.

[8] B. Vermeulen, J. Dielissen, K. Goossens, and C. Ciordas, "Bringing communication networks on chip: Test and verification implications," *IEEE Communications Magazine*, 41(September 2003) (9): 74–81.

[9] M. Cuviello, S. Dey, X. Bai, and Y. Zhao, "Fault modeling and simulation for crosstalk in system-on-chip interconnects." In *Proc. of 1999 IEEE/ACM International Conference on Computer-Aided Design (ICCAD'99)*, San Jose, CA, November 7–11, 1999, 297–303.

[10] X. Bai and S. Dey, "High-level crosstalk defect simulation methodology for system-on-chip interconnects," *IEEE Transactions on Computer-Aided Design of Integrated Circuits and Systems*: 23 September 2004 (9): 1355–1361.

[11] A. J. V. de Goor, I. Schanstra, and Y. Zorian, "Functional test for shifting-type FIFOs." In *Proc. of 1995 European Design and Test Conference (ED&TC 1995)*, Paris, France, March 6–9, 1995, 133–138.

[12] W. J. Dally and B. Towles, "Route packets, not wires: On-chip interconnection networks." In *Proc. of 38th Design Automation Conference (DAC'01)*, Las Vegas, NV, June 18–22, 2001, 683–689.

[13] P. P. Pande, C. Grecu, M. Jones, A. Ivanov, and R. Saleh, "Performance evaluation and design trade-offs for network-on-chip interconnect architectures," *IEEE Transactions on Computers* 54 (August 2005) (8): 1025–1040.

[14] J. Duato, S. Yalamanchili, and L. Ni, *Interconnection Networks—An Engineering Approach*. San Francisco, CA: Morgan Kaufmann Publishers, 2002.

[15] C. Grecu, A. Ivanov, R. Saleh, and P. P. Pande, "Testing network-on-chip communication fabrics," *IEEE Transactions on Computer-Aided Design of Integrated Circuits and Systems* 26 (Dec. 2007) (12): 2201–2214.

[16] Intel IXP2400 datasheet. [Online]. Available: http://www.intel.com/design/network/products/npfamily/ixp2400.htm.

[17] A. Radulescu, J. Dielissen, K. Goossens, E. Rijpkema, and P. Wielage, "An efficient on-chip network interface offering guaranteed services, shared-memory abstraction, and flexible network configuration." In *Proc. of Design, Automation and Test in Europe Conference and Exhibition (DATE'04)*, 2, Paris, France, Feb. 16–20, 2004, 878–883.

[18] P. P. Pande, C. Grecu, A. Ivanov, and R. Saleh, "Switch-based interconnect architecture for future systems on chip." In *Proc. of SPIE, VLSI Circuits and Systems, Gran Canaria, Spain*, 5117, 228–237, 2003.

[19] K. Stewart and S. Tragoudas, "Interconnect testing for networks on chips." In *Proc. of 24th IEEE VLSI Test Symposium (IEEE VTS'06)*, Berkeley, CA, May 1–4, 2005, 100–107.

[20] E. Dupont, M. Nicolaidis, and P. Rohr, "Embedded robustness IPs for transient-error-free ICs," *IEEE Design and Test of Computers* 19(2002) (3): 54–68.

[21] S. Mitra, N. Seifert, M. Zhang, Q. Shi, and K. Kim, "Robust system design with built-in soft error resilience," *IEEE Computer* 38 (Feb. 2005) (2): 43–52.

[22] H. Tseng, L. Scheffer, and C. Sechen, "Timing-and crosstalk-driven area routing," *IEEE Transactions on Computer-Aided Design of Integrated Circuits and Systems*, 20 (Apr. 2001) (4): 528–544.

[23] P. P. Pande, A. Ganguly, H. Zhu, and C. Grecu. "Energy Reduction throuth Crosstalk Avoidance Coding in Networks on Chip." *Journal of System Architecture (JSA)* 54 (3-4) (March-April 2008): 441–451.

[24] S. R. Sridhara and N. R. Shanbhag, "Coding for reliable on-chip buses: Fundamental limits and practical codes." In *Proc. of 19th International Conference on VLSI Design (VLSID'05)*, Kolkata, India, January 3–7, 2005, 417–422.

[25] P. P. Sotiriadis and A. P. Chandrakasan, "A bus energy model for deep submicron technology," *IEEE Transactions on Very Large Scale Integration (VLSI) Systems*, 10 (June 2002) (3): 341–350.

[26] A. Ganguly, P. P. Pande, B. Belzer, and C. Grecu, "Design of low power and reliable networks on chip through joint crosstalk avoidance and multiple error correction coding," *Journal of Electronic Testing: Theory and Applications (JETTA)*, Special Issue on Defect and Fault Tolerance (June 2008): 67–81.

[27] P. P. Pande, A. Ganguly, B. Feero, B. Belzer, and C. Grecu, "Design of low power and reliable networks on chip through joint crosstalk avoidance and forward error correction coding." In *Proc. of 21st IEEE International Symposium on Defect and Fault Tolerance in VLSI Systems (DFT 06)*, Arlington, VA, October 4–6, 2006, 466–476.

[28] S. Murali, T. Theocharides, N. Vijaykrishnan, M. J. Irwin, L. Benini, and G. D. Micheli, "Analysis of error recovery schemes for networks on chips," *IEEE Design and Test of Computers*: 22 (Sept. 2005) (5): 434–442.

[29] D. Bertozzi, L. Benini, and G. D. Micheli, "Error control schemes for on-chip communication links: The energy-reliability tradeoff," *IEEE Transactions on Computer-Aided Design of Integrated Circuits and Systems* 24 (June 2005) (6): 818–831.

[30] S. R. Sridhara and N. R. Shanbhag, "Coding for system-on-chip networks: A unified framework," *IEEE Transactions on Very Large Scale Integration (VLSI) Systems* 13 (June 2005) (6): 655–667.

[31] D. Rossi, A. K. N. S. V. E. van Dijk, R. P. Kleihorst, and C. Metra, "Power consumption of fault tolerant codes: The active elements." In *Proc. of Ninth IEEE International On-Line Testing Symposium (IOLTS 2003)*, Kos Island, Greece, July 7–9, 2003, 61–67.

[32] K. N. Patel and I. L. Markov, "Error-correction and crosstalk avoidance in DSM busses," *IEEE Transactions on Very Large Scale Integration (VLSI) Systems* 12 (Oct. 2004) (10): 1076–1080.

[33] D. Rossi, C. Metra, A. K. Nieuwland, and A. Katoch, "New ECC for crosstalk effect minimization," *IEEE Design and Test of Computers* 22 (2005) (4): 340–348.

[34] A. Ganguly, P. P. Pande, B. Belzer, and C. Grecu, "Addressing signal integrity in networks on chip interconnects through crosstalk-aware double error correction coding." In *Proc. of IEEE Computer Society Annual Symposium on VLSI (ISVLSI 2007)*, Porto Alegre, Brazil, May 9–11, 2007, 317–322.

[35] T. Theocharides, G. Link, N. Vijaykrishnan, and M. Irwin, "Implementing LDPC decoding on network-on-chip." In *Proc. of 19th International Conference on VLSI Design (VLSID'05)*, Kolkata, India, Jan. 3–7, 2005, 134–137.

[36] C. Grecu, P. P. Pande, A. Ivanov, and R. Saleh, "Timing analysis of network on chip architectures for MP-SoC platforms," *Elsevier Microelectronics Journal* 36 (Sept. 2005) (9): 833–845.

[37] L. Benini and D. Bertozzi, "Xpipes: A network-on-chip architecture for gigascale systems-on-chip," *IEEE Circuits and Systems Magazine* 4 (2003) (2): 18–31.

[38] D. Park, C. Nicopoulos, J. Kim, N. Vijaykrishnan, and C. R. Das, "A distributed multi-point network interface for low-latency, deadlock-free on-chip interconnects." In *Proc. of First International Conference on Nano-Networks (Nano-Net 2006)*, Lausanne, Switzerland, September 14–16, 2006, 1–6.

[39] A. Kumar, P. Kundu, A. Singh, L.-S. Peh, and N. Jha, "A 4.6Tbits/s 3.6GHz single-cycle NoC router with a novel switch allocator in 65nm CMOS." In *Proc. of 25th International Conference on Computer Design (ICCD 2007)*, LakeTahoe, CA, October 7–10, 2007.

[40] R. Mullins, A. West, and S. Moore, "Low-latency virtual-channel routers for on-chip networks." In *Proc. of 31st Annual International Symposium on Computer Architecture (ISCA'04)*, Munich, Germany, June 19–23, 2004, 188–197.

8

Monitoring Services for Networks-on-Chips

George Kornaros, Ioannis Papaeystathiou, and Dionysios Pnevmatikatos

CONTENTS

8.1   Introduction

Network monitoring is the process of extracting information regarding the operation of a network for purposes that range from management functions to debugging and diagnostics. Originally started in bus-based systems for the most basic and critical purpose of debugging, monitoring consisted of probes that could relay bus transactions to an external observer (be it a human or a circuit). The observability is crucial for debugging so that the behavior of the system is recorded and can be analyzed, either on- or off-line. When the behavior is recorded into a trace, the run-time evolution of the system can be replayed, facilitating the debugging process. Robustness in time- or life-critical applications also requires monitoring of the system and real-time reaction upon false or misbehaving operation.

Research has already produced valuable results in providing observability for bus-based systems, such as ARM's Coresight technology [1]. Also First Silicon's on-chip instrumentation technology (OCI) provides on-chip logic analyzers for AMBA AHB, OCP, and Sonics SiliconBackplane bus systems [2]. These solutions allow the user to capture bus activity at run-time, and can be combined in a multicore-embedded debug system with in-system analyzers for cores, for example, for MIPS cores.

Because buses offer limited bandwidth, these simple bus-based systems at first evolved using hierarchies of multiple interconnected buses. This solution offered the required increase in bandwidth but made the design more complex and ad hoc, and proved difficult to scale. As systems increase in number of interconnected components, communication complexity, and bandwidth requirements, we see a shift toward the use of generic networks (Networks-on-Chips) that can meet the communication requirements of recent and future complex Systems-on-Chips (SoC). Figure 8.1 shows the use of a regular topology for the creation of a heterogeneous SoC. An example of how a heterogeneous application can be mapped on this SoC is also depicted. Of course the topology does not have to be regular, as shown in Figure 8.2.

However, this change dramatically increased the complexity of monitoring compared to the simpler systems for several reasons. First, the sheer increase in communication bandwidth of each component increases the amount of information that needs to be monitored or traced. Second, the structure of the system does not provide the single, convenient central-monitoring location any more. As communication in most cases is conducted in a point-to-point,

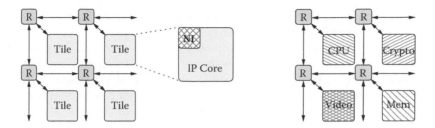

FIGURE 8.1
Network-on-Chip based on a regular topology, and an example with a heterogeneous application. Each node (or tile) is connected to a router, and the routers are interconnected to form the network. The nodes can be identical creating a homogeneous system (i.e, CPUs), or can differ leading to a heterogeneous system.

not broadcast, fashion, monitoring recent and future systems is a distributed operation.

Despite all the difficulties that must be overcome for successful monitoring, the complexity of SoCs offers additional opportunities as well to deal with many challenges such as: short time to market, increasing fabrication (mask) cost, incomplete specifications at design time, and changing customer requirements. These challenges increased the complexity of SoC designs, which are designed with increased versatility so that they can cover more application space and increase the product lifetime. These two factors, increased complexity and increased flexibility, lead to a dynamic system behavior that cannot be known in advance at design time. This opens the possibility for dynamic system management, where application behavior is monitored and adjustments of the system and its operation can be made either to improve the application function (e.g., provide better QoS) or to optimize the system's operation (e.g., consume less energy). Exploiting these opportunities depends on knowing

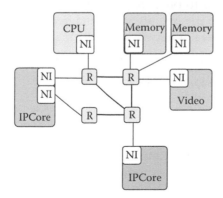

FIGURE 8.2
Network-on-Chip based on an irregular topology. The nodes are again connected to (possibly multiple) routers, but the routers are interconnected on an ad hoc basis to customize the network to the application demands and achieve better cost–performance ratio.

the run-time characteristics of the system's operation. For a network-based design, this can be achieved using network monitoring.

Furthermore, new opportunities appear as we move toward deep submicron implementation technologies. In these future technologies, device reliability is an issue as they are susceptible to a range of postmanufacturing reliability failures. The consequence is that the designer has to deal not only with static faults, but also transient faults, wear-out of devices, etc. To address this challenge, future systems need to support redundant paths and resources, and the ability to rearrange their operation to isolate and bypass failures when they occur. This operation requires a quick identification of the existence of a problem and its location to determine the correct repair action. Regarding the NoC resources, both functions can be achieved by network monitoring.

In the next two sections of this chapter, we first discuss in detail the objectives and the opportunities of network monitoring and their applications. Then, we cover the type of information that needs to be monitored and the required interfaces to extract this information from the distributed monitor points. Following this, we describe the overall NoC monitoring architecture and discuss implementation issues of monitoring in NoCs such as cost and the effects on the design process. In the last two sections, we present a case study where we discuss several approaches to provide complete NoC monitoring services, followed by open issues for future research and our conclusions.

8.2   Monitoring Objectives and Opportunities

Monitoring can be used to provide information for many different applications that are related to the overall SoC management. The following subsections detail the main uses of network monitoring.

8.2.1   Verification and Debugging

Traditional monitoring is achieved by adding observability into internal points of a complex system. The observability is crucial to enable the designer to track the system's operation and determine if and when something goes wrong. To this end, tracking the maximum possible amount of information is desired as it provides the best flexibility to the user, who is then responsible to focus on the exact needed bits and pieces.

For testing the nodes of the SoC, one approach is to provide a Test Access Mechanism (TAM), which reuses the network resources [3] to minimize the cost and improve the speed of testing probes. The key observation is that due to its role, NoC is a central piece of the SoC. TAM interfaces with test wrappers, built around the cores, to apply test vectors to the cores under test, and also collects and delivers the possible responses. However, this type of operation is intrusive and useful only for off-line testing.

Another important benefit of monitoring is to use it for debugging purposes. When the system is in operation and we want to extract information, we can track the system progress without affecting its operation (i.e., in a nonintrusive manner). To achieve this goal, the testing wrappers should provide the necessary information to the monitoring infrastructure, which can then deliver it to the tester without affecting the application's behavior.

8.2.2 Network Parameters Adaptation

Monitoring can be applied in a parameterized network to provide information for the update of the configuration or run-time parameters. For example, when the NoC uses adaptive routing, updates to the routing tables can better distribute the load, reduce the latency variation that is caused by congestion, and improve the overall system's performance. NoC monitoring can provide the necessary input to a decision-making algorithm that updates the routing tables in the network.

A similar application is to detect permanent network link malfunctions and errors (either permanent or even transient) and readjust the routing tables to avoid these defective links. This notion can be carried out even at the node level, where monitoring can detect defective processing nodes and provide feedback to the run-time system. Depending on the application, the run-time system should thus avoid the use of the defective node and migrate the processing to other functional nodes. The mechanisms to support the isolation of defective links or nodes are basically the same as the ones used in adaptive routing (i.e., updates in the routing tables), and, in a way, we can think of defective links or nodes as permanently saturated areas that need to be avoided.

8.2.3 Application Profiling

For applications with dynamically changing behavior, monitoring their network patterns can offer an insight to their overall operation. This can be useful for the purpose of application profiling, a process to collect information about its run-time behavior. Profiling information can then be used to map the application on the existing resources in a better way.

A similar application is when the network supports Quality of Service (QoS) guarantees and can allocate specific portions of link bandwidth to certain communication pairs. Monitoring can detect when the QoS contract is violated, and this information can be used as feedback to either simply detect the problem or take actions (i.e., adjust the QoS parameters) and fix it, if possible.

Obtaining information regarding an NoC-based application can be used to enable intelligent power management of the NoC resources. NoC monitoring can detect statistically important changes in the communication patterns, and can readjust the speed of uncongested portions of the NoC to save power. When links and routers do not support multiple voltage and corresponding speed levels, the identified routers can be shutdown, and their (presumably noncritical) traffic can be rerouted via other low utilization routers.

8.2.4   Run-Time Reconfigurability

Similar to readjusting the parameters is the use of run-time reconfigurable NoC systems. This approach has been explored as a promising way to over- come the potential performance bottlenecks, because communication pa- rameters cannot be estimated beforehand as communication patterns vary dynamically and arbitration performs poorly. As a result, dynamic recon- figuration is used to change the key parameters of the NoC and eventually the communication characteristics can be tuned to better meet the current requirements at any given time. Such run-time reconfigurable NoC systems have been proposed [4–8].

Moreover, as the silicon devices are getting more and more complex, the testing of the NoC structure is becoming more difficult. Different Design-for- Testability (DfT) approaches have been proposed, to provide the means for NoC testing [9]. However, a recent trend which is very promising is the use of certain run-time reconfigurable structures that can be used for ordinary oper- ations as well as for testing of the NoC structures such as by Möller et al. [10].

For all run-time reconfigurable NoCs to adapt their structures on run-time, an efficient online monitoring system is required (such as the one by Mouhoub and Hammami [11]). This system can mainly be based on reconfigurable net- working interfaces. These networking interfaces will route the traffic, which is coming from the IPs that are connected to them, and will only keep statis- tics regarding the different characteristics of the traffic. The main difference between those monitoring systems and the others that are described in this chapter is that the former can take advantage of the run-time reconfiguration aspect in the following manner: Whenever scheduled, the interfaces will be altered in run-time and instead of sending usual data, they will be connected to a separate network infrastructure over which the monitoring data will be transferred to the main monitoring and reconfiguration module. Those run- time reconfigurable monitoring interfaces have the advantage of utilizing the same hardware resources with the standard NoC interfaces, and thus reduce the overall overhead of the monitoring schemes.

8.3   Monitoring Information in Networks-on-Chips

8.3.1   A High-Level Model of NoC Monitoring

As in every monitoring system, an NoC monitoring scheme should collect samples that may range from simple bit-level events to whole messages. The system designer or ultimately the real-time service may need to trace fine grain information such as interrupt notifications, or even protocol messages and data. Testing a multiprocessor SoC obviously calls for a verification strategy, which needs to consider the inherent parallelism: the on-chip network structure and the task attributes. Only a high-level approach can tackle such issues. Abstraction via filtering of a large amount of traced messages can be the key approach for a realistic monitoring service.

8.3.1.1 Events

In the high-level schemes, the data collected are modeled in the forms that are called *events* [12]. Based on this approach, all the events have specific predefined formats and are most frequently categorized because they may have different meanings. According to Mansouri-Samani and Sloman [13], "an event is a happening of interest, which occurs instantaneously at a certain time." Therefore information characterizing an event consists of (a) a timestamp giving the exact time the event occurred, (b) a source id that defines the source of the event, (c) a special identifier specifying the category that the event belongs to, and (d) the information that this event carries. The information of the events are called *attributes* of the events, and consist of an attribute identifier and a value. The exact attributes as well as the number of them depend on the category to which the event belongs.

Regarding the classification of the events, Ciordas et al. have grouped them in five main classes: user configuration events, user data events, NoC configuration events, NoC alert events, and monitoring service internal events [12].

- The user configuration events are initiated by the IP modules that are connected to the NoC to configure the different NoC monitoring components accordingly. They are formatted in such a way that they present a system-level view of the requested information and hide NoC implementation details. The information contained in such events can be large, when many details are needed for the configuration action, or small, when the subsystem to be configured does not need any specific information except probably the timing of the communication and the communication modules.

- The user data events carry the monitored data from the NoC. Collecting the data can be through sniffing from either the various NoC interfaces or from the actual NoC links.

- The NoC configuration events are employed in the programming or configuration, statically or dynamically, in a centralized or distributed way, of the underlying NoC. They are usually employed in NoC debugging and optimization, because they carry all the requested information regarding the configuration of the NoC. For example, such events are produced whenever there is a change in the actual routing protocol or routing state.

- The NoC alert events are generated whenever emergency situations are triggered. Such situations include buffer overflows, internal or

edge congestion, or even missing a hard deadline in a real-time system. Through these events and based on the statistics of the utilization of various NoC resources, the NoC monitoring/administration
system can be alerted on an abnormal situation about to occur (or
already occurred).

- The monitoring service internal events are issued by the monitoring
  service mechanism itself for various reasons such as synchronization, ordering, and data losses.

8.3.1.2 Programming Model

Another critical constituent of the high-level description of an NoC monitoring system is the programming model of the system. Such a model describes in
detail the procedure needed for setting up the various monitoring services. In
general, it consists of a sequence of basic tasks for configuring the NoC monitoring subsystems as well as a detailed reference description of implementing
those tasks. For example, Radulescu et al. have proposed a memory-mapped
I/O programming model for configuring the different submodules of an NoC
monitoring system [14].

The programming model should address the critical issue of NoC monitoring configuration time. In general, an NoC monitoring system can be
configured at three possible points in time: (a) at NoC's initialization time, (b)
at NoC's reconfiguration time, or (c) at run-time. Furthermore, the programming model should define the events that would be generated, the categories
of the events that would be supported, the attributes of those events as well as
a global timing/synchronization scheme and ways to start/freeze/stop the
monitoring system.

The programming model also defines whether the NoC monitoring system
will be centralized or distributed. In a centralized monitoring service, all
the monitoring information is collected at a central point; this approach is
simple yet efficient for small NoCs. However, in case of SoCs with hundreds
of different submodules, the collection of all the monitoring information at a
central point may lead to the bottleneck of the NoC monitoring system. On
the other hand, in a distributed monitoring service, the monitoring data are
collected by concentrating components, which are interconnected together
to be able to take a decision based on the global state of the NoC. Although
this approach is more complicated than the centralized one, it removes the
possible bottlenecks of the centralized approach and is also significantly more
scalable.

8.3.1.3 Traffic Management

Traffic management is another component of most of the abstract models of
NoC monitoring systems. In general, it is divided into two subsystems: the
first manages the *configuration traffic* and the second covers the *event traffic*.

The configuration traffic includes all the messages/events required to set
up and configure the monitoring scheme, such as the events to configure the

NoC monitoring hardware subsystems and the traffic for setting up connections for the transport of data from the actual NoC to the NoC-monitoring processing system. On the other hand, the event traffic management system deals with the traffic generated after the NoC has been thoroughly configured.

8.3.1.4 *NoC Monitoring Communication Infrastructure*

NoC traffic monitoring system can use either the existing NoC intercommunication infrastructure or an added network that is implemented only to cover the requirements of the NoC monitoring systems. The former has the advantage that no extra interconnection system is needed but, on the other hand, it introduces additional traffic to the actual NoC. If this traffic causes performance problems, a dedicated NoC monitoring interconnection infrastructure is to be employed. Based on the selection of the desired interconnection scheme for the transmission of the actual measurements in an NoC monitoring system, the NoC data can be categorized as follows:

- **In-band traffic**. In this case, the NoC traffic is transmitted over the NoC links either by using time division multiplexing (TDM) techniques or by sharing a network interface (NI).
- **Out-of-band traffic**. When hard real-time diagnostic services are needed or when the NoC capacity is limited by communication-bounded applications, a separate interconnection scheme is used and the NoC monitoring traffic is considered *out-of-band*.

Considering that the employed monitoring services are used for debugging, performance optimization purposes, or power management, it is clear that the choice of the appropriate interconnection structure is very critical because it may affect the overall efficiency of the NoC toward the opposite direction of the desired objective.

A self-adapting monitor service could encompass programmable mechanisms to adjust the generated monitoring traffic in a dynamic manner. Using a hybrid methodology, the distributed NoC-monitoring subsystems or the central-monitor controller can support an efficient traffic management scheme and regulate the traffic from the NoC to the central diagnostic manager. However, placing extra functionality increases the overhead of the monitoring probes, in terms of area or energy consumption.

8.3.2 Measurement Methods

One of the main problems in NoC monitoring is that processing the entire contents of every packet imposes high demands on packet probes and their hardware resources. For this reason, probes usually capture only the initial part of the packet that contains valuable information. Even then, and because the current NoCs work at extremely high speeds, the amount of data collected is huge. One way for reducing the volume of the data is by utilizing certain techniques for filtering, aggregation, and sampling just as it is done in the case of telecommunication network monitoring [15].

When sampling is employed and within an NoC monitoring infrastructure, a number of different sampling mechanisms can be utilized as described by Jurga [16]. The most important algorithms are the following:

- Systematic packet sampling, which involves the selection of packets according to a deterministic function. This function can either be count-based, in which every kth packet is saved for monitoring purposes, or time-based, where a packet is selected at every constant time interval. As described by He and Hou [17], the count-based approach gives more accurate results in terms of the estimation of traffic parameters than the time-based one.

- Random sampling, in which the selection of packets is triggered in accordance with a random process. Based on the simple algorithm, n samples are selected out of N packets; hence, it is sometimes called n-out-of-N sampling. A certain algorithm of random sampling is what is called probabilistic sampling. In this technique, the samples are chosen in accordance with a predefined selection probability. When each packet is selected independently with a fixed probability p, the sampling scheme is called uniform probabilistic sampling, whereas when the probability p depends on the input (i.e., packet content) then this is nonuniform probabilistic sampling.

- The adaptive sampling schemes employ either a special heuristic for performing the sample process or certain prediction mechanisms for predicting future traffic and adjusting the sampling rate. The schemes have some inevitable disadvantages. There is always some latency in the adaptation process and, in case of unanticipated NoC traffic bursts, the saturation of the monitoring module will be possible. To avoid it, the NoC monitoring designer would have to allow a certain safety margin by employing systematic undersampling (obviously, at the cost of lower accuracy).

Another way of reducing the traffic recorded by the network monitoring scheme is the use of filtering. In a formal definition, filtering is the deterministic selection of packets based on their content. In practice, the packets are selected if their content matches a specified mask. In the general case, this selection decision is not biased by the packet position in the packet stream. This approach may also require a relatively complex packet content inspection, because, depending on the NoC protocol employed, packets can have different formats and thus a fixed length mask cannot be applied.

Finally, there are the hybrid techniques, which are based on combining a number of packet selection approaches. For instance, Scholler proposes to add packet sampling into the packet and create a scheme combining certain advantages of filtering and sampling [18]. Another example is Stratified Random Sampling. In this approach, packets are grouped into subsets according to a set of specific characteristics. Then, the number of samples is drawn randomly from each group.

Regarding the advantages and the disadvantages of each scheme, it is obvious that systematic sampling can be easily implemented in hardware. As demonstrated by Schöller et al. [18], and Harangsri et al. [19] in general networking environments systematic sampling often performs better than random sampling. The main disadvantage is that "it is vulnerable to bias if the metric being measured itself exhibits a period which is rationally related to the sampling interval" [20]. This problem can be overcome when random sampling is utilized.

Depending on the specific SoC that an NoC is employed in, the hybrid sampling schemes can trigger the optimal point in the trade-off between the amount of data collected and the accuracy of the monitoring scheme. Such schemes can allow building accurate time series of different parameters and improving the accuracy of the classification of NoC traffic into flows, groups, etc.

On the other hand, the filtering methods can be tuned to collect whatever NoC data are needed in each specific case, with the drawback of being relatively complex to be implemented mainly because of the extremely high speed of today's NoCs.

There are mainly two approaches to the actual transmission of the sampled/filtered data to the monitoring management processing system.

- **Raw data transfers**. Here the probing components need to be simple enough, and a centralized monitoring management unit is employed.

- **Filtered data transfers**. These incorporate some manipulation on the sniffed data on-the-spot. If the system design can afford extra area, statistic functions that measure traffic can be used, such as average, min/max, stddev.

8.3.3 NoC Metrics

There are several metrics which can be used to characterize the NoC monitoring systems in terms of a number of parameters. The most important of those metrics are the following:

- BW—bandwidth requirements of each NoC monitoring module, because the traffic characteristics of the NoC's individual links vary as well as the information generated by each NoC monitoring subsystem. Additionally, even if two such systems collecting the monitoring data are identical, they may require different networks due to their different configurations. For instance, a particular module can be configured to generate coarse grain statistics for diagnostic services, whereas another identical one used for debugging may supply large amounts of data.

- NP—path coverage, which represents the number of paths in an NoC under monitoring when the NoC monitoring subsystems are placed in specific locations in the NoC architecture.
- RH—resource history, which denotes the time duration in which the monitoring information will be saved. To deduce a valid result at the transaction level or at a higher abstraction level, the required storage at the different NoC monitoring subsystems may be significant, depending on the NoC monitoring scheme.
- RTR—real-time response requirements, which is the time between issuing a trigger event and getting back the requested information. For fault tolerant systems or hard real-time environments, this is a very critical metric.
- ND—NoC dependencies, which are determined in terms of interfacing and protocol compatibility of the monitoring traffic and the standard on-chip NoC traffic.
- AD—application dependencies, which cover the mapped application on the NoC and the associated NoC monitoring requirements in terms of resilience to faults, performance, power consumption, etc.

8.4   NoC Monitoring Architecture

To aid the development of large complex SoCs, it is necessary to efficiently assist the realtime debugging of the complete design. This can be achieved by embedding hardware components that are usually application- or circuit-specific, as they are tailored to the verification needs of a particular circuit. One common approach is to attach monitoring components to a bus (as by ARM [1]). Monitoring at this level consists of signal observation or event identification; events may be restricted to be activated only under prespecified conditions. Several research and commercial solutions are built around this principle to achieve realtime debugging of designs; using in-circuit emulators is also a methodology that is gaining popularity.

System-on-Chip designs are growing larger and larger by integrating a number of IP cores connected using an NoC that consists of small routers and dedicated communication links. In this environment, data are transferred in the form of messages or packets. Packet switching in turn is commonly based on the wormhole approach, where a packet is broken up into flow control units that are called flits, the smallest unit over which the flow control is performed, and the flits follow the header in a pipeline way.

The monitoring messages described in the last section should not block or even interfere with the typical user traffic, and should follow the on-chip communication protocol. In terms of interfacing, the monitoring component in NoC should also comply with the router and NIs. The aim of the monitoring services ranges in goals and functionality, and correspondingly the type of data extracted to support meaningful conclusions varies. The type of data can be categorized as follows:

- Measurements, for example, statistics using counters that are usually event-triggered
- Content-based measurements performed by sniffing the contents of the data units (flits, packets, messages) transferred on the NoC
- Filtered data extracted from the contents of the data units
- Headers from the packets
- Payload data meeting particular conditions (filters)
- Configuration data exchanged between routers

The use of monitoring units to collect NI statistics to assist the operating system controlling the NoC has been proposed by Nollet et al. [21]. Such performance monitoring can be used to optimize communication resource usage to control the interaction between concurrent applications. On the other hand, router-attached performance monitors are used by Pastrnak et al. [22] to keep track of the network utilization. This information is used by the network manager to adjust the QoS levels of the running applications.

The probing method is associated with the type of required monitoring information and can be categorized as *cycle-level* or *transaction-level* probing. A probing component with the capability of filtering the sniffed data and identifying messages up to transaction level is called *transactor*, throughout the subsequent sections. Independent of the type of sniffed data, a cycle-level transactor operates on every clock cycle, and its bandwidth requirements depends on the number of signals to observe. At transaction level, the transactor completes each observed datum at the end of each transaction. Thus, processing and possibly increased storage may be required on the spot to allow each probe/transactor to communicate only with the higher level of an interconnect link. A cycle-level probe needs to use either high bandwidth links or a compression scheme to reduce the rate of data monitoring. A monitoring component can potentially include both options as shown in Figure 8.3, unless the area cost cannot be afforded.

At a higher level, all these strategies are derived from the event model. The monitored information can be modeled in the form of events, with an event model to specify the event format, for example, time-stamped events or not. Event taxonomy helps to distinguish different classes of events and to present their meaning.

A general-purpose monitoring for on-chip interconnects is usually based on a layered structured approach, similar to the one depicted in Figure 8.4. The first step to embed a monitor-sniffer with dummy manipulation of data is to include only the most basic functionality, such as only copying the observed data on an on-chip link. However, observing all NoC transfers even for a single link may generate large amounts of data that require further analysis. This is

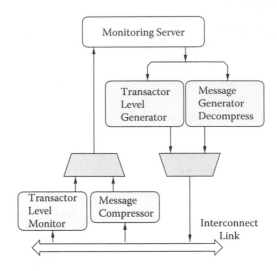

FIGURE 8.3
Monitoring component combining the two alternatives: sniffing data and filtering up to transaction level, and streaming the messages using compression to reduce transferring large amounts data.

a significant issue because the collected data has to be processed at a different location, either on-chip or off-chip. The complexity increases dramatically based on the number of monitors and the amount of required captured data.

Second, the postprocessing of the sniffed data should be done by a more advanced *transaction* level analyzer, to deduce useful conclusions. In between the two layers, it is necessary to place another layer to filter and classify the data. In particular, because each message conveyed from end to end consists of header and payload, the monitoring component must perform decapsulation in the same way as a network interface. The flits transferred over a communication link may be interleaved by the network interface or the routers. The monitors must therefore collect and construct messages from the right flits.

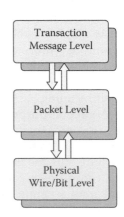

FIGURE 8.4
Layered organization of a monitoring component.

In addition, connections with different bandwidth demands should be identi-fied and treated accordingly. The connections of interest are usually a subset, but in the worst case all may be monitored.

Hence, the monitoring service inside an NoC-based system must deal with the following main issues:

- Conditioning/filtering of sniffed data
- Prioritizing critical vs. noncritical sniffed data (i.e., for statistical purposes)
- On-the-spot analysis and, possibly, reactions based on the analyzed data.

Filtering is based on prior knowledge of the type and the format of data, and in some cases also on the timing of the data of interest. Dimensioning the filters though is a trade-off between flexibility and increased area cost. It is feasible to use masks and even more intelligent event-based filters as long as the total overhead is affordable. The benefits of appropriate conditioning focus on reducing the traced data to only the critically needed pieces of information.

Monitoring services can be characterized as best effort (BE) assuming either that the probing of a link is done periodically or that the messages sent and henceforth the reaction to them is not strictly real-time. This type of service is useful for observing liveness of a core or of a link. Moreover, because pri-oritization of on-chip connections is a usual mechanism to differentiate and ensure QoS for on-chip user traffic, the same prioritization should be ensured also for the monitoring services.

Meanwhile, monitoring services that need guaranteed accuracy (GA) might be required when an exact piece of information is needed; for example, to cal-culate throughput based on bytes sent over a link or for debuging purposes. In addition, hard real-time performance necessitates the quality of GA ser-vices in terms of low latency and complete view of the traffic or capacity to sustain monitoring traffic at full throughput. Guarantees are obtained by means of separate physical links or by means of TDMA slot reservations in NoC interfaces.

Although this is a modular approach, it may suffer from transferring and possibly keeping in memory large amounts of data. If the memory refers to on-chip memory resources, then the issue that can be raised is the amount of available memory, although if off-chip storage is used the issues shift to band-width needs, pad limitations, or augmented system complexity. Additionally, multiplexing with already used memory interfaces affects the available band-width and may raise redesign considerations.

If on-the-fly analysis is desired, then such a monitoring must be application-specific (i.e., hard monitor). Alternatively, an embedded software solution can also perform such analysis provided that software latencies can be tolerated. This is a viable option in low bandwidth configurations of an NoC or for a monitoring application that is not critical in a real-time environment.

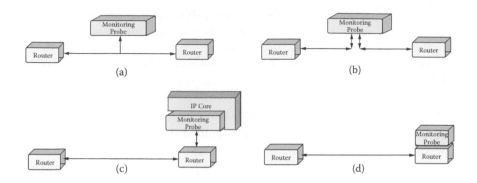

FIGURE 8.5
Attachment options of a monitoring component: (a) sniffing packets from a link, (b) operating as a bridge observing and even injecting packets, (c) collecting data also from the core of an IP, (d) accessing also the internal status of a router.

Although attaching a monitoring probe to a link serves for real-time observation, as shown in Figure 8.5a, the alternative organization shown in Figure 8.5b can handle total failures of nodes. If the monitoring probe is also attached to another link, then the system liveness is ensured in case of a dead router by replacing the router functionality with the monitoring services, as long as it can be supported. In the case shown in Figure 8.5c, the monitoring probe embedded at the *edge* of an IP core can also collect information from the link connected to the router and from the IP core. Finally, in Figure 8.5d the monitoring probe also has access to the core functional part of a router. This configuration has the potential to also allow the observation of the internal state of the router and its port interfaces.

8.5 Implementation Issues

This section describes various monitoring implementation alternatives and the impact on the total NoC-based SoC. The communication requirements of the monitoring probes are sometimes not known beforehand. The monitoring probes are designed only after the NoC has been monitored, or at least after some steps in the NoC design flow are performed so that it is convenient to make decisions about the location and functionality of the probes. The mapping of the application to the NoC nodes is first completed, so that designers can have an overall view of the needs of the system and evaluate the requirements and costs of each option. These are closely related problems. Once an NoC architecture is near the completion of the development phase, some issues have to be evaluated regarding the integration of the monitoring probes: the location and the number of the monitoring probes, the communication requirements among them, and their impact on the total area and energy consumption.

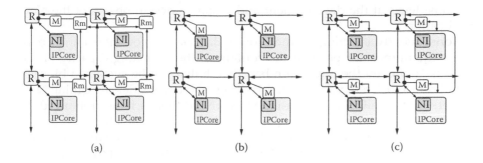

(a)   (b)   (c)

FIGURE 8.6
Monitoring architectural options: (a) use a separate monitoring NoC for transfering monitor traffic, (b) share the user NoC also for monitoring, (c) use separate bus-based interconnect.

8.5.1 Separate Physical Communication Links

Using a nonshared interconnect scheme allows a nonintrusive independent possibility to transfer the monitoring data at high speeds. Thus, the original NoC remains intact and the user data is not affected in any way with respect to latency or blocking side-effect. This solution comes though at the expense of increased area. Although any interconnect may be used, depending on the area cost margins of the system, the bandwidth requirements, and the scalability potential, a separate physical NoC may be adopted due to its scalable attributes. This option is depicted in Figure 8.6a. One probe (M) and a monitoring router (Rm) is added per router in the NoC. In a simple scenario, this monitoring probe is attached to the router; however, it could be attached to the NI of each IP core or to the communication link. If a probe is connected to the NI, whole messages can be collected from the IP core itself, whereas attaching the probe to the router permits only collecting data in the form of flits.

Another costly yet high-speed solution that also uses individual nonshared interconnect is to employ point-to-point links to the monitoring manager of the SoC. Thus, the multihop latencies are avoided and the system monitor manager can instantly send and retrieve data from the probes.

On the other hand, a shared bus or set of buses can be used and still the nonintrusive integration of the probes could be achieved because no interference is possible via the physically disjoint interconnects. Moreover, a separate shared bus amortizes the cost of extra area.

8.5.2 Shared Physical Communication Links

An alternative solution is to use the existing NoC infrastructure for both data traffic and monitoring traffic. The NoC resources can be completely shared with the integrated monitoring units. Otherwise, a subnetwork or virtual flow for the monitoring traffic could be implemented. Of course, the user requirements for bandwidth must be respected and they should be considered at the development and mapping design phases of the NoC application.

Consequently, injecting the monitoring data via the NoC interconnect may affect the number of ports of the routers, or the switching capacity of the routers, and even the maximum bandwidth that can be sustained by each link. From a different viewpoint, this design choice considers designing NoC routers with monitoring probe capabilities. Nevertheless, if the probe needs to operate at transaction level, a separate probe block attached to the NI would be a more structured approach. Although when using shared NoC resources, the interconnect cost is amortized in terms of area, the cost of the probes still remains the same as in the previous alternative.

8.5.3 The Impact of Programmability on Implementation

A monitoring probe is usually a modular component that consists of the NoC interface, the event generator, and the interface to a centralized monitoring manager either on- or off-chip. By identifying packets, messages, and local filtering, a classification of messages per connection is possible when raising the filtering to transaction level. Then, a detailed inspection of transactions can be achieved. Apparently, each level increases the complexity of the monitoring component probe. Along with the needed logic, storage is also required, which is proportional to the message size, the number of simultaneous connections, and the depth level of inspection. If the headers of each message is the only part of interest and examination of the payload is not required, then the necessary storage can be minimized.

Figure 8.7 shows a modular approach of the transaction monitoring probe as described by Ciordas et al. [23]. The monitoring data path starts at the sniffer that captures the raw data from the router links and provides them to the transaction monitor. The link of interest can be selected at run-time by configuring the first filter. The flits can be further filtered as BE or GT in the second filtering block. Further filtering of flits is done by identifying a single connection from the set of connections sharing the same link in the next filter. Transactions are composed of messages. Message identification allows viewing from within the NoC, when a write or a read message has been issued and from where or to which IP or memory. Messages are packed in payload packets. Therefore, message identification requires depacketization, a procedure usually done at the NI. Finally, the monitoring server must be notified by the transaction monitor according to the preconfigured format. Regarding the implementation impact of such a modular probe, it is reported that in a 0.13 μm CMOS technology, the implementation of a transaction monitor supporting the first four filtering stages costs 0.026 mm^2 in silicon. Assuming that no filtering/abstraction is done locally at the monitor, the bandwidth requirements of the transaction monitors are comparable with the bandwidth of the monitored connection. This example demonstrates the potential of providing intelligent services to the system designer to help him monitor an NoC-based design at run-time.

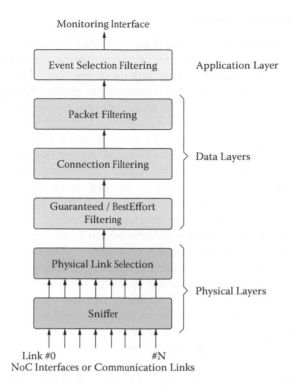

FIGURE 8.7
Layered organization of a transaction monitor: Each filter layer can be configured at run-time via a command-based interface. The required functionality defines the number of layers of a monitoring probe.

8.5.4 Cost Optimizations

If the area budget of the SoC is limited, then sharing a monitoring probe between two or even more routers of the user NoC may be an option. A transaction-level monitoring probe as described by Ciordas et al. [23] requires significant processing that is performed in layers so as to depacketize a message, for example. Hence, assuming the bandwidth requirements of the user and of the monitoring services are tolerable, the same probe with this processing engine could be shared among many routers, thus saving the useful silicon area. In the same direction, a monitoring component may collect information from the IP core as well. Thus, by acquiring a view of the future transactions to be performed by the IP core, the monitoring probe can reserve space or do a more intelligent conditioning of the events such as tracking the messages passing via an NoC router depending on the status of the IP core. The benefits and the total impact of every architectural option have to be evaluated in the environment of real applications mapped on the NoC.

8.5.5   Monitor-NoC Codesign

Integrating monitoring units efficiently into an NoC is not a straightforward task. Depending on the type and extent of monitoring, the monitoring units and their traffic will affect the performance of the NoC. Hence, the simple approach of designing the NoC and subsequently adding the necessary monitoring infrastructure will often lead to suboptimal results. There are several steps in the design process where monitoring support is interweaved and affects the traditional NoC design. The most obvious is the network dimensioning. If the monitors use the same network to transfer their information, the offered load at the links will be higher depending on the monitoring demands. This will in turn affect the decision of choosing the link speed, or the dimensions of the network, so that the total traffic can be supported and the desired maximum and average latencies are met.

Even if monitoring uses a different network for its communication, the two networks coexist and share the same area resources. Hence, the physical distance between nodes is increased even when monitoring uses a dedicated network. This increase leads to longer point-to-point latencies, affecting the NoC clock cycle. The optimal NoC topology depends partly on the relative position of its nodes as well as on the availability of wires to connect the corresponding routers, so even when a separate dedicated monitoring network is used, the regular NoC design is affected.

When the monitoring units are fully deployed at every node (or more generally at a regular topology with respect to the rest of the nodes or network), placing the monitors largely depends on the physical location of the nodes, and the corresponding requirements are known in advance. When the monitoring units are placed irregularly close to nodes that are deemed important for a particular application (i.e., the monitors are customized for that application), the effect of their placement cannot be determined in advance. Further complications may occur when the application itself is mapped into SoC nodes in many different ways. In such a context, Application Aware Monitor Mapping [23,24] is required to achieve good results.

A design flow for monitor placement should also attempt to optimize the required number of monitors according to the exact placement of the initially prescribed monitors. In certain cases, the functionality of multiple monitors may be merged into a single one. There are many conditions that must be met to achieve such a merger. For example, the total aggregate monitoring traffic should not exceed the monitoring link (or channel when the network is shared) capacity, the merged monitor should have access to all the monitored information (being transactions, events, etc.), and the monitor programmable resources (if any) should be sufficient to satisfy the tasks of all the merged monitors. Once merging becomes achievable, there are direct or indirect benefits. For instance, smaller overall area is a direct consequence, but this also indirectly improves latencies and clock rates.

Figure 8.8 shows an integrated design flow similar to the one proposed by Ciordas et al. [23,24] for shared network use between the monitors and regular

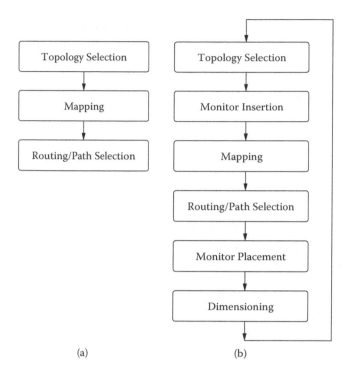

FIGURE 8.8
Integrated NoC-monitor design flow. Part (a) shows a simplified flow for simple NoCs, although part (b) shows how it is changed to integrate monitor placement and optimization.

communications. The input to these flows is the description of the application network requirements (i.e., communication flows, bandwidth, possible latency requirements for QoS, etc.), and, for the case with monitors, the monitoring requirements. Although the main steps that determine the structure of the network remain the same, they are intermixed with a step that determines the monitor insertion and mapping. The second step in the flow, monitor insertion, places virtual monitors in the locations specified by the user. These virtual monitors are later materialized in the monitor placement step that can optimize both the number and the location of the monitors, ensuring that the overall monitoring functionality is preserved. Finally, the dimensioning of the network is shown separated from the topology selection in the traditional NoCs because the overall network requirements are modified (increased) by the number and placement of the monitors. Finally, the process is iterated when the initial parameters (topology, etc.) do not lead to a feasible system that meets all the requirements.

If the monitors use a separate network, the NoC design is simpler and consists of designing an NoC for communications [Figure 8.8(a)] and a separate network for the monitors using a simplified version of the integrated flow of Figure 8.8(b). In this case, convergence is much easier as the only interaction in the design of the regular NoC design and monitoring, and its network is through the increased area to accommodate the monitors and the monitoring network.

8.6 A Case Study

In this section, we discuss several existing approaches that deal with the design and implementation of network monitoring. This section is meant to motivate the range of applications that can use monitoring functionality.

8.6.1 Software Assisted Monitoring Services

Kornaros et al. propose a hybrid monitoring scheme for NoCs that feature the flexibility of a software manager inside a customized embedded CPU. The proposed monitoring scheme is enhanced by small hardware agent components to guarantee a very high response time. These components reside at the "edge" of the NoC [25].

The proposed system consists of the following subsystems:

- Hardware agents, which are responsible for providing information to the embedded CPU and performing the reconfiguration operations commanded by this CPU.

- An interrupt controller, which communicates with the agents, arbitrates the access of the agents to the corresponding CPUs data cache, and interrupts the CPU when necessary.

- A specialized RISC-CPU core optimized for the monitoring application. The RISC-CPU supports high performance applications, covers very small silicon area, and has a very low power consumption.

Figure 8.9 shows an architecture of the hybrid monitoring system of a software monitoring manager assisted by hardware interface accelerators. In this

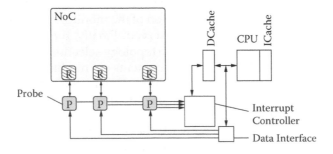

FIGURE 8.9
Architecture of the hybrid monitoring system of a software monitoring manager assisted by hardware interface accelerators.

architecture, a centralized scheme is adopted to manage the traffic of the on-chip interconnect by controlling the limits of guaranteed throughput and best effort priority classes. Nevertheless, special hardware mechanisms are employed to offload the centralized CPU from complex calculations. A hardware data structure is located at each NI, which logs the activity of the flits and the calculated statistic measurements. Programmable event generators are assisting to support fine-grain interrupts if this is desirable, or to mask out selected events. A master block implements additional timers with a varying resolution. The desired objective for the management system is to react quickly enough when fluctuations in traffic are identified. Even when the NoC is scaled to a larger number of routers, the added complexity is shifted to the special hardware components that interface the CPU. The proposed monitoring system is implemented in a Xilinx Virtex4 device (FX100) occupying 12K slices and operating at 100 MHz. A rather large crossbar is used as an interconnect under monitoring; an 8-by-8 64-bit wide buffered crossbar was used as an interconnect, which operates at 120 MHz on the same device without monitors. The simulations show that interrupt handling is achieved in less than 100 cycles for all the monitoring tasks.

Finally, this monitoring system can additionally discover and overcome defects in the NoC. Diagnostic and failure analysis may be periodically performed by the software to ensure operational integrity.

8.6.2 Monitoring Services Interacting with OS

The general problem of mapping a set of communicating tasks onto the heterogeneous resources of a platform on a chip, while dynamically managing the communication between the tiles, is an extremely challenging task. Nollet et al. describe a system where each node of a packet-switched NoC includes a traffic statistic monitor probe, a simple interface to the packet-switched data NoC called data network interface component (dNIC), and a control network interface component (cNIC) that assists the operating system (OS) to control the NoC [21]. The OS is able to control the interprocessor communication in the NoC environment, matching the communication needs to provide the required quality of service. The OS can optimize communication resource allocation and thus minimize interaction between concurrent applications. The three main OS tools provided by the NoC are as follows:

1. The ability to collect statistics of data traces.
2. The ability to limit the time interval in which a processing element (PE) is allowed to send (this time interval is called injection rate control).
3. The ability to dynamically adapt the routing in the NoC.

Although the experimental setup presented by Nollet et al. [21] includes only a few nodes, the layered organization of the cNIC provides a well-structured approach. The main role of the cNIC is to provide the OS with a unified view of the communication resources. For instance, the message statistics collected in the dNIC are processed and supplied to the core OS by the cNIC. Additionally, it allows the core OS to dynamically set the routing table in the data router or to manage the injection rate control mechanism of the dNIC. Another role of the cNIC is to provide the core OS with an abstract view of the distributed processing elements. Hence, it can be considered as a distributed part of the OS.

8.6.3 Monitoring Services at Transaction Level and Monitor-Aware Design Flow

Ciordas et al. [23] presents, in detail, how monitoring services can be taken into account at design time and how designers can integrate the monitoring functionality and placement in the NoC through the system design flow. The proposed solution directs designers toward a unified approach by automating the insertion of the monitors whenever their communication requirements are known, thus leading to a monitoring-aware NoC design flow. The proposed flow is exemplified with the concrete case of transaction monitoring, in the context of the Æthereal NoC and UMARS design flow. The objective in the methodology presented is the mapping of transaction monitors to routers in a way that a full coverage of user channels is achieved. Hence, they extended the coupling of mapping, path selection, and time-slot allocation from the original flow to also include the monitoring probes.

In addition, the cost of the complete monitoring solution is quantified [23]; this cost includes the monitors, the extra NIs, NI ports or enlarged topology needed to support monitoring in addition to the original communication infrastructure. Results show an area-efficient solution for integrating monitoring in NoC designs. Monitors alone do not add much to the overall area numbers as the designs remain dominated by the area of NIs. Reconfiguration of the monitoring system is also considered, showing acceptable reconfiguration times.

It is worthwhile to explore both approaches, that is, using a separate NoC for the transportation of the monitoring messages and on the other end sharing the monitoring NoC with the application NoC. In the first option, assuming the bandwidth requirements are known, it is usually more expansive in area; however, it allows more degrees of freedom for the location and topology of the monitoring interconnect. In the shared case, the combined communication requirements may not fit on the existing application NoC. In this case, it is clear that a new NoC must be generated, for example, by increasing the topology and repeating the process. By increasing the topology, the number of NoC routers increases and in turn the number of required transaction monitors may increase (e.g., if probing all routers is required). This leads to the recomputing of the monitoring communication requirements and monitoring IPs. This means that the NoC monitoring flow must be revised. The reason for investigating this option is that the developed NoC using this approach has the minimum cost.

These options are evaluated by experiments using a 0.13 μm CMOS technology. Results show that in the case of choosing a separate physical interconnect for monitoring, the total NoC area cost of 3.82 mm^2 (2.35 mm^2 original + 1.47 mm^2 extra) was determined based on the addition of (1) seven NIs for the six probes, (2) one monitoring service access (MSA) point, and (3) six routers. When the application NoC was shared with the monitoring components, the total cost of the NoC area was 2.75 mm^2 (2.35 mm^2 original + 0.4 mm^2 extra). This was based on the addition of seven network interface ports (six for connecting the probes and one for the MSA). The added monitoring traffic fitted completely in the original network.

The evaluation of the proposed monitoring methodology is done by benchmarks based on the Æthereal NoC [26]. The Æthereal NoC runs at 500 MHz and offers a raw link bandwidth of 2 GB/s in a 0.13 μm CMOS technology. Æthereal offers transport layer communication services to IPs, in the form of connections, comprising BE and GT services. Guarantees are obtained by means of TDMA slot reservations in NIs. The main objective is to investigate how the monitors affect the mapping, routing, slot allocation in the design flow, and the resulting area implications.

Two real applications, mpeg and audio, were tested. *Mpeg* is an mpeg2 encoder/decoder using the main profile (4:2:0 chroma sampling) at main level (720 × 480 resolution with 15 Mb/s), supporting interlaced video up to 30 frames per second. This application consists of 15 processing cores and an external SDRAM, and has 42 channels (with an aggregated bandwidth of 3 GB). *Audio* is an application that performs sample rate conversion, MP3 decoding, audio postprocessing and radio. The application consists of 18 cores and has 66 channels all configured to use guaranteed throughput. They have also combined these two applications into four cases to be used as examples: mpeg (Design1), mpeg + audio (Design2), 2 mpeg + audio (Design3), 4 mpeg + audio (Design4).

The authors also generated synthetic application benchmarks for testing the proposed design flow. These benchmarks are structured to follow the application patterns of real SoCs. The following two benchmarks were created:

1. Spread communication benchmarks (spread), where each core communicates to a few other cores. These benchmarks characterize designs such as the TV processor that has many small local memories with communication evenly spread in the design.

2. Bottleneck communication benchmarks (bottleneck), where there are one or multiple bottleneck vertices to which the core communication takes place. These benchmarks resemble designs using shared memory/external devices such as the set-top boxes.

For the synthetic benchmarks, the average area cost is almost 15 percent, although for the real examples, the total area increase ranges from 2.2 to 16.1 percent. The concluding result is that, in all cases the area of the transaction monitors is insignificant relative to the total area of the designs, dominated by the area of the NIs. In the Æthereal NoC, the number of cores connected affects the number of NIs and the associated channels and not the routers. Thus, full coverage requires a large number of transaction monitors attached to the NIs. In other NoCs with cores attached to the same channel, a lower number of transaction monitors will be required. From the real examples, the area-efficient solutions were achieved when all routers were probed. Finally, in all designs, the area of the monitors is several times lower than the area of the routers involved.

It must be noted that in the case of bottleneck designs, the number of routers was inevitably increased. The situation might be even worse assuming that an irregular topology might be in use, or in the case where TDMA was not employed. The benchmarks showed a dependence between the slot table size and the NoC topology; a mesh comprised of fewer routers required larger slot table size. Even most important, it is noticeable that the monitoring service itself is not considered in the design stage. It could dynamically affect and ultimately alter the application, which is mapped on the NoC, so as to discover and avoid bottleneck situations or hotspots at run-time.

There is also very little research done regarding other synchronization, arbitration mechanisms in NoCs, and the impact of monitoring traffic to it. Additionally, the transaction monitors in all these studies follow a centralized organization. The MSA, for example, configures the monitor function layers and collects the sniffed data. A distributed control monitoring scheme will obviously deviate from the previous conclusions and needs investigation.

8.6.4 Hardware Support for Testing NoC

Correa [27] and Cota [3] analyze methods to test a packet-switched network model named SOCIN (System-on-Chip Interconnection Network) by reusing the NoC access channels to avoid the inclusion of extra hardware at system level. Originating from test strategies for on-chip multiprocessor architecture, where the processors are connected in a network-based model, the test of the routers exploits the similarity of those blocks by using broadcast messages throughout the network, showing that the test time can be minimized. Particularly, they focus in the test of NoC wrappers, and the strategy to shorten its design time, based on the available network parallelism. In this case, the wrappers are homogeneous, but the cores may be heterogeneous. NoC switching is based on the wormhole approach, where a packet is broken up into flits. With their methodology, the externally generated vectors are transformed into messages to be sent through the network so that each wrapper is tested separately. The area overhead is minimal; however, in this strategy, the objective is to reuse the NoC for system testing although it is not in normal operation.

8.6.5 Monitoring for Cost-Effective NoC Design

Kim et al. propose the use of reconfigurable prototypes to achieve optimal NoC design [28]. Faced with the need to determine all the NoC architectural design parameters such as IP mapping, network topology, routing, etc., they find that many of these choices are affected by the actual on-chip traffic patterns. Thus, to achieve good results, the NoC design requires refinement steps based on knowledge of traffic patterns. To obtain these traffic patterns the designers can use analytical evaluation, simulation, or actual execution.

An analytical approach is very quick but assumes that an accurate theoretical model exists for the application and its traffic pattern. In most cases, in NoC design, this assumption is not valid. A simulation-based approach provides accurate internal traffic observation at the cost of very long simulation time for detailed evaluation. This problem becomes worse as the SoC complexity increases in terms of interconnected nodes and processing power. Another alternative is the use of HW emulation, that is, executing the actual application on hardware, but emulators do not provide the observability required to measure the various NoC parameters.

The final option considered by Kim et al. [28] is executing the application on hardware coupled with the use of monitors to capture all the necessary information. This approach provides an accurate NoC evaluation and enables the determination of design parameters based on real traffic patterns. Because this approach is many orders of magnitude faster than simulation, iterative design refinement is feasible and can be used to achieve better results. They constructed a system that allows the measurement of the following traffic parameters: end-to-end latency, backlog, output conflict status, total execution time, bandwidth between IPs, and link/switch/buffer utilization. Using this system, they investigated the best settings for buffer-sized-assignment, network frequency selection, and run-time routing path modifications, although additional applications such as IP mapping and routing path selection, etc., are also possible.

They also applied the NoC run-time traffic monitoring system and the collection of dynamic statistics on a portable multimedia system running a 3-D graphics application, and found that through more accurate determination of the application needs, they can reduce the NoC buffer size by 42%. They also found that using adaptive routing based on the run-time monitoring results can reduce the path latency by 28 percent. They also discussed using monitoring to choose the lowest frequency that meets the desired processing and communication bandwidth.

8.6.6 Monitoring for Time-Triggered Architecture Diagnostics

El Salloum et al. studied the integration of diagnostic mechanisms for embedded applications (e.g., automotive, avionics, consumer electronics) and SoCs built around the Time-Triggered Architecture (TTA) [29]. The desired property of these systems is to achieve predictable execution for component-based design.

The goals of the diagnostic service is the identification of faulty IP blocks and to distinguish between transient and permanent faults. TTA uses a slotted approach for communication using global time base for the time-triggered

NoC, allowing a diagnostic unit to easily pinpoint the faulty components. The diagnostic unit collects messages with failure indications of other components at the application level and at the architecture level. Failure detection messages are sent on the same TT NoC. Each message is a tuple < type, timestamp, location >, which provides information concerning the type of the occurred failure (e.g., crash failure of a micro component, illegal resource allocation requests), the time of detection with respect to the global time base, and the location within the SoC.

To provide full coverage, failures within the diagnostic unit itself must be detected and all the failure notifications analyzed by correlating failure indications along time and space. The diagnostic unit can distinguish permanent and transient failures, and determine the severity of the action whether to restart a component or to take the component off-line and call for maintenance action. The authors conclude that the determinism inherent in the TTA facilitates the detection of out-of-norm behavior and also find that their encapsulation mechanisms were successful in preventing error propagation.

8.7 Future Research

Future research in the field of NoC monitoring is needed to offer more monitoring flexibility at a smaller cost. As the cost of processing logic becomes lower than communication, intelligent information preprocessing and compression can reduce the amount of data transferred. Also the mechanisms used in profiling at the processing nodes and the monitoring of network resources encourage designers to use a common mechanism. In particular, regarding the future work on the monitoring systems for NoCs, there are numerous specific challenges that have not been addressed by the existing systems.

First, the programmability aspect of the monitoring system has not been covered by the existing approaches. To efficiently and widely use the proposed monitoring systems, the operating and run-time systems should be able to seamlessly support them; this requires the development of efficient and, if possible, standardized high-level interfaces and special modules that support certain OS attributes. The complexity of this task is augmented due to the fact that numerous NoC monitoring systems are highly distributed.

NoC monitoring systems that will not only measure the performance of the interconnection infrastructure but also the power consumption and even some thermal issues may also prove to be highly useful. Such systems will utilize certain heuristics for evaluating the power consumption, the operating temperature, and the thermal gradient of the hardware structure that is monitored.

Another interesting issue that has not been addressed yet is how the monitoring system can be utilized in conjunction with the partial real-time reconfigurable features of the state-of-the-art Field Programmable Gate

Arrays (FPGAs). In such a future system, the monitoring modules will decide when and how the NoC infrastructure will be reconfigured based on a number of different criteria such as the ones presented in the last paragraph. Because the real-time reconfiguration can take a significant amount of time, the relevant issues that should be covered are how the traffic will be routed during the reconfiguration and how the different SoC interfaces connected to the NoC will be updated after the reconfiguration is completed. This feature will not only be employed in FPGAs but can also be used in standard ASIC SoCs, because numerous field-programmable embedded cores are available, which can be utilized within an SoC and offer the ability to be real-time reconfigured in a partial manner.

The monitoring systems can also be utilized, in the future, to change the encoding schemes employed by the NoC. For example, when a certain power consumption level is reached, the monitoring system may close down some of the NoC individual links and adapt the encoding scheme to reduce the power consumption at the cost of reduced performance. To have such an efficient system, the monitoring module should be able to communicate and alter all the NoC interfaces to be aware of the updated data encoding system.

It would also be beneficial if the future monitoring systems are very modular and are combined with a relevant efficient design flow to offer flexibility to the designer to instantiate only the modules needed for her or his specific device. For example, in a low-cost, low-power multiprocessor system only the basic modules will be utilized, which will allow the processors to have full access directly to the monitoring statistics that would be collected in the most power-efficient manner. On the other hand, in a heterogeneous system consisting of numerous high-end cores, the monitoring system will include the majority of the provided modules as well as one or more processors, which will collect numerous different detailed statistics that will be further analyzed and processed by the monitoring CPU(s).

8.8 Conclusions

Network monitoring is a very useful service that can be integrated in future NoCs. Its benefits are expected to increase as the demand for short time to market forces designers to create their SoCs with an incomplete list of features, and rely on programmability to complete the feature list during the product lifetime instead of before-the-product creation. SoC reuse for multiple applications or even a simple application's extensions may lead to a product behavior that is vastly different than the one originally imagined during the design phase.

Monitoring the system operation is a vehicle to capture the changes in the behavior of the system and enable mechanisms to adapt to these changes. Network monitoring is a systematic and flexible approach and can be integrated into the NoC design flow and process. When the monitored information can be abstracted at higher-level constructs, such as complex events, and when monitoring is sharing resources with the regular SoC communication, the cost of supporting monitoring can be much higher. However, given the potential benefits of monitoring during the SoC lifetime, supporting a more detailed (lower level) monitoring abstraction can be acceptable, especially when the monitoring resources are reused for traditional testing and verification purposes.

References

[1] "Coresight," ARM. [Online]. Available: http://www.arm.com/products/solutions/CoreSight.html.

[2] R. Leatherman, "On-chip instrumentation approach to system-on-chip development," First Silicon Solutions, 1997. Available: http://www.fs2.com/pdfs/OCI_Whitepaper.pdf.

[3] Érika Cota, L. Carro, and M. Lubaszewski, "Reusing an on-chip network for the test of core-based systems," *ACM Transactions on Design Automation of Electronic Systems (TOADES)* 9 (2004) (4): 471–499.

[4] A. Ahmadinia, C. Bobda, J. Ding, M. Majer, J. Teich, S. Fekete, and J. van der Veen, "A practical approach for circuit routing on dynamic reconfigurable devices," *Rapid System Prototyping, 2005. (RSP 2005).* In *Proc. of the 16th IEEE International Workshop,* June 2005, 8–10, 84–90.

[5] T. Bartic., J.-Y. Mignolet, V. Nollet, T. Marescaux, D. Verkest, S. Vernalde, and R. Lauwereins, "Topology adaptive network-on-chip design and implementation," In *Proc. of the IEEE Proceedings on Computers and Digital Techniques,* 152 (July 2005) (4): 467–472.

[6] C. Zeferino, M. E. Kreutz, and A. A. Susin, "Rasoc: A router soft-core for networks-on-chip." In *Proc. of Design, Automation and Test in Europe conference (DATE'04),* February 2004, 198–203.

[7] B. Sethuraman, P. Bhattacharya, J. Khan, and R. Vemuri, "Lipar: A light-weight parallel router for FPGA-based networks-on-chip." In *GLSVSLI '05: Proc. of 15th ACM Great Lakes symposium on VLSI.* New York: ACM, 2005, 452–457.

[8] S. Vassiliadis and I. Sourdis, "Flux interconnection networks on demand," *Journal of Systems Architecture* 53 (2007) (10): 777–793.

[9] A. Amory, E. Briao, E. Cota, M. Lubaszewski, and F. Moraes, "A scalable test strategy for network-on-chip routers." In *Proc. of IEEE International Test Conference (ITC 2005),* November 2005, 9.

[10] L. Möller, I. Grehs, E. Carvalho, R. Soares, N. Calazans, and F. Moraes, "A NoC-based infrastructure to enable dynamic self reconfigurable systems." In *Proc. of 3rd International Workshop on Reconfigurable Communication-Centric Systems-on-Chip (ReCoSoC),* 2007, 23–30.

[11] R. Mouhoub and O. Hammami, "NoC monitoring hardware support for fast NoC design space exploration and potential NoC partial dynamic reconfiguration." In *Proc. of International Symposium on Industrial Embedded Systems (IES '06),* October 2006, 1–10.

[12] C. Ciordas, T. Basten, A. Rădulescu, K. Goossens, and J. V. Meerbergen, "An event-based monitoring service for networks on chip," *ACM Transactions on Design Automation of Electronic Systems (TOADES)* 10 (2005) (4): 702–723.

[13] M. Mansouri-Samani and M. Sloman, "A configurable event service for distributed systems." In *Proc. of Third International Conference on Configurable Distributed Systems*, 1996, 210–217.

[14] A. Radulescu, J. Dielissen, S. Pestana, O. Gangwal, E. Rijpkema, P. Wielage, and K. Goossens, "An efficient on-chip NI offering guaranteed services, shared-memory abstraction, and flexible network configuration," *IEEE Transactions on Computer-Aided Design of Integrated Circuits and Systems* 24 (January 2005) (1): 4–17.

[15] P. Amer and L. Cassel, "Management of sampled real-time network measurements." In *Proc. of 14th Conference on Local Computer Networks*, Oct. 10–12, 1989 62–68.

[16] M. H. R. Jurga, "Packet sampling for network monitoring," CERN Technical Report, Dec. 2007. [Online]. Available: http://cern.ch/openlab.

[17] G. He and J. C. Hou, "An in-depth, analytical study of sampling techniques for self-similar internet traffic." In *ICDCS '05: Proc. of 25th IEEE International Conference on Distributed Computing Systems*, 2005, 404–413.

[18] M. Schöller, T. Gamer, R. Bless, and M. Zitterbart, "An extension to packet filtering of programmable networks." In *Proc. of the 7th International Working Conference on Active Networking (IWAN)*, Sophia Antipolis, France, November 2005.

[19] B. Harangsri, J. Shepherd, and A. Ngu, "Selectivity estimation for joins using systematic sampling." In *Proc. of Eighth International Workshop on Database and Expert Systems Applications*, 1–2 September 1997, 384–389.

[20] B.-Y. Choi and S. Bhattacharrya, "On the accuracy and overhead of cisco sampled netflow." In *Proc. of ACM Sigmetrics Workshop on Large-Scale Network Inference (LSNI)*, Banff, Canada, June 2005.

[21] V. Nollet, T. Marescaux, and D. Verkest, "Operating-system controlled network on chip." In *Proc. of 41st Design Automation Conference*, 2004, 256–259.

[22] M. Pastrnak, P. H. N. de With, C. Ciordas, J. van Meerbergen, and K. Goossens, "Mixed adaptation and fixed-reservation QoS for improving picture quality and resource usage of multimedia (NoC) chips." In *Proc. of 10th IEEE International Symposium on Consumer Electronics (ISCE)*, June 2006, 1–6.

[23] C. Ciordas, A. Hansson, K. Goossens, and T. Basten, "A monitoring-aware network-on-chip design flow." *Journal of Systems Architecture*, (2008) 54(3–4): 397–410. http://dx.doi.org/10.1016/j.SYSARC.2007.10.003.

[24] C. Ciordas, A. Hansson, K. Goossens, and T. Basten, "A monitoring-aware network-on-chip design flow." In *DSD '06: Proc. of 9th EUROMICRO Conference on Digital System Design*. Washington, DC: IEEE Computer Society, 2006, 97–106.

[25] G. Kornaros, Y. Papaefstathiou, and D. Pnevmatikatos, "Dynamic software-assisted monitoring of on-chip interconnects." In *Proc. of DATE'07 Workshop on Diagnostic Services in Network-on-Chips*, April 2007.

[26] K. Goossens, J. Dielissen, and A. Radulescu, "Aethereal network on chip: Concepts, architectures, and implementations," *Design and Test of Computers, IEEE* 22 (September–October 2005) (5): 414–421.

[27] E. Correa, R. Cardozo, E. Cota, A. Beck, F. Wagner, L. Carro, A. Susin, and M. Lubaszewski, "Testing the wrappers of a network on chip: A case study." In *Proc. of Natal, Brazil*, 2003, 159–163.

[28] K. Kim, D. Kim, K. Lee, and H. Yoo, "Cost-efficient network-on-chip design using traffic monitoring system." In *Proc. of DATE'07 Workshop on Diagnostic Services in Network-on-Chips*, April 2007.

[29] C. E. Salloum, R. Obermaisser, B. Huber, H. Paulitsch, and H. Kopetz, "A time-triggered system-on-a-chip architecture with integrated support for diagnosis." In *Proc. of DATE'07 Workshop on Diagnostic Services in Network-on-Chips*, April 2007.

9

Energy and Power Issues
in Networks-on-Chips

Seung Eun Lee and Nader Bagherzadeh

CONTENTS

NoC is emerging as a solution for an on-chip interconnection network. Most optimizations considered so far have focused on performance, area, and complexity of implementation of NoC. Another substantial challenge facing designers of high-performance computing processors is the need for significant reduction in energy and power consumption. Although today's processors are much faster and far more versatile than their predecessors using high-speed operation and parallelism, they consume a lot of power. The International Technology Roadmap for Semiconductors highlights system power consumption as the limiting factor in developing systems below the 50-nm technology point. Moreover, an interconnection network dissipates a significant fraction of the total system power budget, which is expected to grow in the future.

A power-aware design methodology emphasizes the graceful scalability of power consumption with factors such as circuit design, technology scaling, architecture, and desired performance, at all levels of the system hierarchy. The energy scalable design methodologies are specifically geared toward mobile applications. At the hardware level, the redundant energy consumption is effected by the low-traffic activity of a link. This design adapts to varying active workload conditions with dynamic voltage scaling (DVS) or on-off links techniques. At the software level, energy agile algorithms for topology selection or application mapping provide energy-performance trade-offs. Energy aware NoC design encompasses the entire system hierarchy, coupling software that considers the energy-performance trade-offs with respect to the hardware that scales its own energy consumption accordingly.

This chapter covers energy and power issues in NoC. Power sources, including dynamic and static power consumptions, and the energy model for NoC are described. The techniques for managing power and energy consumption on NoC are discussed, starting with microarchitectural-level techniques, followed by system-level power and energy optimizations. Power reduction methodologies at the microarchitectural level are highlighted, based on the power model for CMOS technology, such as low-swing signaling, link encoding, RTL optimization, multithreshold voltage, buffer allocation, and performance enhancement of a switch. System-level approaches, such as DVS, on-off links, topology selection, and application mapping, are addressed. For each technique, recent efforts to solve the power problem in NoC are presented. It is desirable to get detailed trade-offs for power and performance early in the design flow, preferably at the system level. To evaluate the dissipation of communication energy in NoC, energy models based on each NoC components are used. Methodologies for power modeling, which are capable of providing a cycle accurate power profile and enable power exploration at the system level, are introduced. The power models enable designers to simulate various system-level power reduction technologies and observe their impact on power consumption, which is not feasible with gate-level simulation. The chapter concludes with a summary of power management strategies.

9.1   Energy and Power

Energy and power are commonly defined in terms of the work that a system performs. Energy is the total electrical energy consumed over time, while performing the work. Power is the rate at which the system consumes electrical energy while performing the work.

$$P = W/T \tag{9.1}$$

$$E = P \cdot T \tag{9.2}$$

where P is power, E energy, T a time interval, and W the total work performed in that interval.

The concepts of energy and power are important because techniques that reduce power do not necessarily reduce energy. For instance, the power consumed by a network can be reduced by halving the operating clock frequency, but if it takes twice as long to forward the same amount of data, the total energy consumed will be similar.

9.1.1   Power Sources

Silicon CMOS (Complementary Metal Oxide Semiconductor) has emerged as the dominant semiconductor technology. Relative to other semiconductor technologies, CMOS is cheap, more easily processed and scaled, and has higher performance-power ratio. For CMOS technology, total power consumption is the combination of dynamic and static sources (Figure 9.1).

$$P = P_{dynamic} + P_{static} \tag{9.3}$$

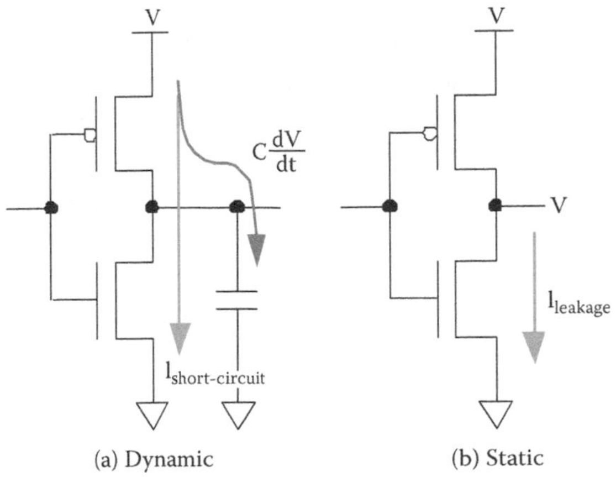

(a) Dynamic   (b) Static

FIGURE 9.1
(a) Dynamic and (b) static power dissipation mechanisms in CMOS.

Equation (9.3) defines power consumption P as the sum of dynamic and static components, $P_{dynamic}$ and P_{static}, respectively.

9.1.1.1 Dynamic Power Consumption

Dynamic power dissipation is the result of switching activity and is ideally the only mode of power dissipation in CMOS circuitry. It is primarily due to charging of capacitative load associated with output wires and gates of subsequent transistors ($C\frac{dV}{dt}$). A smaller component of dynamic power arises from the short-circuit current that flows momentarily when complementary types of transistors switch current. There is an instant when they are simultaneously on, thereby creating a short circuit. In this chapter, power dissipation caused by short-circuit current will not be discussed further because it is a small fraction of the total power and researchers have not found a way to reduce it without sacrificing performance.

As the following equation shows, dynamic power depends on four parameters, namely, a switching activity factor (α), physical capacitance (C), supply voltage (V), and the clock frequency (f).

$$P_{dynamic} = \frac{1}{2}\alpha C V^2 f \tag{9.4}$$

$$f_{max} = \eta\frac{(V - V_{th})^\beta}{V} \tag{9.5}$$

Equation (9.5) establishes the relationship between the supply voltage V and the maximum operating frequency f_{max}, where V_{th} is the threshold voltage, and η and β are experimentally derived constants.

Architectural efforts to control power dissipation have been directed primarily at the dynamic component of power dissipation. There are four ways to control dynamic power dissipation:

(1) Reduce switching activity: This reduces α in Equation (9.4).
(2) Reduce physical capacitance or stored electrical charge of a circuit: The physical capacitance depends on lower-level design parameters such as transistor size and wire length.
(3) Reduce supply voltage: Lowering supply voltage requires reducing clock frequency accordingly to compensate for additional gate delay due to lower voltage.
(4) Reduce operating frequency: This worsens performance and does not always reduce the total energy consumed.

9.1.1.2 Static Power Consumption

As transistors become smaller and faster, another mode of power dissipation has become important, that is, static power dissipation, or the power due to leakage current of the MOS transistor in the absence of any switching activity. As the following equation illustrates, it is the product of the supply voltage (V) and leakage current (I_{leak}):

$$P_{static} = V \cdot I_{leak} \tag{9.6}$$

Technology scaling is increasing both the absolute and relative contribution of static power dissipation. Although there are many different leakage modes, subthreshold and gate-oxide leakages dominate the total leakage current [1].

9.1.1.2.1 Subthreshold Leakage

Subthreshold leakage flows between the drain and source of a transistor. It depends on a number of parameters that constitute the following equation:

$$I_{sub} = K_1 W e^{-V_{th}/nV_\theta} (1 - e^{-V/V_\theta}) \tag{9.7}$$

K_1 and n are experimentally derived constants. W is the gate width and V_θ the thermal voltage.

9.1.1.2.2 Gate-Oxide Leakage

Gate-oxide leakage flows from the gate of a transistor into the substrate.

$$I_{ox} = K_2 W \left(\frac{V}{T_{ox}} \right) e^{-\gamma T_{ox}/V} \tag{9.8}$$

K_2 and γ are experimentally derived. The gate-oxide leakage I_{ox} decreases exponentially as the thickness T_{ox} of the gate's oxide material increases. Unfortunately, it also degrades the transistor's effectiveness because T_{ox} should decrease proportionally with process scaling to avoid short channel effects [2].

Equations (9.6) and (9.7) highlight several avenues that can be targeted for reducing leakage power consumption:

- Turn off supply voltage: This sets V to zero in Equation (9.7) so that the factor in parentheses also becomes zero.
- Increase the threshold voltage: As Equation (9.7) shows, this can have a dramatic effect on even small increments, because V_{th} appears as a negative exponent. However, it reduces performance of the circuit.
- Cool the system: This reduces subthreshold leakage. As a side effect, it also allows a circuit to work faster and eliminates some negative effects from high temperatures.
- Reduce size of a circuit: The total leakage is proportional to the leakage dissipated in all transistors. One way of doing this is to eliminate the obvious redundancy. Another method to reduce size without actually removing the circuit is to turn them off when they are unused.

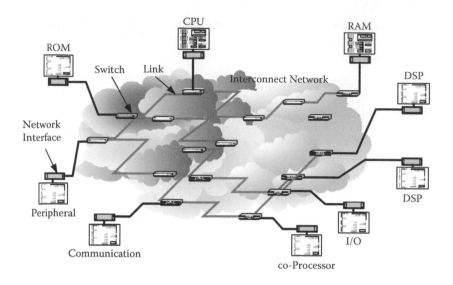

FIGURE 9.2
Example of Network-on-Chip architecture.

9.1.2 Energy Model for NoC

As shown in Figure 9.2, NoC consists of switches that direct packets from source to destination node, links between adjacent switches, and network interfaces that translate packet-based communication into a higher level protocol. Total NoC energy consumption, E_{NoC}, is represented as

$$E_{NoC} = E_{network} + E_{network\ interface} \qquad (9.9)$$

where $E_{network}$ and $E_{network\ interface}$ are energy sources consumed by the network, including link and switch, and network interface, respectively.

When a flit travels on the interconnection network, both links and switches toggle. We use an approach similar to the one presented by Eisley and Peh [3] to estimate the energy consumption for a network. $E_{network}$ can be further decomposed as

$$E_{network} = H \cdot E_{switch} + (H - 1) \cdot E_{link} \qquad (9.10)$$

where E_{link} is the energy consumed by a flit when traversing a link between adjacent switches, E_{switch} is the energy consumed by each flit within the switch, and H is the number of hops a flit traverses. A typical switch consists of several microarchitectural components: buffers that house flits at input ports, routing logic that steers flits toward appropriate output ports along its way to destination, arbiter that regulates access to the crossbar, and a crossbar that transports flits from input to output ports. E_{switch} is the summation of energy consumed on the internal buffer E_{buffer}, arbitration logic $E_{arbiter}$, and crossbar $E_{crossbar}$.

$$E_{switch} = E_{buffer} + E_{crossbar} + E_{arbiter} \qquad (9.11)$$

A network consumes approximately the same amount of energy to transport a flit to its destination independently of the switching technique used. The power consumption can be readily obtained from the energy used in a finite amount of time.

9.2 Energy and Power Reduction Technologies in NoC

Based on the basic power and energy equations for NoC in the previous section, we now discuss energy and power reduction techniques: (1) microarchitectural level and (2) system level optimizations.

Many microarchitectural techniques have been proposed: reducing link power by using low swing signaling and link-encoding schemes; reducing each component power using RTL and buffer optimization; reducing leakage power with multithreshold circuit; and enhancing the throughput of a link. Moving on to system-level techniques, DVS, on-off links, topology optimization, and application mapping algorithms have been introduced. However, any power reduction technique suffers from certain limitations. For example, power management circuitry itself has power and area overheads, which cannot be applied at the lowest granularity. First, each technique's impact on power and energy is analyzed in-depth by using a power model, and previously published results are addressed.

9.2.1 Microarchitecture Level Techniques

9.2.1.1 Low-Swing Signaling

Low-swing signaling alleviates power consumption, obtaining quadratic power savings. As shown in Figure 9.3, binary logic is encoded using lower voltage (V_{swing}), which is smaller than V_{dd}. Typically, these schemes are implemented using differential signaling where a signal is split into two signals

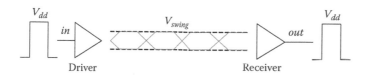

FIGURE 9.3
Low-swing differential signaling.

of opposite polarity bounded by V_{swing}. The receiver is a differential sense amplifier, which restores the signal swing to its full-swing voltage V_{dd} level.

Zhang and Rabaey [4] investigated a number of low-swing on-chip interconnection schemes and presented an analysis of their effectiveness and limitation, especially on energy efficiency and signal integrity. Svensson [5] demonstrated the existence of an optimum voltage swing for minimum power consumption for on-chip and off-chip interconnection. Lee et al. [6] applied a differential low-swing signaling scheme to NoC and found out the optimum voltage swing at which the energy and delay product has the smallest value.

Low-swing differential signaling has several advantages in addition to reduced power consumption. It is immune to crosstalk and electromagnetic radiation effect [7], but supply voltage reduction contributes to a decrease of noise immunity of the interconnection network implementation. Additional complexity is the extra power supply, distributed to both the driver and the receiver.

9.2.1.2 Link Encoding

NoC communication is done through links that transmit bits between adjacent switches. For every packet forwarding, the number of wires that switch depends on the current and previous values forwarded. Link-encoding schemes (Figure 9.4) attempt to reduce switching activity in links through intelligent coding, where a value is encoded and then transmitted, such as bus inversion [8]. Bus inversion ensures that at most half of the link wires switch during a transaction by transmitting the inverse of the intended value and asserting a control signal, which indicates recipients of the inversion when the Hamming distance between current and previous values is more than half the number of wires.

For deep submicron technology, the cross-talk effect between adjacent wires has become another source of power consumption. One way of reducing cross talk is to insert a shield wire between adjacent wires, but this method doubles the number of wires [9]. Another way to prevent cross talk is the use of an encoding scheme. Victor and Keutzer [10] introduced self-shielding codes to prevent cross talk, and Patel and Markov [11] adopted encoding that simultaneously addresses error-correction requirements and cross-talk noise avoidance. Hsieh et al. [12] proposed a de-assembler/assembler structure to eliminate undesirable cross-talk effect on bus transmission. Lee et al. proposed the SiLENT [13] coding method to reduce the transmission power of communication by minimizing the number of transitions on a serial wire. These encoding schemes usually have additional logic gates for data encoding and decoding.

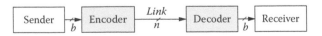

FIGURE 9.4
Model of link encoding.

9.2.1.3   *RTL Power Optimization*

Clock gating, operand isolation, and resource hibernation are well-known techniques for RTL power optimization. These techniques can be adopted for the design of network components such as switches, network interfaces, and FIFO buffers. Clock gating stops the clock to registers, which are not in use during specific clock cycles. Power saving is achieved by reducing switching activity on the clock signal to synchronous registers and the capacitive load on the clock tree. Operand isolation identifies redundant computations of datapath components and isolates such components using specific circuitry, preventing unnecessary switching activity. Determination of how operations can be identified and instantiation of isolating logic are key issues in operand isolation. Resource hibernation is a coarse-grained technique which powers down modules with sleep modes, where each mode represents a trade-off between wake-up latency and power savings.

9.2.1.4   *Multithreshold (V_{th}) Circuits*

As Equation (9.7) shows, increasing the threshold voltage reduces the subthreshold leakage exponentially, which also reduces the circuit's performance. Modern CMOS technology, referred to as MTCMOS (multithreshold CMOS), produces devices with different threshold voltage, allowing for an even better trade-off between static power and performance. For a network component logic circuit, such as switches and FIFO buffers, a higher threshold voltage can be assigned to those transistors in the noncritical paths, so as to reduce leakage current, although the performance is maintained due to the low-threshold transistors in critical paths.

An MTCMOS circuit structure was analyzed by Kao et al. [14]. Algorithms for selecting and assigning an optimal high-threshold voltage transistor were investigated to reduce leakage power under performance constraints [15–18].

9.2.1.5   *Buffer Allocation*

Design of buffers in NoC influences power consumption, area overhead, and performance of the entire network. Buffers are a key component of a majority of network switches. Buffers have been estimated to be the single largest power consumer for a typical switch for an NoC. Application-specific buffer management schemes that allocate buffer depth for each input channel depending on the traffic pattern have been studied [19,20]. Kodi et al. [21] achieved power and chip area savings by reducing the number of buffers with static and dynamic buffer allocation, enabling the repeaters to adaptively function as buffers during congestion. Banerjee and Dutt [22] investigated energy-power characteristics of FIFOs to reduce buffer energy consumption in the context of an on-chip network. Power-aware buffers, which place idle buffers in an inactive mode, based on actual utilization, were proposed as an architectural technique for leakage power optimization in interconnection networks [23].

9.2.1.6 Performance Enhancement

Improving network performance has power saving potential for an NoC. Express Cube [24] lowers network latency by reducing average hop counts. The main idea is to add extra channels between nonadjacent nodes, so that packets spanning long source-destination distances can shorten their network delay by traveling mainly along these express channels, reducing the average hop counts. Besides its performance benefit, an Express Cube can also reduce network power, because it reduces hop counts effectively removing intermediate switch energy completely [25].

The enhanced throughput of a switch can result in power saving by reducing the operating frequency of a switch for certain communication bandwidth requirements that are usually defined by an application. A speculative virtual-channel router [26] optimistically arbitrates the crossbar switch operation in parallel with allocating an output-virtual channel. This speculative architecture largely eliminates the latency penalty of using virtual-channel flow control, having the same per-hop router latency as a wormhole router, although improving throughput of the router. A clock boosting router [27] increases the throughput of an adaptive wormhole router. The key idea is the use of different clocks for head and body flits, because body flits can continue advancing along the reserved path that is already established by the head flit, while the head flit requires the support of complex logic. This method reduces latency and increases throughput of a router by applying faster clock frequency to a boosting clock to forward body flits. Express virtual channels [28], which use virtual lanes in the network to allow packets to bypass nodes along their path in a nonspeculative fashion, reduced delay and energy consumption. In this case, the performance enhancement of a router results in design complexity, which increases energy consumption of a switch (E_{switch}).

9.2.1.7 Miscellaneous

The crossbar is one of the most power-consuming components in NoC. Wang et al. [25] investigated power efficiency of different microarchitectures: segmented crossbar, cut-through crossbar, write-through buffer, and Express Cube, evaluating their power-performance-area impact with power modeling and probabilistic analysis. Kim et al. [29] reduced the number of crossbar ports, and Lee et al. [6] proposed a partially activated crossbar reducing effective capacitive loads.

Different types of interconnect wire have different trade-offs for power consumption and area cost. Power consumption of RC wires with repeated buffers increases linearly with the total wire length. Increasing the spacing between wires can reduce power consumption, but result in additional on-chip area. Using a transmission line is appropriate for long-distance high frequency on-chip interconnection networks, but has complicated transmitter and receiver circuits that may add to the overhead cost. Hu et al. [30] utilized a variety of interconnect wire styles to achieve high-performance, low-power, and on-chip communication.

9.2.2 System-Level Techniques

9.2.2.1 Dynamic Voltage Scaling

A communication link in NoC is capable of scaling energy consumption gracefully, commensurate with traffic workload. This scalability allows for efficient execution of energy-agile algorithms. Suppose that a link can be clocked at any nominal rate up to a certain maximum value. This implies that different levels of power will be consumed for different clock frequencies. One option would be to clock all the links at the same rate to meet the throughput requirements. However, if there is only one link in the design that requires to be clocked at a high rate, the other links could be evaluated by a slower link, consuming less power.

Dynamic power consumption can be reduced by lowering the supply voltage. This requires reducing the clock frequency accordingly to compensate for the additional gate delay due to lower voltage. The use of this approach in run-time, which is called dynamic voltage scaling (DVS), addresses the problem of how to adjust the supply voltage and clock frequency of the link according to the traffic level. The basic idea is that because of high variance in network traffic, when a link is underutilized, the link can be slowed down without affecting performance.

Dynamic power for a single wire is estimated to be

$$P_{wire} = \frac{1}{2}\alpha C_L V_{link}^2 f_{link} \tag{9.12}$$

where P_{wire} is the power consumed by a wire, C_L the load capacitance, V_{link} the link voltage, and f_{link} the link frequency.

By assuming that no coupling capacitance exists between two adjacent wires due to shielding, the total link power becomes

$$P_{link} = \sum_{i=1}^{N} P_{wire_i} \tag{9.13}$$

with N, the number of wires per link.

The energy consumed during voltage transition from V_a to V_b is discussed by Burd and Brodersen [31].

$$E_{link-transition} = (1 - \eta)C_{filter}|V_a^2 - V_b^2| \tag{9.14}$$

where η is the efficiency of the DC-DC converter and C_{filter} is the filter capacitance of the power supply regulator.

Therefore, the total link energy with DVS is represented as

$$E_{link} = \sum_{i=1}^{M} T_{f_i} P_{link_{f_i}} + n E_{link-transition} \tag{9.15}$$

where M is the number of different frequency levels, T_{f_i} is the time occupied by the frequency level i, and n is the number of frequency transitions.

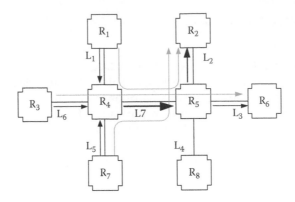

FIGURE 9.5
Network containing three traffics: $R_1 \rightarrow R_2$, $R_3 \rightarrow R_6$, and $R_7 \rightarrow R_2$.

In this scenario, the goal is to find the configuration that maximizes energy and power savings while delivering a prespecified level of performance. The network in Figure 9.5 shows an example of an NoC architecture that consists of eight nodes and seven links. Each node R_i represents a router and solid line L_j represents a link connection. There are three network traffic flows that could occur simultaneously: (1) from node R_1 to R_2; (2) from R_3 to R_6; and (3) from R_7 to R_2. Assuming the same amount of traffic load for three flows, the link traffics ξ_{L_i} on link L_i are ordered as $\xi_{L_7} > \xi_{L_2} > \xi_{L_1}, \xi_{L_3}, \xi_{L_5}$, and $\xi_{L_6} > \xi_{L_4}$. Thus, we can assign the link frequencies as $f_{L_7} > f_{L_2} > f_{L_1}, f_{L_3}, f_{L_5}$, and $f_{L_6} > f_{L_4}$ at that time period, reducing the energy and power of the links.

Wei and Kim proposed chip-to-chip parallel [32] and serial [33] link design techniques where links can operate at different voltage and frequency levels (Figure 9.6). When link frequency is adjusted, supply voltage can track to the lower suitable value. It consists of components of a typical high-speed link: a transmitter to convert digital binary signals into electrical signals; a signaling channel usually modeled as a transmission line; a receiver to convert electrical

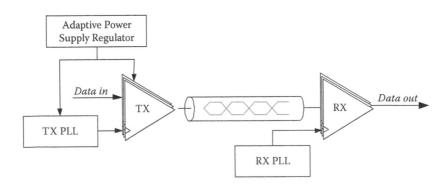

FIGURE 9.6
Example of a DVS link.

signals back to digital data; and a clock recovery block to compensate for delay through the signaling channel. Although this link was not designed for both dynamic voltage and frequency settings, the link architecture can be extended to accommodate DVS [34].

In applying DVS policy to a link, we confront two general problems. One is the estimation of link usage for a given application and the other is the algorithm that adjusts the link frequency according to the time varying workload.

(1) How to predict future workload with reasonable accuracy?

This requires knowing how many packets will traverse a link at any given time. Two issues complicate this problem. First, it is not always possible to accurately predict future traffic activities. Second, a subsystem can be preempted at arbitrary times due to user and I/O device requests, varying traffic beyond what was originally predicted.

There are three kinds of estimation scheme. One is an online scheme that adjusts the link speed dynamically, based on a hardware prediction mechanism by observing the past traffic activity on the link. Based on the variable-frequency links discussed by Kim and Horowitz [33], Shang et al. [35] developed a history-based DVS policy which adjusts the operating voltage and clock frequency of a link according to the utilization of the link and the input buffer. Worm et al. [36] proposed an adaptive low-power transmission scheme for on-chip networks. They minimized the energy required for reliable communication, while satisfying QoS constraints.

One of the potential problems with a hardware prediction scheme is that a misprediction of traffic can be costly from performance and power perspectives. Motivated by this observation, Li et al. [37] proposed a compiler-driven approach where a compiler analyzes the application code and extracts communication patterns among parallel processors. These patterns and the inherent data dependency information of the underlying code help the compiler decide the optimal voltage/frequency to be used for communication links at a given time frame. Shin and Kim [38] proposed an off-line link speed assignment algorithm for energy-efficient NoC. Given the task graph of a periodic real-time application, the algorithm assigns an appropriate communication speed to each link, while guaranteeing the timing constraints of real-time applications.

Combining both online and off-line approaches reduces the misprediction penalty, adjusting links to the run-time traffic based on off-line speculation. Soteriou et al. [39] proposed a software-directed methodology that extends parallel compiler flow to construct a power-aware interconnection network. By using application profiling, it matches DVS link transitions to the expected levels of traffic, generating DVS software directives that are injected into the network along with the application. These DVS instructions are executed at run-time. Concurrently, a hardware online mechanism measures network congestion levels and fine-tunes the execution of these DVS instructions. They reported significantly improved power performance, as compared to prior hardware-based approaches.

(2) How fast to run the link?
Even though the link utilization estimator predicts the workload correctly, determining how fast to run the network is nontrivial. Intuitively, if a link traffic is going to be high, the link frequency can be increased. On the contrary, when a link traffic falls below the threshold value, the link frequency can drop to save power. Shang et al. [35] used link utilization level, which is a direct measure of traffic workload over some interval. Worm et al. [36] introduced residual error rates and transmission delay for clock speed optimization. For each indicator, threshold value is an input to the link control policy and is specified by the user in design-time or optimized in run-time. Shin and Kim [38] adopted the energy gradient $\Delta E(\tau_i)$, that is, the energy gain when the time slot for τ_i is increased by Δt [40]. The clock speed selection algorithm first estimates the slack time of each link and calculates the energy gradient. After increasing the time slot with the largest $\Delta E(\tau_i)$, by a time increment Δt, it repeats the same sequence of steps until there is no task with slack time.

9.2.2.2 On-Off Links

An alternative to save link energy is to add hardware such that a link can be powered down when it is not used heavily. By assuming that a link consumes constant power regardless of the link utilization, the power dissipation of a link that is turned on can be represented by a constant P_{on}. Similarly, when a link is turned off, its power dissipation is assumed to be P_{off}. Thus, the energy consumption of total links, E_{link}, is estimated as follow:

$$E_{link} = \sum_{i=1}^{L} (P_{on} T_{on_i} + P_{off} T_{off_i} + n_i E_P) \qquad (9.16)$$

where T_{on_i} and T_{off_i} are the length of total power on and power off time periods for link i, n_i is the number of times link i has been reactivated, E_P is an energy penalty during the transition period, and L is the total number of links in the network. By assuming $P_{off} \simeq 0$, the energy consumption of links can be reduced to $E_{link} \simeq \sum_{i=1}^{L} (P_{on} T_{on_i} + n_i E_P)$. There can be trade-offs based on the values of n_i and E_P. For instance, link L_4 in Figure 9.5 can be turned off to reduce the energy and power consumption for the network.

Dynamic link shutdown (DLS) [41] powers down links intelligently when their utilizations are below a certain threshold level and a subset of highly used links can provide connectivity in the network. An adaptive routing strategy that intelligently uses a subset of links for communication was proposed, thereby facilitating DLS for minimizing energy consumption. Soteriou and Peh [42] explored the design space for communication channel turn-on/off based on a dynamic power management technique depending on hardware counter measurement obtained from the network during run-time.

Compiler-directed approaches have benefits as compared to hardware-based approaches. Based on high-level communication analysis, these techniques determine the point at which a given communication link is idle and can be turned off to save power, without waiting for a certain period of time

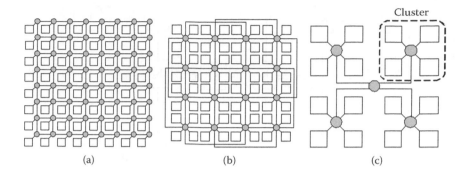

FIGURE 9.7
Network topologies: (a) Mesh, (b) CMesh, and (c) hierarchical star.

to be certain that the link has truly become idle. Similarly, the reactivation point which was identified automatically eliminates the turn on performance penalty. Chen et al. [43] introduced a compiler-directed approach, which increases the idle periods of communication channels by reusing the same set of channels for as many communication messages as possible. Li et al. [44] proposed a compiler-directed technique to turn off the communication channels to reduce NoC energy consumption.

9.2.2.3 Topology Optimization

The connection pattern of nodes defines the network topology that can be tailored and optimized for an application. More specifically, network topologies determine the number of hops and the wire length involved in each data transmission, both critically influencing the energy cost per transmission (Figure 9.7).

Equation (9.10) expressed as

$$E_{network} = H \cdot E_{switch} + D \cdot E_{avg} \tag{9.17}$$

where D is the distance from source to destination and E_{avg} is the average link traversal energy per unit length. Among these factors, H and D are strongly influenced by the topology. For instance, the topology in Figure 9.5 can be changed to Figure 9.8(a), by adding additional links, while reducing the number of hop counts. The power trade-offs are determined by interaction of all factors dynamically, and the variation of one factor will clearly impact other factors. For example, topology optimization can effectively reduce the hop count, but it might inevitably increase router complexity, which increases the switch energy (E_{switch}).

Energy efficiency of different topologies was derived and compared based on the network size and architecture parameters for technology scaling [45]. Based on the model, Lee et al. [6] showed that hierarchical star topology has the lowest energy and area cost for their application. For any given average point-to-point communication latency requirement, an algorithm which

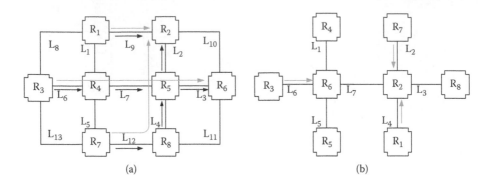

FIGURE 9.8
(a) Topology optimization, (b) application mapping.

finds the optimal NoC topology from a given topology library (mesh, torus, and hypercube) was proposed, balancing between NoC power efficiency and communication latency [46]. Balfour and Dally [47] developed area and energy models for an on-chip interconnection network and described trade-offs in a tiled CMP. Using these models, they investigated how aspects of the network architecture including topology, channel width, routing strategy, and buffer size affect performance, and impact area and energy efficiency. Among the different topologies, the Concentrated Mesh (CMeshX2) network was substantially the most efficient. Krishnan et al. [48] presented an MILP formulation that addresses both wire and router energy by splitting the topology generation problem into two distinct subproblems: (1) system-level floor planning and (2) topology and route generation. A prohibitive greedy iterative improvement strategy was used to generate an energy optimized application specific NoC topology which supports both point-to-point and packet switched networks [49].

9.2.2.4 *Application Mapping*

As shown in Equation (9.10), optimizing the mapping and routing path allocation reduces energy consumption, by reducing the number of hop counts H. For instance, the traffic flows shown in Figure 9.5 can be transformed to Figure 9.8(b) by choosing different application mapping, while balancing the amount of traffic in links. It also increases the number of idle links, enabling the opportunity for on-off link control.

Hu and Marculescu [50] proposed a branch and bound algorithm to map the processing cores onto a tile-based NoC mesh architecture to satisfy bandwidth constraints and minimize total energy consumption. Murali et al. [51] considered the topology mapping problem together with the possibility of splitting traffic among various paths to minimize the average communication delay while satisfying bandwidth constraints. Morad et al. [52] placed clusters of different cores on a single die. The larger and faster cores execute single-threaded programs and the serial phase of multithreaded programs for high energy per instruction, whereas the smaller and slower cores execute the parallel phase for lower energy per instruction, reducing power consumption for similar performance.

9.2.2.5 Globally Asynchronous Locally Synchronous (GALS)

Another challenge for low-power design is the globally asynchronous locally synchronous (GALS) system. GALS architecture is composed of large synchronous blocks which communicate with each other on an asynchronous basis. Working in the asynchronous domain has advantages in terms of performance, robustness, and power. As each synchronous block operates asynchronously with respect to each other, the operating frequency of each synchronous block is tailored to the local demand, reducing the average frequency and the overall power consumption.

Hemani et al. [53] analyzed power savings in GALS with respect to its overheads, such as communication and local clock generation, to use for partitioning the system into an optimal number of synchronous blocks. NoC architectures based on the GALS scheme, providing low latency for QoS, were proposed by researchers [54–56]. On-chip and off-chip interfaces, which not only handle the resynchronization between the synchronous and asynchronous NoC domains but also implement NoC communication priorities, were designed for GALS implementation [57,58].

Systematic comparison between GALS and fully asynchronous NoCs is discussed by Sheibanyrad et al. [59]. In a typical shared memory multiprocessor system using a best effort micronetwork, the fully asynchronous router consumed less power than GALS due to the less idle power consumption, even though the energy required for packet transmission is larger in the asynchronous router than in GALS.

9.3   Power Modeling Methodology for NoC

Communication architectures have a significant impact on the performance and power consumption of NoC. Customization of such architectures for an application requires the exploration of a large design space. Accurate estimates of the power consumption of the implementation must be made early in the design process. This requires power models for NoC components. Power models are classified based on different levels of abstraction. The lowest level of abstraction is the gate level, which represents the model at transistor level and is more accurate than any of the higher levels. These models are extremely time-consuming and intractable as far as power profile for complicated multiprocessors are concerned. The next level in the abstraction is the register transfer level, which considers the transfer of data at register and wire levels. The highest level of abstraction is the system level which emulates the functionalities performed without going into the hardware details of components. This level is less accurate but requires less simulation time. Power models for NoC are targeted for power optimization, system performance, and power trade-offs.

9.3.1   Analytical Model

In NoC, bits of data are forwarded through links from a source node to a destination node via intermediate switches. The power consumed is the sum of the power consumed by links and intermediate switches, which includes the power consumed by internal components such as FIFO buffers, arbiters, and crossbars during switching activity.

One way to model power consumption of an NoC is to derive detailed capacitance equations for various switch and link components, assuming specific circuit designs for each components. These equations are then plugged into a cycle-accurate simulator so that actual network activity triggers specific capacitance calculations and derives dynamic power estimates. The capacitance for each network component is derived, based on architectural parameters. The other approach is to evaluate the energy and power consumption of each component by using gate-level simulation with technology libraries. Overall energy consumption in NoC is estimated with the energy model that was described in Section 9.1.2.

There have been several power estimation approaches for network components in NoC. Patel et al. [60] first noted the need to consider power constraints in interconnection network design, and proposed an analytical power model of switch and link. Wang et al. [61] presented the architectural-level parameterized power model named Orion by combining parameterized capacitance equations and switching activity estimations for network components. These analytical models are based on evaluation of switching capacitance and estimate dynamic power consumption. Chen and Peh [23] extended the Orion by adding the leakage power model, which was based on empirical characterization of some frequently used circuit components. Ye et al. [62] analyzed the power consumption of switch fabric in network routers and proposed the bit-energy models to estimate the power consumption. Average energy consumption on each bit was precalculated from *Synopsys^{TM} Power Compiler* simulation. Eisley and Peh [3] approximated NoC power consumption with just link utilizations, which is the percentage of cycles that link has used. Xi and Zhong [63] presented a transaction-level power model for switch and link in SystemC, providing both temporal and spatial power profiles.

9.3.2   Statistical Model

The basic idea is to measure power consumption using a series of different input data patterns where a generally valid power model from the obtained results is rendered. The problem is to find a regression curve that best approximates the dependence of power on variables from the sampled data.

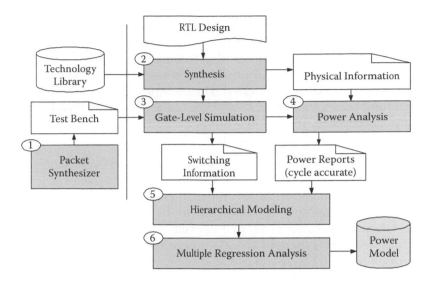

FIGURE 9.9
Power model generation methodology.

A power macro model consists of variables, which represent factors influencing power consumption, and regression coefficients that reflect contributions of the variables for power consumption. A general power macro model for a component is expressed as

$$\hat{P} = \alpha_0 + A \cdot \Psi \tag{9.18}$$

where α_0 is the power term that is independent of the variables and $A = [\alpha_1 \, \alpha_2 \ldots \alpha_k]$ is the regression coefficients for the variables $\Psi = [\psi_1 \, \psi_2 \ldots \psi_k]^T$.

Power macro modeling is to find out the regression coefficients for the variables that provide the minimum mean square error. Figure 9.9 illustrates a procedure to create a power macromodel for NoC components.

- Step 1: A packet synthesizer generates traffic patterns that exercise the network under different conditions.
- Step 2: The RTL description is synthesized to the gate-level net-list using a technology library. As part of this step, physical information is also generated to be used for gate-level power analysis.
- Step 3: The gate-level simulation extracts the switching information of the variables Ψ for modeling.
- Step 4: The gate-level power analysis creates nanosecond detailed power waveform using switching and physical information. To develop a cycle accurate model, the extracted waveform is modified to a cycle-level granularity power waveform.
- Step 5: For the hierarchical power model, the power consumption of the network is analyzed to estimate the power contributions of

various parts of the network. Based on the analysis, the levels of hierarchical model are defined.

- Step 6: The switching information is compared with cycle accurate power reports and macro-model templates for each node in the hierarchical model are generated. These templates consist of variables (Ψ) and power values for every cycle of the test bench. Finally, multiple regression analysis to correlate the effect of each variable to power consumption is performed to find coefficients for variables.

A statistical approach based on multiple linear regression analysis was adopted to generate cycle accurate power estimation. Palermo et al. [64] proposed automatic generation of an analytical power model of network elements based on design space parameters and the traffic information derived from simulation. Wolkotte et al. [65] derived an energy model for packet- and circuit-switched routers by calculating the average energy per bit to traverse on a single router based on possible scenarios empirically. They insisted that the power consumption of a single router depends on four parameters: (1) the average load of every data stream; (2) the amount of bit-flips in the data stream; (3) the number of concurrent data streams; and (4) the amount of control overhead in a router. Penolazzi et al. [66] presented an empirical formulation to estimate the power consumption of the Nostrum NoC. They chose reference power with static input, number of total switching bits, number of static logic one bits, and total number of static bits as parameters for this analysis. The accuracy of their models demonstrated the average difference with respect to gate-level simulation to be about 5 percent.

9.4 Summary

The key feature of on-chip interconnection network is the capability to provide required communication bandwidth and low power and energy consumption in the network. With the continuing progress in VLSI technology where billions of transistors are available to the designer, power awareness becomes the dominant enabler for a practical energy-efficient on-chip interconnection network. This chapter discussed a few of the power and energy management techniques for NoC. Ways to minimize the power consumption were covered starting with microarchitectural-level techniques followed by system-level approaches.

The microarchitectural-level power savings were presented by reducing supply voltage and switching activity. RTL optimization enables circuit-level power savings and multithreshold circuit reduces the static power consumption for NoC components. There are trade-offs in performance and power dissipation for buffer allocation and throughput of switches. System-level power management is to allow the system power scale with changing conditions and performance requirements. Energy savings are achieved by DVS for an interconnection network using an adjustable voltage and frequency link. Energy scalable algorithms running based on this implementation consume less energy with DVS than with a fixed supply voltage. The on-off links technique enables a link to be powered down when it is not used heavily. Topology and task mapping can be tailored and optimized for the application while reducing power and energy consumption on NoC. GALS architecture reduces power dissipation by using different clock frequencies according to the local demand. In general, these techniques have different trade-offs and not all of them reduce the total energy consumed. For power management, accurate estimates of the power consumption of NoC must be made early in the design process, allowing for the exploration of the design space.

References

[1] A. P. Chandrakasan, W. J. Bowhill, and F. Fox, *Design of High-Performance Microprocessor Circuits*. Hoboken, NJ: Wiley-IEEE Press, 2000.

[2] N. S. Kim, T. Austin, D. Blaauw, T. Mudge, K. Flautner, J. S. Hu, M. J. Irwin, et al., "Leakage current: Moore's law meets static power," *Computer* 36 (2003) (12): 68–75.

[3] N. Eisley and L.-S. Peh, "High-level power analysis for on-chip networks." In *CASES '04: Proc. of 2004 International Conference on Compilers, Architecture, and Synthesis for Embedded Systems*. New York: ACM, 2004, 104–115.

[4] H. Zhang and J. Rabaey, "Low-swing interconnect interface circuits." In *ISLPED '98: Proc. of 1998 International Symposium on Low Power Electronics and Design*. New York: ACM, 1998, 161–166.

[5] C. Svensson, "Optimum voltage swing on on-chip and off-chip interconnect," *IEEE Journal of Solid-State Circuits* 36 (Jul. 2001) (7): 1108–1112.

[6] K. Lee, S.-J. Lee, and H.-J. Yoo, "Low-power network-on-chip for high-performance SoC design," *IEEE Transactions on Very Large Scale Integration Systems* 14 (2006) (2): 148–160.

[7] V. Venkatachalam and M. Franz, "Power reduction techniques for microprocessor systems," *ACM Computing Surveys* 37 (2005) (3): 195–237.

[8] M. R. Stan and W. P. Burleson, "Bus-invert coding for low-power I/O," *IEEE Transactions on Very Large Scale Integration Systems* 3 (1995) (1): 49–58.

[9] C. N. Taylor, S. Dey, and Y. Zhao, "Modeling and minimization of interconnect energy dissipation in nanometer technologies." In *DAC '01: Proc. of 38th Conference on Design Automation*. New York: ACM, 2001, 754–757.

[10] B. Victor and K. Keutzer, "Bus encoding to prevent crosstalk delay." In *ICCAD '01: Proc. of 2001 IEEE/ACM International Conference on Computer-Aided Design*. Piscataway, NJ: IEEE Press, 2001, 57–63.

[11] K. N. Patel and I. L. Markov, "Error-correction and crosstalk avoidance in DSM busses," *IEEE Transactions on Very Large Scale Integration Systems* 12 (2004) (10): 1076–1080.

[12] W.-W. Hsieh, P.-Y. Chen, and T. Hwang, "A bus architecture for crosstalk elimination in high performance processor design." In *CODES+ISSS '06: Proc. of 4th*

International Conference on Hardware/Software Codesign and System Synthesis. New York: ACM, 2006, 247–252.

[13] K. Lee, S.-J. Lee, and H.-J. Yoo, "Silent: serialized low energy transmission coding for on-chip interconnection networks." In *Computer Aided Design, 2004. ICCAD-2004. IEEE/ACM International Conference,* 7–11 November 2004, 448–451.

[14] J. Kao, A. Chandrakasan, and D. Antoniadis, "Transistor sizing issues and tool for multi-threshold CMOS technology." In *DAC '97: Proc. of 34th Annual Conference on Design Automation.* New York: ACM, 1997, 409–414.

[15] L. Wei, Z. Chen, M. Johnson, K. Roy, and V. De, "Design and optimization of low voltage high performance dual threshold CMOS circuits." In *DAC '98: Proc. of 35th Annual Conference on Design Automation.* New York: ACM, 1998, 489–494.

[16] K. Roy, "Leakage power reduction in low-voltage CMOS design." In *Proc. of IEEE International Conference on Circuits and Systems,* Lisboa, Portugal, 1998, 167–173.

[17] Q. Wang and S. B. K. Vrudhula, "Static power optimization of deep submicron CMOS circuits for dual VT technology." In *ICCAD '98: Proc. of 1998 IEEE/ACM International Conference on Computer-Aided Design.* New York: ACM, 1998, 490–496.

[18] M. Liu, W.-S. Wang, and M. Orshansky, "Leakage power reduction by dual-VTH designs under probabilistic analysis of VTH variation." In *ISLPED '04: Proc. of 2004 International Symposium on Low Power Electronics and Design.* New York: ACM, 2004, 2–7.

[19] J. Hu and R. Marculescu, "Application-specific buffer space allocation for networks-on-chip router design." In *ICCAD '04: Proc. of 2004 IEEE/ACM International Conference on Computer-Aided Design.* Washington, DC: IEEE Computer Society, 2004, 354–361.

[20] C. A. Nicopoulos, D. Park, J. Kim, N. Vijaykrishnan, M. S. Yousif, and C. R. Das, "Vichar: A dynamic virtual channel regulator for network-on-chip routers." In *MICRO 39: Proc. of 39th Annual IEEE/ACM International Symposium on Microarchitecture.* Washington, DC: IEEE Computer Society, 2006, 333–346.

[21] A. Kodi, A. Sarathy, and A. Louri, "Design of adaptive communication channel buffers for low-power area-efficient network-on-chip architecture." In *ANCS '07: Proc. of 3rd ACM/IEEE Symposium on Architecture for Networking and Communications Systems.* New York: ACM, 2007, 47–56.

[22] S. Banerjee and N. Dutt, "FIFO power optimization for on-chip networks." In *GLSVLSI '04: Proc. of 14th ACM Great Lakes Symposium on VLSI.* New York: ACM, 2004, 187–191.

[23] X. Chen and L.-S. Peh, "Leakage power modeling and optimization in interconnection networks." In *ISLPED '03: Proc. of 2003 International Symposium on Low Power Electronics and Design.* New York: ACM, 2003, 90–95.

[24] W. Dally, "Express cubes: Improving the performance of k-ary n-cube interconnection networks," *Computers, IEEE Transactions* 40 (September 1991) 9: 1016–1023.

[25] H. Wang, L.-S. Peh, and S. Malik, "Power-driven design of router microarchitectures in on-chip networks." In *MICRO 36: Proc. of 36th Annual IEEE/ACM International Symposium on Microarchitecture.* Washington, DC: IEEE Computer Society, 2003, 105.

[26] L.-S. Peh and W. J. Dally, "A delay model and speculative architecture for pipelined routers." In *HPCA '01: Proc. of 7th International Symposium on High-Performance Computer Architecture.* Washington, DC: IEEE Computer Society, 2001, 255.

[27] S. E. Lee and N. Bagherzadeh, "Increasing the throughput of an adaptive router in network-on-chip (NoC)." In *CODES+ISSS'06: Proc. of 4th International Conference on Hardware/Software Codesign and System Synthesis*, 2006, 82–87.

[28] A. Kumar, L.-S. Peh, P. Kundu, and N. K. Jha, "Express virtual channels: Towards the ideal interconnection fabric." In *ISCA '07: Proc. of 34th Annual International Symposium on Computer Architecture*. New York: ACM, 2007, 150–161.

[29] J. Kim, C. Nicopoulos, and D. Park, "A gracefully degrading and energy-efficient modular router architecture for on-chip networks," *SIGARCH Computer Architecture News* 34 (2006) (2): 4–15.

[30] Y. Hu, H. Chen, Y. Zhu, A. A. Chien, and C.-K. Cheng, "Physical synthesis of energy-efficient networks-on-chip through topology exploration and wire style optimizations." In *ICCD '05: Proc. of 2005 International Conference on Computer Design*. Washington, DC: IEEE Computer Society, 2005, 111–118.

[31] T. Burd and R. Brodersen, "Design issues for dynamic voltage scaling," *Low Power Electronics and Design, 2000. In ISLPED '00. Proc. of 2000 International Symposium*, 9–14, 2000.

[32] G. Wei, J. Kim, D. Liu, S. Sidiropoulos, and M. A. Horowitz, "A variable-frequency parallel I/O interface with adaptive power-supply regulation," *IEEE Journal of Solid-State Circuits* 35 (2000) (11): 1600–1610.

[33] J. Kim and M. A. Horowitz, "Adaptive supply serial links with sub-1v operation and per-pin clock recovery," *IEEE Journal of Solid-State Circuits*, 37 (2002) (11): 1403–1413.

[34] L. Shang, L.-S. Peh, and N. K. Jha, "Power-efficient interconnection networks: Dynamic voltage scaling with links," *IEEE Computer Architecture Letters*, 1 (2006) (1): 6.

[35] L. Shang, L.-S. Peh, and N. K. Jha, "Dynamic voltage scaling with links for power optimization of interconnection networks." In *HPCA'03: Proc. of 9th International Symposium on High-Performance Computer Architecture*, Anaheim, CA, 2003, 91–102.

[36] F. Worm, P. Ienne, P. Thiran, and G. de Micheli, "An adaptive low-power transmission scheme for on-chip networks." In *ISSS'02: Proc. of 15th International Symposium on System Synthesis*, Kyoto, Japan, 2002, 92–100.

[37] F. Li, G. Chen, and M. Kandemir, "Compiler-directed voltage scaling on communication links for reducing power consumption." In *ICCAD '05: Proc. of 2005 IEEE/ACM International Conference on Computer-Aided Design*. Washington, DC: IEEE Computer Society, 2005, 456–460.

[38] D. Shin and J. Kim, "Power-aware communication optimization for networks-on-chips with voltage scalable links." In *CODES+ISSS '04: Proc. of International Conference on Hardware/Software Codesign and System Synthesis*. Washington, DC: IEEE Computer Society, 2004, 170–175.

[39] V. Soteriou, N. Eisley, and L.-S. Peh, "Software-directed power-aware interconnection networks," *ACM Transactions on Architecture and Code Optimization* 4 (2007) (1): 5.

[40] M. T. Schmitz and B. M. Al-Hashimi, "Considering power variations of DVS processing elements for energy minimisation in distributed systems." In *ISSS '01: Proc. of 14th International Symposium on Systems Synthesis*. New York: ACM, 2001, 250–255.

[41] E. J. Kim, K. H. Yum, G. M. Link, N. Vijaykrishnan, M. Kandemir, M. J. Irwin, M. Yousif, and C. R. Das, "Energy optimization techniques in cluster

interconnects." In *ISLPED '03: Proc. of 2003 International Symposium on Low Power Electronics and Design*. New York: ACM, 2003, 459–464.

[42] V. Soteriou and L.-S. Peh, "Design-space exploration of power-aware on/off interconnection networks." In *ICCD'04: Proc. of IEEE International Conference on Computer Design*, 2004, 510–517.

[43] G. Chen, F. Li, and M. Kandemir, "Compiler-directed channel allocation for saving power in on-chip networks," *SIGPLAN Notices* 41 (2006) (1): 194–205.

[44] F. Li, G. Chen, M. Kandemir, and M. J. Irwin, "Compiler-directed proactive power management for networks." In *CASES '05: Proc. of 2005 International Conference on Compilers, Architectures and Synthesis for Embedded Systems*. New York: ACM, 2005, 137–146.

[45] H. Wang, L.-S. Peh, and S. Malik, "A technology-aware and energy-oriented topology exploration for on-chip networks." In *DATE '05: Proc. of Conference on Design, Automation and Test in Europe*. Washington, DC: IEEE Computer Society, 2005, 1238–1243.

[46] Y. Hu, Y. Zhu, H. Chen, R. Graham, and C.-K. Cheng, "Communication latency aware low power NoC synthesis." In *DAC '06: Proc. of 43rd Annual Conference on Design Automation*. New York: ACM, 2006, 574–579.

[47] J. Balfour and W. J. Dally, "Design tradeoffs for tiled CMP on-chip networks." In *ICS '06: Proc. of 20th Annual International Conference on Supercomputing*. New York: ACM, 2006, 187–198.

[48] K. Srinivasan, K. Chatha, and G. Konjevod, "Linear-programming-based techniques for synthesis of network-on-chip architectures," *IEEE Transactions on Very Large Scale Integration (VLSI) Systems*, 14 (April 2006) (4): 407–420.

[49] J. Chan and S. Parameswaran, "Nocout: NoC topology generation with mixed packet-switched and point-to-point networks." In *ASP-DAC '08: Proc. of 2007 Conference on Asia South Pacific Design Automation*. Washington, DC: IEEE Computer Society, 2008, 265–270.

[50] J. Hu and R. Marculescu, "Exploiting the routing flexibility for energy/performance aware mapping of regular NoC architectures." In *DATE '03: Proc. of Conference on Design, Automation and Test in Europe*. Washington, DC: IEEE Computer Society, 2003, 10688.

[51] S. Murali and G. D. Micheli, "Bandwidth-constrained mapping of cores onto NoC architectures." In *DATE '04: Proc. of Conference on Design, Automation and Test in Europe*. Washington, DC: IEEE Computer Society, 2004, 20896.

[52] T. Y. Morad, U. C. Weiser, A. Kolodny, M. Valero, and E. Ayguade, "Performance, power efficiency and scalability of asymmetric cluster chip multiprocessors," *IEEE Computer Architecture Letters* 5 (2006) (1): 4.

[53] A. Hemani, T. Meincke, S. Kumar, A. Postula, T. Olsson, P. Nilsson, J. Oberg, P. Ellervee, and D. Lundqvist, "Lowering power consumption in clock by using globally asynchronous locally synchronous design style." In *DAC '99: Proc. of 36th ACM/IEEE Conference on Design Automation*. New York: ACM, 1999, 873–878.

[54] T. Bjerregaard and J. Sparso, "A scheduling discipline for latency and bandwidth guarantees in asynchronous network-on-chip." In *ASYNC '05: Proc. of 11th IEEE International Symposium on Asynchronous Circuits and Systems*. Washington, DC: IEEE Computer Society, 2005, 34–43.

[55] D. R. Rostislav, V. Vishnyakov, E. Friedman, and R. Ginosar, "An asynchronous router for multiple service levels networks on chip." In *ASYNC '05: Proc. of 11th*

IEEE International Symposium on Asynchronous Circuits and Systems. Washington, DC: IEEE Computer Society, 2005, 44–53.

[56] E. Beigne, F. Clermidy, P. Vivet, A. Clouard, and M. Renaudin, "An asynchronous NoC architecture providing low latency service and its multi-level design framework." In *ASYNC '05: Proc. of 11th IEEE International Symposium on Asynchronous Circuits and Systems*. Washington, DC: IEEE Computer Society, 2005, 54–63.

[57] E. Beigne and P. Vivet, "Design of on-chip and off-chip interfaces for a GALS NoC architecture." In *ASYNC '06: Proc. of 12th IEEE International Symposium on Asynchronous Circuits and Systems*. Washington, DC: IEEE Computer Society, 2006, 172.

[58] D. Lattard, E. Beigne, C. Bernard, C. Bour, F. Clermidy, Y. Durand, et al., "A telecom baseband circuit based on an asynchronous network-on-chip." In *Solid-State Circuits Conference, 2007. ISSCC 2007. Digest of Technical Papers. IEEE International*, San Francisco, CA, February 11–15, 2007, 258–601.

[59] A. Sheibanyrad, I. M. Panades, and A. Greiner, "Systematic comparison between the asynchronous and the multi-synchronous implementations of a network on chip architecture." In *DATE '07: Proc. of Conference on Design, Automation and Test in Europe* Nice, France, 2007, 1090–1095.

[60] C. Patel, S. Chai, S. Yalamanchili, and D. Schimmel, "Power constrained design of multiprocessor interconnection networks." In *ICCD '97: Proc. of 1997 International Conference on Computer Design (ICCD '97)*, Austin, Texas, 1997, 408–416.

[61] H.-S. Wang, X. Zhu, L.-S. Peh, and S. Malik, "Orion: A power-performance simulator for interconnection networks." In *Proc. 35th Annual IEEE/ACM International Symposium on Microarchitecture (MICRO-35)*, Istanbul, Trukey, 2002, 294–305.

[62] T. T. Ye, G. D. Micheli, and L. Benini, "Analysis of power consumption on switch fabrics in network routers." In *DAC '02: Proc. of 39th Conference on Design Automation*, New Orleans, LA, 2002, 524–529.

[63] J. Xi and P. Zhong, "A transaction-level NoC simulation platform with architecture-level dynamic and leakage energy models." In *GLSVLSI '06: Proc. of 16th ACM Great Lakes Symposium on VLSI*. New York: ACM, 2006, 341–344.

[64] G. Palermo and C. Silvano, "Pirate: A framework for power/performance exploration of network-on-chip architectures," *Lecture Notes in Computer Science* 3254 (2004): 521–531.

[65] P. Wolkotte, G. Smit, N. Kavaldjiev, J. Becker, and J. Becker, "Energy model of networks-on-chip and a bus." In *Proc. of International Symposium on System-on-Chip*, Tampere, France, 82–85, November 17, 2005.

[66] S. Penolazzi and A. Jantsch, "A high level power model for the Nostrum NoC." In *DSD '06: Proc. of 9th EUROMICRO Conference on Digital System Design*, Dubrovnik, Crotia, 2006, 673–676.

10

The CHAIN®Works Tool Suite: A Complete Industrial Design Flow for Networks-on-Chips

John Bainbridge

CONTENTS

The challenge of today's multimillion gate System-on-Chip (SoC) designs is to deal with system-level complexity in the presence of deep submicron (DSM) effects on the physical design while coping with extreme development schedule pressures and rapidly escalating development costs. Previously, performance and functionality were limited by transistor switching delays, but in today's shrinking feature sizes, interconnect and signal integrity have become predominant factors. As the number of gates per square millimeter has increased, tying them together and achieving system-level timing closure has also become increasingly challenging. In this environment, conventional bus interconnects and their derivatives are particularly problematic and are being replaced by Networks-on-Chip (NoC).

Achieving maximum advantage from moving to an NoC rather than conventional bus hierarchies requires the use of two new approaches. First, synthesis tools are required to provision the NoC to achieve the best architecture to meet the specific requirements of the SoC. Second, the implementation should encompass clockless logic to avoid the pollution of the interconnect-centric and interface-based design approach with the need to distribute global clocks as part of the NoC physical implementation. When such techniques are combined, the user benefits not only from the NoC scalability that enables the construction of ever more complex systems, but also from reliable and improved predictability of power, performance, area, and their trade-offs at the early architectural stage of the design process.

10.1 CHAIN®Works

To address the complexity issues of combining these techniques and deploying NoC technology, Silistix has introduced CHAINworks, a suite of software tools and clockless NoC IP blocks that fit into the existing ASIC or COT flows and are used for the design and synthesis of CHAIN networks that meet the critical challenges in complex devices. CHAINworks consists of

- CHAIN®architect—used to architect the specific implementation of the Silistix interconnect to meet the needs of the system being designed. Trade-off analysis of network topology, link widths, pipelining depth, and protocol options is performed by CHAIN architect that uses a language-based approach to specify the requirements of the system.
- CHAIN®compiler—processes the architecture synthesized by CHAINarchitect to configure and connect the components used to construct the interconnect. It produces verilog netlists, manufacturing test vectors, timing constraints, validation code, behavioral models at a variety of abstraction levels in both SystemC and verilog,

and a variety of scripts to ensure seamless integration of the Silistix clockless logic into a standard EDA flow.

- CHAIN®library—contains and manages the underlying technology information and hard macro views used by the CHAINworks tool suite.

10.2 Chapter Contents

This chapter takes the user on a guided tour through the steps involved in the creation of an NoC using the CHAINworks tool suite, and its use in an SoC design flow. As part of this process, aspects of the vast range of trade-offs possible in building an NoC will become apparent as will the increased capabilities beyond what bus-based design can achieve and the need for tools like CHAINarchitect to automate the trade-off exploration. Also highlighted in this chapter are some of the additional challenges and benefits of using a self-timed NoC to achieve a true top-level asynchrony between endpoint blocks—as is predicted by the International Technology Roadmap for Semiconductors (ITRS) [1] to become much more mainstream. Topics discussed include the following:

- Requirements capture—introducing the C-language-like Connection Specification Language (CSL) used as input to CHAINarchitect for describing the pertinent aspects of the endpoint blocks to be connected together by the interconnect, and the traffic requirements between them.
- NoC building blocks—introducing the range of basic function blocks and hardware concepts available in the CHAINlibrary. These units include a mixture of clocked and self-timed logic blocks for routing, queuing, protocol conversion, timing-domain crossing, serialization, deserialization, etc.
- Topology exploration—explaining the basics of the algorithms, choices and calculations used by CHAINarchitect to find a good-fit network that meets the requirements for the system.
- Design-for-Test (DFT)—explaining how DFT is performed for the mixture of clocked and self-timed logic of the NoC, and looking at the impact such an NoC has on the overall SoC DFT flow.
- Physical implementation—exploring the interaction of the CHAINworks tools with the floor-planning and place-and-route steps of physical implementation. The flexibility provided by the clockless logic at this stage of the design process is one of the key contributors to the much improved predictability of power, performance,

and area achievable with the CHAINworks flow when compared to other approaches.

- Validation—considering some of the issues involved in validation of the logic of an NoC and of systems constructed using one. Particular focus is given to simulation at multiple stages down the ASIC design flow including SystemC, prelayout RTL, structural, and post-layout modeling and the support that the NoC hardware and the CHAINcompiler synthesis tool can provide.

10.3   CHAIN® NoC Building Blocks and Operation

Many of the unique predictability, power, and performance properties of the NoCs constructed by the CHAINworks tools are attributable to the asynchronous implementation of the transport layer components whose operating frequency (and consequently the available bandwidth) is determined entirely by the sum of the logic and wire-delays in each asynchronous flow-control loop. To better understand the entire design philosophy of building a system around a CHAIN NoC [2], one has to grasp the following implications:

- High frequency communication can be implemented without using fast clocks.
- Pipelining can be used to tune for bandwidth, with negligible impact on latency.
- Low-cost serialization and rate-matching FIFOs can be used to change link widths.
- Protocol conversion and data transport between endpoints are treated as two separate operations, resulting in a two-layer communication model.

These concepts are somewhat different from the design principles to which a clocked interconnect designer is accustomed and warrant further explanation.

10.3.1   Differences in Operation as Compared to Clocked Interconnect

The first major difference between clocked interconnect and the self-timed approach used by CHAIN is the idea that using higher frequencies does not make achieving timing closure more difficult. This is because with the self-timed operation there is no matching high-speed clock to distribute. It is possible because the timing is implicit in the signaling protocol and is only determined by the length of the wires—shorter wires allow faster operation, and typically such circuits operate at much higher frequencies than the surrounding IP blocks that are being connected together.

The second major difference between clocked interconnect and the self-timed approach used by CHAINworks is in the ability to use pipelining to tune for bandwidth without having to consider latency. This stems from the fact that the C-element* based half-buffer pipeline latch has only a single gate-delay propagation time. This is very different from the use of clocked registers where insertion of each extra register adds an additional whole clock-cycle of latency to a communication. Clocked designers are thus accustomed to having to use registers sparingly, requiring the P&R tool to perform substantial buffering and struggle to meet timing. However, exactly the opposite approach is best in the design of a CHAINworks system—copious use of pipelining results in shorter wires and provides extra bandwidth slack, facilitating easier timing closure.

Finally, the combination of low-cost serializers and rate-matching FIFOs enables simple bandwidth aggregation where narrow links merge to deliver traffic onto a wider link [3]. Typically this is difficult to achieve in a clocked environment and can only be performed with time-division multiplexing requiring complex management of time-slots and global synchrony across the system.

These differences from clocked interconnect implementation impact on the architectural, logical and physical design of the CHAIN NoC, but are largely hidden from the user, with just the benefits visible through the use of the CHAINworks tools.

10.3.2 Two-Layer Abstraction Model

The CHAIN NoC uses a two-layer approach to provide seamless connection of many IP blocks speaking disparate protocols. The low level serves as a transport layer abstracting the switching, arbitration, and serialization to provide a simple interface to the layer above. The upper layer handles protocol mapping to tunnel and bridge transactions from one endpoint to another, even when they speak different protocols. The interface between these two layers is a simple clocked proprietary protocol known as the CHAIN Gateway Protocol (CGP). The two-level logical hierarchy is facilitated through a variety of different components each parameterized across a range of widths and other options.

- Protocol adapters that bridge between the endpoint bus-protocol interface and the CGP interface presented by the Silistix transport layer.

- Transmit and receive gateway units that handle packet encoding, static-route symbol calculation, clock-domain crossing, and serialization or deserialization.

- Link-level serialization and deserialization units that allow width conversion within the transport layer such that different regions

---

*The C-element is a stateholding gate often encountered in asynchronous design with logic function Q=A.B+Q.(A+B) such that if the inputs are the same then the output takes that level, otherwise when the inputs differ the output holds its previous state.

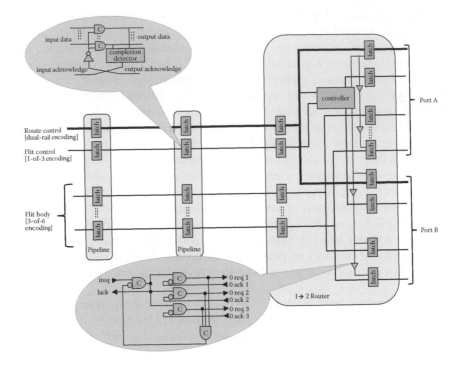

FIGURE 10.1
Link and logic structure showing two pipeline latches and a router.

of the transport layer network fabric can operate at different widths.

- 1-to-N route components, N-to-1 merge components, and N-to-N switch components used to provide the steering of traffic between endpoints. All routing decisions are precalculated by the transmit gateway units.

- Pipeline latches used to break long hops into shorter ones to maintain the high-bandwidth operation of the asynchronous connections between the other transport-layer modules.

- FIFOs used to buffer traffic for the management and avoidance of congestion, and also to provide rate-matching that is sometimes required as an artifact of the asynchronous serializers and deserializers.

Some of these units are visible in the figures used later in this chapter, for example, in the CHAINarchitect floorplan view in Figure 10.5 and the pipeline and routing components shown in slightly more detail in Figure 10.1.

10.3.3   Link-Level Operation

The operation and architecture of the current CHAIN NoC circuits at the link level has advanced since earlier descriptions to provide improved latency and better scalability to wider datapaths [4]. These improved components operate using a mixture of quasi delay-insensitive (QDI) asynchronous protocols [5], which operate such that their signaling encodes both the data validity and data value in a manner that is tolerant of delays on wires. The link structure is as shown in Figure 10.1 with three different sections.

- Routing information—A set of wires dedicated to carrying the routing information for each packet. This is a fixed number of bits throughout a design, the same on all links. Typically this has been 8 bits of data encoded using dual-rail* signaling on recent designs. The route is flow-controlled, separately, with its own dedicated acknowledge.

- Control—A set of three wires operating with 1-hot signaling and their own dedicated acknowledge flow-control signal. The CHAIN transport layer passes messages between the protocol adapters, and multiple such messages can be sent in a contiguous stream by wrapping them in the same packet so that they share the same routing header. These wires indicate for each fragment (or flit) in the transfer whether it is the final flit in a message, the final flit in a packet, or neither (i.e., there is a further flit coming immediately afterwards).

- Payload—Multiple sets of an m-of-n code group, each with its own dedicated acknowledge. The current release of CHAINworks uses the 3-of-6 encoding to give the most efficient performance, power, and area trade-off possible, while retaining the delay-insensitive signaling. This encoding is explained in greater detail by Bainbridge et al. [6] and provides a substantial wire-area saving when compared to the use of conceptually simpler 1-of-4 or 1-of-2 (dual-rail) codes.

The CHAINworks, concept uses fine grained acknowledges [4] (here, each grouping of 4 bits of the payload has its own acknowledge) to achieve high frequency operation, and partition the wide datapath into small bundles to ease placement and routing. Such use of fine-grained acknowledges also allows for skew tolerance between the route, control, and payload parts of the flits and packets passing through the network. Skew of this nature is to be expected as a result of the early-propagation techniques used in the switches when they open a new route. Consideration of the steps involved in opening a new route in a 1-to-2 port router with internal structure, as illustrated in Figure 10.1, helps to explain how this happens.

- The bottom bits are tapped off the routing control signals.
- The tapped-off route bits are used to select which output port to use.

---

* Dual-rail codes use two wires to convey a single bit by representing the logic level 0 using signaling activity on one wire and logic level 1 by signaling activity on the other wire.

- The flit-control symbols are steered to the selected output port.
- Concurrently, the route is rotated (simple wiring) and output to the selected port.
- Finally, the select-signal fans out to switch the datapath.

The first steps are performed with low latency and the flit-control, and updated route symbols are output approximately together. Then, to perform the final step, significant skew is introduced as a result of the C-element-based pipeline tree (represented as simple buffers in the diagram) used to achieve the fanout while maintaining the high frequency operation. The latency introduced is of the order of $\log_3(\text{width}/4)$ gate delays for a datapath of width bits implemented using 4-bit 3-of-6 code groups.

All of the transport layer components are designed using this early propagation technique to pass the performance-critical route control information through to the next stage as quickly as possible. Consequently, all the components can accommodate this variable skew, up to a maximum of half of a four-phase return-to-zero handshake, and the receive gateway realigns the data wavefront as part of the process of bridging data back into the receiving clock domain.

10.3.4 Transmit and Receive Gateways and the CHAIN Gateway Protocol

The CGP interface presented by the transport layer upwards to the protocol layer uses a simple format and carries the following few fields:

- Sender identifier (SID)—uniquely identifying the originating endpoint block or function
- Receiver identifier (RID)—uniquely identifying the destination endpoint block or function
- Payload data—the data to be transported, arranged in groups of 4 bits each with their own nibble-enable signal
- Sequence control—used to indicate temporal relationships between transfers
- Status signals—local queue fullness/readiness signals

The interface uses conventional clocked signaling and has bidirectional flow control. All local formatting of data is abstracted (by the protocol adapter) by the time it is passed through to the transport layer, which means that the transport layer never needs to inspect the contents of the payload as all the information required to route the traffic is accessible from the SID, RID, sequence and status signals.

Current instantiations of the CHAIN NoC have all used static source-based wormhole routing where an RID/route lookup is performed by the transmitting gateway, the lookup tables being constructed as part of the synthesis performed by CHAINcompiler. For simple networks with suitable assignment of routing bits, this can often be a direct mapping although more complex networks require a few levels of logic to implement this function. This static routing approach is easy to implement in self-timed logic allowing extraction of the routing symbols required by a switch using a simple wiring tap-off of the bottom bits of the route. Each switch rotates the route by the number of bits it uses so that the bottom bits then contain the symbols that will be used by the downstream switch.

The majority of the logic within the transmit and receive units is involved in implementing the serialization and clock-boundary crossing required to bridge between the slow clock domains of the NoC endpoint IP blocks and the high frequency asynchronous NoC fabric. One of the difficulties in provisioning these blocks is in achieving the minimum area while maintaining a good bridge between the slow, wide, parallel, and clocked CGP port and the fast, narrow, and serial asynchronous NoC fabric port. There are many variables to consider including the following:

- Number of stages of asynchronous serialization
- Number of stages of clocked serialization
- Effect of synchronization on throughput
- Frequency ratio required between the two domains
- Traffic bandwidth ratio required
- Width ratio required

If one tries to perform the provisioning manually, it rapidly becomes apparent that automation is required!

10.3.5 The Protocol Layer Adapters

The protocol layer handles the mapping and encapsulation of information contained in the transactions on the IP block interface onto the CGP interface. At this level there is a need to handle support for a variety of protocols including being able to bridge between endpoints that communicate using different protocols. Currently support is provided for the AMBA and OCP standards. The implementation of these blocks is done using conventional clocked RTL design and a protocol mapping format that is used as a standard base-set of features onto which all adapters map their local protocol-specific actions. A command transaction transfer in flowing from an initiator to a target thus goes through the following steps:

- Protocol adapter: mapping AMBA/OCP/proprietary protocol to the base format
- Transmit gateway: static route lookup and message formatting
- Transmit gateway: clocked to asynchronous conversion

- Transmit gateway: wide message serialized to multiple narrower flits
- Switches: packet is wormhole routed to the receiver
- Receive gateway: deserialization reconstructs the wide message from narrow flits
- Receive gateway: asynchronous to clocked synchronization
- Protocol adapter: mapping base format to AMBA/OCP/proprietary protocol

The process is symmetrical for returning the response part of a transaction from target to initiator, although the network route may follow a different topology.

10.4   Architecture Exploration

One of the first steps in the design of a complex SoC is the determination of the interconnect architecture of the chip. This important step is performed early in the design process, and many of the problems of a conventional approach stem from the fact that decisions at this stage are taken based on sketchy information of requirements, separated from the physical floorplan design, which happens later in the flow and are performed manually with no automated support. Combined, these factors often result in an interconnect architecture that is grossly over-provisioned while at the same time being very difficult for the backend team to implement.

Silistix CHAINworks automates the synthesis of such top-level interconnect, removing the guesswork and ad hoc nature of the process, and the need for extensive high-level simulation to ensure that the chosen architecture will provide the necessary capabilities. Using an automated approach provides many benefits including eliminating the risk of errors in the process, but the key improvement is that it allows many more trade-offs to be analyzed in the architectural stage of the design process than are feasible if a human is doing this manually using spreadsheets and high-level models.

CHAINarchitect is the CHAINworks graphical framework for the capture of system requirements using CSL, and for processing such requirements to synthesize suitable interconnect architectures that can meet the specification. The CHAINarchitect graphical user interface

- Allows capture and editing of CSL specifications with syntax high-lighting
- Serves as a cockpit to manage the runs of the CSL compiler topology synthesis engine and CHAINcompiler
- Displays report files generated by the CSL compiler for synthesized architectures

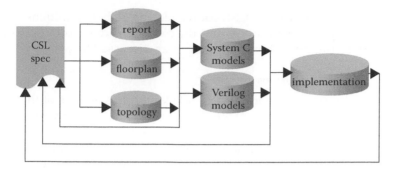

FIGURE 10.2
CHAINarchitect architecture exploration flow.

- Displays connectivity graphs showing the topology of the synthe-sized interconnect
- Displays suggested and estimated floorplan information for the design

Figure 10.2 shows the use model that can be used for this frontend of the Silistix CHAINworks tools.

Using CHAINarchitect, it is possible to iterate over many variations on a design, exploring the impact of different requirements and implementations in a very short period of time. The tightness of this iterative loop is provided by the fact that the CSL language is used to capture all of the requirements of the interconnect such as the interface types and properties, and the com-munication requirements between each of these interfaces. This formalized approach leads to a more robust design process by requiring the system ar-chitect to consider all of the IP blocks in the system rather than focusing only on the interesting or high-performance blocks. When all blocks are consid-ered, a much more complete picture of the actual traffic in the system results in being able to use reduced design margins, that is, the interconnect can be more closely matched to the needs of the system.

Once the architect is satisfied with the results predicted by the CSL compiler for his synthesized architecture, he can proceed to transaction- or cycle-level modeling of the system with the SystemC and verilog models, and finally to the implementation.

10.4.1 CSL Language

The CSL language serves as an executable specification of the requirements of the system, and as such has to be written and used for exploration of suitable architectures by the system architect. But it must also be easily understood by others involved in later stages of the design or validation of the SoC. In an effort to achieve this wide-spectrum comprehension, CSL uses C-language syntax and preprocessor directives.

There are four sections in a CSL source file: the global definitions, the address maps, the port descriptions, and the connection descriptions. Explanations and code-fragments for each of these are discussed below.

10.4.1.1 Global Definitions

The global definitions section of the CSL file captures aspects of the design such as the total estimated area of the chip, the cell library, and process technology to be used and the global relative importance of optimizing for minimum area or power once the performance has been met. The cell library and process information are the most important of the global parameters as these configure the internal tables used in the CSL compiler synthesis algorithms. A range of commonly available combinations such as cell libraries from ARM (formerly Artisan) and processes from TSMC or UMC are supported. Proprietary cell libraries have been used with this tool also.

10.4.1.2 Endpoints and Ports

The endpoints and ports section of the specification is where the properties of each interface of each IP block are captured. All aspects of the port are captured here from the course-grained detail of the clock frequency at which the port operates (and the clock domain it is in, if multiple blocks or ports use the same clock) or the type of protocol it uses, for example, OCP or AMBA-AHB to the fine-grained detail of the protocols such as the interleaving depth allowed across the interface or the burst-lengths and widths used. Properties of the endpoint such as its response time between transaction commands and responses must also be specified if they are nonzero to facilitate correct synthesis.

Some of the properties captured at this stage are used directly in the synthesis algorithms to determine and sanity check the traffic provisioning required, for example, the data field width and the clock frequency, combined with the knowledge of the bus protocol, which determine the peak traffic that could theoretically be generated by an initiator. Other information captured at this stage such as the supported transfer widths, burst lengths, and interleaving depths are fed through the flow to configure the implementation, for example, optimizing state-machines in the RTL and provisioning appropriate reordering and FIFO facilities to allow for any interleaving mismatches between initiators and targets that require communication with each other.

Although the properties are protocol specific, there are many commonalities between protocols that have to be captured. A substantial (but incomplete) set of properties that CHAINarchitect currently uses includes the following:

- Protocol variant—for example, APB v2 or v3 affects that signals are in the interface.
- Address width—the number of address wires in the interface. Bridging between different sized interfaces may require truncating or expanding the address and allows potential optimizations to reduce the total NoC traffic.

- Read- and write-data width—which can be different for some interface types and must be known to allow bridging between interfaces and transactions using different widths.
- Supported/used read-data sizes—only supporting the sizes necessary for each port that allows minimization of the logic in the implementation of the NoC protocol adapters, thus reducing area costs.
- Interleaving capabilities—when there is a mismatch between interleaving capabilities and traffic expected across an interface, the tools can provision the NoC to restrict traffic to avoid congestion at that interface, or can provision the protocol adapters to be able to queue and possibly reorder operations to better match the requirements with the capabilities.
- Burst-lengths and addressing modes used—restricting protocol adapter support to only the lengths and addressing modes required and used by the IP block, rather than providing adapters that support the full capabilities of the protocol specification it allows substantial savings on the area of the adapters.
- Caching/buffering signal usage—these are transported by the NoC between endpoints but often have to be bridged between differing models for different protocols, and sometimes have implications for the transport layer operation.
- Endianness—protocol adapters can convert endianness when intermixing big and little-endian IP blocks.
- Data alignment needs
- Interface handshaking properties—some interfaces support optional flow-control signals
- Atomic operation/locking capabilities—most complex protocols allow read-modify-write operations, but their implementation adds complexity to NoC protocol adapters. Many initiators do not use these capabilities allowing area optimizations in such cases.
- Error signaling model—some endpoints signal errors using imprecise methods, such as interrupts, but others use exact, in-order methods provided by the bus-interface. Typically such exact methods introduce additional overheads into the adapter mapping of high-level protocols onto low-level NoC transport layer primitives.
- Ordering model and tagging used—necessary to configure and provision the reordering capabilities of NoC gateways and protocol adapters to accommodate mismatches.
- Rejection capabilities supported/used—rejection is a technique encountered in older bus-interfaces to handle congestion and can provoke complex interactions and unpredictable traffic if supported in an NoC environment. Typically many rejection scenarios can be avoided through suitable provisioning.

In each case, the exact options available and syntax used to capture the specification are protocol specific, and many of the attributes have default values. A typical CSL fragment for an endpoint port description is shown below.

```
power_domain pd0 {                    //power domain (shutoff/voltage grouping)
   domain pd0a (500 MHz)              //clock domain (frequency/skew grouping)
     { // Embedded Processor:
      ACPU { // eg ARM 1176
         protocol = "AXI";            //the interface type
         outstanding = 4;             //maximum number of outstanding commands
         initiator i_port {           //initiator port - single port in this example
            address = 32 bits;        //AXI address bus width
            data = 64 bits;           //AXI data bus width
            peak = 200 MBs;           //upper limit on burst-traffic generation
            nominal = 100 MBs;        //average total traffic generated
            burstsize = 32 bytes;     //longest bursts used by endpoint
            address_map = global_map; //which address map this endpoint uses
            register_cgp_cmd = 0;     //specific protocol adapter attribute
   } } } }
```

10.4.1.3 Address Maps

Address maps are required for each initiator in the system to describe the segmenting of their address space into regions used to access different targets. CSL provides the ability to describe address maps that are specific to each initiator or shared across a group of initiators. An example of address map specification is

```
address map address_map1 {   // Address map
  range range_target0 0x00000000 .. 0x0000ffff;
  range range_target1 0x0001ffff .. 0x0004ffff;
}
```

Any number of address ranges can be supported in an address map, although they are not allowed to overlap because a one-to-one mapping to unique targets is required. This would then be associated with an initiator using the statement

```
address map = address_map;
```

as part of the initiator port specification. The target entries from each independent address map are then bound in the descriptions of the targets using statements such as

```
address range = {address_map1.range_t1,
  address_map2.range_t2,
  address_map3.range_t1};
```

in the port specification of the target. This indicates that range_t1 in address_map1 and in address_map3 both cause transactions to this target as do operations to addresses in range_t2 of address_map2.

10.4.1.4 Connectivity Specification

The final section of the CSL specification involves capturing the connectivity of the system showing which initiators communicate with which targets. In addition to the basic connection matrix, this section of the specification also captures the bandwidth and worst-case latency requirements of each communication path. The bandwidth is specified from the viewpoint of the bus-protocol data fields of the transaction that the typical user is accustomed to measuring. Understanding of this issue is significant because, for example, a specification of 100 MB per second means there is 100 MB per second of payload data to be transported but for the purposes of the NoC provisioning algorithms, the tools must inflate these numbers to accommodate the need to also transport control information such as addresses and byte enables. Requirements are captured separately for read and write transactions.

The CSL grammar allows for partitioning the connection requirements into sections, each called a mode. Modes can be nested to cater for describing the many different operating situations of a typical SoC. A good example is a modern smartphone, which would have many different functions embodied on the same chip including a camera, a color LCD display, and voice speech processing hardware/software. Even a simplistic consideration of partial functionality of such a chip highlights the wide ranges of tasks that may have to be performed. For example, consider the camera found on a typical smartphone: taking a still image with the camera involves many actions such as charging the flash, focusing, adjusting exposure apertures, reading the image from the CCD, and then processing, displaying, and storing the image. Some of these actions are obviously sequenced—capturing, processing, and then displaying the image—but others are continually, concurrently occurring with these such as adjusting the focus and the aperture when the camera function is enabled. Each of these suboperations has its own traffic profile, and CSL mode descriptions allow these to be captured and the complex interrelationships to be specified also.

An example of a connectivity specification for a set of connections is shown below.

```
//Read connections:
cpu.i_port <= dram.t_port(bandwidth=200 Mbs, latency = 120ns);
mpeg.i_port <= dram.t_port(bandwidth=800 Mbs);
dma.i_port <= dram.t_port(bandwidth=200 Mbs);
dma.i_port <= eth.t_port(bandwidth=50 Mbs);

//Write connections:
cpu.i_port => dram.t_port(bandwidth=200 Mbs);
mpeg.i_port => dram.t_port(bandwidth=200 Mbs);
dma.i_port => dram.t_port(bandwidth=200 Mbs);
dma.i_port => eth.t_port(bandwidth=100 Mbs);
```

10.4.2   NoC Architecture Exploration Using CHAIN®architect

With a specification described in CSL, the user can proceed through the process of analyzing the CSL for consistency, synthesizing an NoC to meet the requirements, and viewing reports and structure for the proposed implementation. If the results are considered unsatisfactory, the user can refine the specification and the synthesis directives, and iterate through the process multiple times. Fragments of an example report file created using the synthesis defaults for a small example are shown below.

```
CSL Compiler Version 2008.0227 report run on Sun Feb 24 02 20:39:36 2008
command line = "-or:.\rep\silistix_training_demo_fpe_master.rep -nl -ga "


System Statistics
-----------------
Initiators: 3
Targets: 2
Adaptors: 5 (0.088 mm2, 38.3 kgates - 1.0%)
TX: 5 (0.063 mm2, 27.5 kgates - 0.7%)
Route: 1 (0.001 mm2, 0.6 kgates - 0.0%)
Serdes: 3 (0.003 mm2, 1.5 kgates - 0.0%)
.....       .....       .....
Total fabric area: 0.242 mm2, 105.7 kgates (2.7%)
Fabric nominal power: 62.918457 mWatt


Network Bill of Materials (truncated)
-------------------------
domain: dram_block
    1 tahb                   0.025 mm2, 11.1 kgates
domain: cpu_block
    1 iahb                   0.014 mm2, 6.2 kgates
domain: self-timed
    1 tx52x4             0.013 mm2, 5.7 kgates
    3 tx100x4            0.031 mm2, 13.6 kgates
    1 tx52x8             0.019 mm2, 8.3 kgates

    ......               .......
    subtotal         0.159 mm2, 69.2 kgates
Network total area       0.247 mm2, 107.5 kgates


Roundtrip Connections (truncated)
---------------------
cpu_block.cpu.i_port (500.0MHz) <= dram_block.dram.t_port (333.0MHz)
    (Req: outstanding=1, burst=512 bits, oa=2, op=3, ol=1)
    cmd path = {iahb_0 -> tx100x4_17 -> mg4x2x1_18 -> pl4_61 ->
                rx4x100_9 -> tahb_4}
    rsp path = {tahb_4 -> tx52x8_23 -> pl8_69 -> pl8_63 -> pl8_66 ->
                rt8x1x2_24 -> fifo8x8_58 -> serdes8x4_57 ->
                rx4x52_19 -> iahb_0}
    Sustained bandwidth (200.000 Mbs) slack: 1519.552 Mbs
    Network roundtrip latency: 49.252 ns
```

```
System roundtrip latency (120.000 ns) slack: 10.748 ns
Energy per packet: 21.961 uW
Worst case utilization 19.453% at mg4x2x1_18
...            ...              ...
```

The bill of materials shows a list of the Silistix library components required for the design, including the clockless hard-macro components implementing the NoC transport layer and the protocol adapters coupling the endpoint interfaces of those protocols onto the transport layer. The other important data provided in the results is a hop-by-hop breakdown of the path through the network for each communication showing the bandwidth and latency slack.

10.4.3   Synthesis Algorithm

Two key algorithms underpin the synthesis performed by the CSL compiler. The first of these is used to calculate the occupancy of a link based on the traffic flowing through it, and consequently also the worst-case congestion-imposed latency that might be encountered by such traffic. The second is an iterative exploration that monotonically tends toward a solution driven by a set of heuristics that makes stepwise improvements to the bandwidth or latency of a path.

The CSL description captures for each communication connection the length and width of the bursts in the transfer, and also the peak and nominal bandwidth requirements of the connection. With this information, assuming the burst is tightly packed and performed at full peak bandwidth, it is possible to determine a window of time and the portion of that window the communication will be using the link connected to the IP block port, as illustrated in Figure 10.3.

Once this analysis is applied to every port of each IP block, it is easy to see how multiple demands can be serviced by one link—if two such ports have duty cycles that can be interleaved without overlapping and bandwidth requirements, which sum to less than the 100 percent (or lower utilization threshold if specified) available from the link, then they can share the link. Provisioning a wider link increases the available bandwidth, and then the

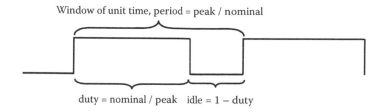

FIGURE 10.3
Trafic time-window usage model.

use of the self-timed transport layer allows simple in-fabric serdes support for changing the aspect ratio (and consequently duty cycle on the link) of transactions as they move from one link width to another. This is an important distinction from clocked implementations where the rigidity of the clock makes changes to the aspect ratio or duty cycle much more challenging to achieve.

The final part of this static time-window traffic model affects the added latency impact of contention and congestion in the system. When considered from an idle situation, some transactions will have to wait because they encounter a link that is busy, the worst case delay being determined by the sum of all of the duty terms of the other communications performed over the same link. However, such "latency from idle" analysis is not representative of a real system, where once a stable operating condition is achieved, which is of course regulated by the traffic-generation rates of the endpoints, the congestion-imposed latency is substantially lower than the theoretical worst case and typically negligible.

Figure 10.4 shows a time-window illustration of two transfer sequences, A and B, which are merged onto a shared link. Initially both sequences encounter added latency, with the first item of sequence B suffering the worst delay (due to the arbiter in the merge unit resolving in favor of A on the first transfer). Then the second transfer on A suffers a small delay waiting for the link to become available. By the third and fourth transfers in the sequences the congestion-imposed delays are minimal. Thus, although the upper limit on the jitter introduced into each transfer sequence is of magnitude equal to the duration of the activity from the other contending sequence, the average jitter is substantially smaller and almost negligible once synchronization delays experienced at the edges of the network are considered, provided the contiguous flit-sequence lengths are short. If wider jitter can be tolerated, as is often the case, then longer sequences can be used.

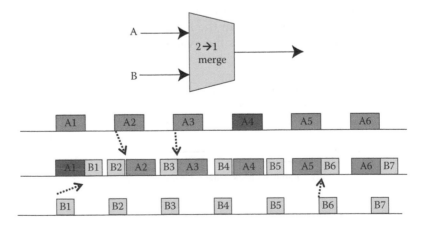

FIGURE 10.4
Arbitration impact on future transfer alignment.

10.4.4   Synthesis Directives

The fundamental objective of any synthesis tool is to implement the specification supplied while achieving lowest cost in the implementation. However, determining the meaning of lowest cost is nontrivial for interconnect synthesis as there are actually many properties of the solution that are all interrelated—lowering the cost on one metric may raise the cost measured in terms of other metrics. To aid the user in obtaining a best-fit that meets their expectations, a set of optimization directives exists in CSL that affect the synthesis process—in effect guiding the heuristics in selecting which of the many implementations that could meet the constraints to use.

The primary, course-grained control that can be applied to influence the synthesis algorithms is the specification of a utilization. This determines the maximum occupancy that a link in the network should be allowed to have; so, for example, the statement

$$\text{utilization} = 80 \text{ percent}$$

would mean that links are provisioned such that the total traffic requirements over each link can be satisfied using only 80 percent of the possible peak bandwidth. In effect, this is similar to specifying a 20 percent design margin to accommodate uncertainty in the traffic specification.

Beyond the utilization, trade-off priority can be controlled across latency, area, and power so that all the sensible solutions that CSL compiler determines would meet the requirements, and the user can coax the heuristics to choose one closest to his preference. For example, the statement

$$\text{optimize latency} = 1, \quad \text{optimize area} = 2, \quad \text{optimize power} = 3$$

selects the implementation that meets the bandwidth and utilization requirements, with any slack being absorbed to provide better latency even if that means the power consumption or area increases, and to then minimize area even if that increases power consumption. Such trade-off specifications can affect, for example, the amount of serialization used in the resulting system—if there is slack, serialization can be used to trade latency for area with little impact on power consumption. As with many aspects of the CSL language, the utilization and area, latency and power trade-off priorities can be specified at a global level and inherited or overridden at more local scope level, for example, per port or per link.

10.5   Physical Implementation: Floorplanning, Placement, and Routing

Historically, asynchronous VLSI design has been associated with difficulties at the physical implementation and validation stages of the design process, but the exact opposite is true with the Silistix CHAINworks tools. All of the transport-layer logic blocks necessary to implement the NoC are delivered as precharacterized hard-macros, which eliminate the problems of achieving timing closure within the logic forming most of the interconnect except the protocol adapters. Timing closure is easy to achieve for these adapters when synthesized, placed, and routed in conjunction with the IP blocks to which they connect.

However, the use of asynchronous logic does not totally avoid the issue of achieving timing closure of the network but instead translates it into the easier-to-solve problem of tailoring the architecture and floorplan together so that a working, predictable physical implementation is always obtained. This means that the synthesis performed using CHAINarchitect has to be floorplan-aware so that it can accommodate the higher latencies of long paths and provision deeper pipelining along such paths accordingly to ensure that they do not impede bandwidth.

CHAINarchitect takes in physical information such as the estimated area (and aspect ratio) of endpoint blocks, and can use this in conjunction with any fixed positioning, clock, and power-domain groupings that are known early in the design process for some blocks to use the ability to juggle the floorplan as one of the trade-offs when configuring a network. It then outputs the resulting floorplan estimate for use further down the flow. A screen-capture showing an example-generated floorplan estimate is shown in Figure 10.5. In this

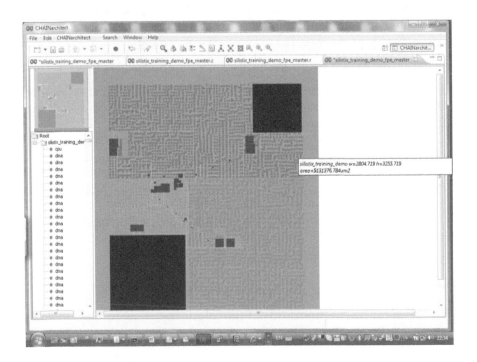

FIGURE 10.5
CHAINarchitect floor-plan estimate.

floorplan estimate, the small black blocks are the Silistix asynchronous logic, and the pipeline latches are just visible as the really small blocks spanning the distance between the other components. Also key, here, is the observation noted earlier about the ease and nondamaging impact of over-provisioning pipelining that is central to the methodology. This means that once CHAINarchitect has settled on a suitable floorplan and topology, it can calculate the link widths and pipelining depths necessary for the implementation and sufficiently over-provision the pipelining to ensure that the system will meet its requirements while allowing for the uncertainty that is inherent in the later steps of the physical design flow in moving from a rough floorplan estimate constructed at an abstract level to a real physical implementation post place-and-route. A basic spring model is used as part of this process to evenly distribute the switching components and pipeline latches of the network fabric across the physical distances to be spanned. For a typical SoC, the runtime of the CHAINworks synthesis and provisioning algorithms is a few minutes thereby allowing the system architect to rapidly iterate through the exploration of a range of system architectures.

Once the floorplan is finalized and the hard-macro components of the network fabric are placed, along with any other hard macros in the design, the placement and routing of other blocks is performed as normal. Timing constraints are output by CHAINcompiler in conventional SDC format for the self-timed wires between the NoC transport-layer components for use with the mainstream timing-driven flows. Using these, the place and route tools perform buffer insertion and routing of the longer wires as appropriate.

10.6   Design-for-Test (DFT)

NoC brings interesting opportunities for improving the DFT methodology of SoCs by providing a high-speed access mechanism to each IP block's primary interfaces. The IEEE 1500 [7] macrocells test access method can be easily facilitated over the NoC and is a good fit with the need to separately and concurrently test many blocks within the NoC architecture. It also provides the modularity needed to separate testing of the self-timed NoC implementation from the usual test procedure of the endpoint logic.

Such separation is required because the self-timed implementation of the CHAIN NoC means that it is incompatible with the conventional scan insertion (and in some cases also the ATPG tools) from mainstream vendors. However, in this sense these components are no different than other blocks that are instantiated as hard macros such as memories and analog blocks, and like those blocks the vendor has to provide a DFT solution.

Multiple test strategies can be used for the CHAIN NoC components. The first, and least intrusive to a normal EDA flow, is to use full scan where the CHAINlibrary components are constructed such that every C-element

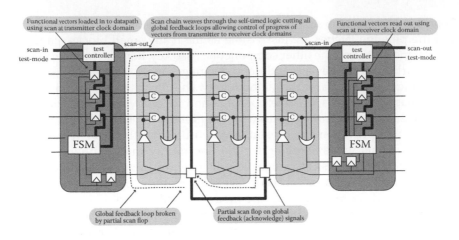

FIGURE 10.6
Scan-latch locations for partial scan.

features an integral scan latch. This approach is compatible with all existing scan-chain manipulation tools and conventional ATPG approaches. The more advanced and lower cost approach relies on a combination of functional patterns and sequential, partial scan. In both cases similar 99.xx percent stuck-at fault coverage is achieved as in regular clocked logic test. Consideration of a pipelined path from a transmitter to a receiver, as shown in Figure 10.6, can illustrate how the partial scan [8] approach works.

Scan flops are placed on targeted nodes, typically feedback loops, state-machine outputs and select lines that intersect the datapath. These are shown explicitly in the simplified pipeline of Figure 10.6, but in reality they are encapsulated in (and placed and routed as part of) the hard-macro components. Testing the network is then a three-stage process.

- The first pass of the test process uses just the conventional scan flops in the transmit and receives hard macros to check the interface with the conventionally clocked logic.

- Second, in transport-layer test mode, the same scan flops are used to shift functional vectors into the datapath at the transmitter. The circuit is switched back into operational mode (where the global-feedback partial-scan latches connect straight through without interrupting the loops) and the vectors are then transmitted through the network at speed and then read out at the receiver using its scan chain. This achieves good coverage of all the datapath nodes and many of the control-path nodes through the network. It is not essential, just more efficient to use this step because all faults can also be detected using the final pass below.

- The final pass uses the partial-scan flops on the global feedback loops in non-bypass mode to break the global loops allowing access to monitor and control the acknowledge signals. This enables the propagation of the test vectors to be single stepped through the pipeline giving further increased coverage and improved ability to isolate the location of any faults that are detected.

The full set of required patterns are generated by the CHAINworks tools in STIL format for use with conventional testers and test pattern processing tools. Achievable coverage is verified using conventional third-party concurrent fault simulators.

Highly efficient delay-fault testing is performed using a variant on the functional-test approach. Patterns injected at a transmit unit are steered through the network to a receiver and the total flight time from the transmitters clocked/asynchronous converter to the receivers asynchronous/clocked converter is measured. Any significant increase above the expected value is indicative of a delay fault somewhere on the path between the two ends. This approach is very efficient for detecting the absence or presence of delay faults, but does not help in the exact location of a delay fault. However, the scan access facilitates such localization of any faults detected.

10.7 Validation and Modeling

The CHAINworks tools provide support for modeling and validation at many stages through the design flow. The most abstract support is provided through Programmer's View SystemC models that provide the system connectivity and protocol mappings but operate with idealized timing. These are intended for supporting high-performance simulations to enable software debug at a very early stage of the design process. Moving down the design flow, the next level of detail is provided in the Architect's View SystemC models. These model the activity within the network at the flit-level, and are cycle-approximate models with cycle-level timing for the clocked components and timing information annotated from the characterized hard macros for the self-timed components. In both the SystemC models, the separation of the transport layer and protocol mapping layer is retained allowing traffic properties to be inspected at both interfaces. Inspection of traffic levels on a hop-by-hop basis in the switching fabric is only possible with the more detailed architect view models. The models support use with a variety of TLMs including the OSCI TLM v1 with posted operations and the CoWare SCML and PV TLMs. The performance of the models is primarily limited by the number of threads required to model the concurrent asynchronous operation. Moving further down the design flow, the next more detailed level of modeling is at the verilog level. Here, CHAINcompiler outputs a top-level verilog netlist structured as shown in Figure 10.7. The protocol adapter and an instantiation of the gateway for each endpoint are output to separate files—they will be put through synthesis, P&R, and validation as a group with the endpoint IP

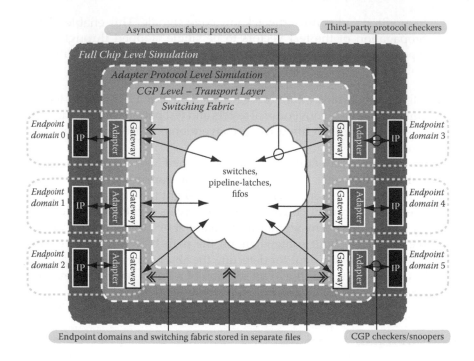

FIGURE 10.7
Output verilog netlist partitioning.

block attached. The clocked logic generated is RTL, ready for synthesis or simulation. For the self-timed logic that will be implemented as hard macros in the realization of the system, behavioral models are provided that simulate substantially faster than a gate-level simulation of the real structural netlist of the asynchronous circuits. These models are built using verilog2001 language constructs and their timing is calibrated against the characterized hard macros allowing realistic time-accurate simulations of the system to be performed.

The final level of detailed accuracy, possible with this flow, is to simulate a combination of RTL (or the synthesized gate-level netlists) of the clocked components with the gate-level netlist of the asynchronous macros using back-annotated timing. This gives very accurate timing simulation but at the expense of substantial run-times.

10.7.1   Metastability and Nondeterminism

Within the CHAIN NoC, there are two types of location where metastability can naturally occur. The first is in the logic used to move between the self-timed and clocked domains. At these boundaries, there is always a synchronizer on the control signal, entering the clock domain. These signals are used with a signaling protocol that ensures every transition on them is observed by the receiving logic, and the spread of arrival times means that there is variability in the time from an asynchronous request event arriving until it is observed in the clock domain. For the typical, default, two-flop synchronizer approach used this means anywhere between just under one to just over two clock cycles of latency for crossing through the synchronizer. This variation in delay is faithfully represented (and can be observed) in the models. Metastability has no impact on it, because failure to resolve will cause an illegal, indeterminate value to be propagated as in any use of synchronizers. The second place where metastability can occur is in the switch components where contention can occur between multiple inputs for the same output. Such contention can occur at any time, albeit with a small probability, and is resolved using the mutual exclusion (mutex) element based on a circuit structure by Seitz [9] to localize the metastability ensuring that its outputs always remain at zero or one logic levels. In noncontended operation the mutex gives a response time of two gate delays, but when contention occurs the response time is determined by the metastability resolution equation meaning that it is statistically extremely rare to experience a substantial delay and that it could resolve in favor of either contender. The SystemC and verilog models correctly represent the noncontended behavior, but for contention they only approximate a round-robin behavior and do not model the variability in delay.

10.7.2   Equivalence Checking

Validation methodologies often revolve around equivalence checking at each level of the design flow. For the CHAIN NoC this cannot yet be performed with current formal equivalence-checking tools but can be performed using simulation. To assist with the validation of one abstraction level versus the next, the CHAINworks flow outputs transactors for stimulating, monitoring, and checking the functionality of each of the whole-system models and the final verilog implementation. Furthermore, stimuli files are generated that create traffic levels matching those specified in the CSL requirements to validate that the network meets the requirements when simulating at the adapter protocol level as illustrated in Figure 10.7. Alternatively, simulation can be performed at the CGP transport-layer level or at the whole chip level to obtain real traffic. For logical equivalence checking of the remainder of the SoC, the asynchronous logic components are black-boxed.

10.8   Summary

This chapter introduces the CHAINworks tools, a commercially available flow for the synthesis and deployment of NOC-style interconnect in SoC designs. A new language for the capture of system-level communication requirements has been presented and some of the implementation challenges that impact the conventional ASIC design flow as a result of moving toward a globally asynchronous, locally synchronous (GALS) approach have been discussed including place-and-route, DFT, and validation; in each case showing that there are workable solutions allowing this exciting new technology to be used today.

References

1. International Technology Roadmap for Semiconductors, 2007 edition, http://www.itrs.net.
2. L. A. Plana, W. J. Bainbridge, and S. B. Furber, "The design and test of a smartcard chip using a CHAIN self-timed network-on-chip." In *Proc. of Design, Automation and Test in Europe Conference and Exhibition*, Paris, France, February 2004.
3. L. A. Plana, J. Bainbridge, S. Furber, S. Salisbury, Y. Shi, and J. Wu, "An on-chip and inter-chip communications network for the SpiNNaker massively-parallel neural network." In *Proc. of 2nd IEEE International Symposium Networks on Chip*, New Castle, United Kingdom, April 2008.
4. W. J. Bainbridge and S. B. Furber, "CHAIN: A delay insensitive CHip area INterconnect," *IEEE Micro Special Issue on Design and Test of System on Chip*," 142 (Sep. 2002) (4): 16–23.
5. T. Verhoeff, "Delay-Insensitive codes—An overview, distributed computing," 3 (1988): 1–8.
6. W. J. Bainbridge, W. B. Toms, D. A. Edwards, and S. B. Furber, "Delay-insensitive, point-to-point interconnect using m-of-n codes." In *Proc. of 9th IEEE International Symposium on Asynchronous Circuits and Systems*, Vancouver, Canada, May 2003, pp. 132–140.
7. IEEE Std 1500–2005, "Standard testability method for embedded core-based integrated circuits," IEEE Press.
8. A. Efthymiou, J. Bainbridge, and D. Edwards, "Test pattern generation and partial-scan methodology for an asynchronous SoC interconnect," *IEEE Transactions on Very Large Scale Integration (VLSI) Systems*, 13, December 2005, (12): 1384–1393.
9. C. L. Seitz, "System timing." In *Introduction to VLSI Systems*, C. A. Mead and L. A. Conway, eds. Reading, MA: Addison-Wesley, 1980.

11

Networks-on-Chip-Based Implementation: MPSoC for Video Coding Applications

Dragomir Milojevic, Anthony Leroy, Frederic Robert,
Philippe Martin, and Diederik Verkest

CONTENTS

11.1   Introduction

In the near future, handheld, mobile, battery-operated electronic devices will integrate different functionalities under the same hood, including mobile telephony and Internet access, personal digital assistants, powerful 3D game engines, and high-speed cameras capable of acquiring and processing high-resolution images at realtime frame rates. All these functionalities will result in huge computational complexity that will require multiple processing cores to be embedded in the same chip, possibly using the Multi-Processor System-on-Chip (MPSoC) computational paradigm. Together with increased computational complexity, the communication requirements will get bigger, with data streams of dozens, hundreds, and even thousands of megabytes per second of data to be transferred. A quick look at the state-of-the art video encoding algorithms, such as AVC/H.264 for high-resolution (HDTV) re-altime image compression applications, for example, already indicate bandwidths of a few gigabytes per second of traffic. Such bandwidth requirements cannot be delivered through traditional communication solutions, therefore Networks-on-Chips (NoC) are used more and more in the development of such MPSoC systems. Finally, for battery-powered devices both processing and communication will have to be low-power to increase the autonomy of a device as much as possible.

In this chapter we will present an MPSoC platform, developed at the Interuniversity Microelectronics Center (IMEC), Leuven, Belgium in partnership with Samsung Electronics and Freescale, using Arteris NoC as communication infrastructure. This MPSoC platform is dedicated to high-performance (HDTV image resolution), low-power (700 mW power budget for processing), and real-time video coding applications (30 frames per second) using state-of-the-art video encoding algorithms such as MPEG-4, AVC/H.264, and Scalable Video Coding (SVC). The proposed MPSoC platform is built using six Coarse Grain Array (CGA) ADRES processors also developed at IMEC, four on-chip memory nodes, one external memory interface, one control processor, one node that handles input and output of the video stream, and Arteris NoC as communication infrastructure. The proposed MPSoC platform is supposed to be flexible, allowing easy implementation of different multimedia applications and scalable to the future evolutions of the video encoding standards and other mobile applications in general.

Although it is obvious that NoCs represent the future of the interconnects in the large, high-performance, and scalable MPSoCs, it is less evident to be convinced of their area and power efficiency. With the NoC, the raw data has to be encapsulated first in packets in the network interface unit on the master side (IP to NoC protocol conversion) and these packets have to travel through a certain number of routers. Depending on the routing strategy, a portion of the packet (or the complete packet) will eventually have to be buffered in some memory before reaching the next router. Finally, on the slave network interface side, the raw data has to be extracted from packets before reaching the target (here we assume a write operation; a read operation would require a similar path for the data request, but would also include a path in the opposite direction with actual data). Therefore all these NoC elements use some logic resources and dissipate power.

In this work we show that in the context of a larger MPSoC system adapted to today's standards (64 mm^2 Die in 90 nm technology with 13 computational and memory nodes) and complex video encoding applications such as MPEG-4 and AVC/H.264 encoders, the NoC accounts for less than three percent of the total chip area and for less than five percent of the total power budget. In absolute terms, this means less than 450 kgates and less than 25 mW of power dissipation for a fully connected NoC mesh composed of 12 routers and for a traffic of about 1 GB per second. Such communication performance, area, and power budget are acceptable even for smaller MPSoC platforms.

The remainder of this chapter is structured as follows: in Section 11.2 we will briefly present a survey of different interconnect solutions. In Section 11.3 we will present in some more details the Arteris NoC. We will first introduce a description of some of the basic NoC components (NoC protocol, network interfaces, and routers) provided within the Arteris Danube NoC IP library. We will also briefly describe the associated EDA tools that will be used for NoC design space exploration, specification, RTL generation, and verification. In Section 11.4, we will describe the architecture of the MPSoC platform in a more detailed way. We will give a description of the ADRES CGA processor architecture, memory subsystem, NoC topology and configuration. Finally, we will present the MPSoC platform synthesis results and power dissipation figures. In Section 11.5 we will present the power models of different NoC components (network interfaces, routers and wires) and will provide the power model of the complete NoC. Such models will be used to derive the power dissipation of the MPEG-4 and AVC/H.264 simple profile encoders for different frame resolutions and different applications mapping scenarios. The results obtained will be compared with some of the state-of-the-art implementations already presented in the literature.

11.2   Short Survey of Existing Interconnect Solutions

In the context of the MPSoC platforms for high-performance computing, simple shared buses based on a simple set of wires interconnected to master devices, slave devices, and an arbiter are no longer sufficiently scalable and cannot provide enough bandwidth for throughput hungry applications. In consequence, advanced on-chip buses are now based on crossbars to reach higher throughput and communication concurrency, such as the ST Microelectronics STBus, IBM Coreconnect, or ARM AMBA. Many research teams propose generic on-chip communication architectures, which can be customized at design time depending on the SoC requirements. The largest projects so far are the Nostrum backbone (KTH) [1–3] and the ×pipes [4,5].

Recent publications present interesting surveys focusing mainly on current academic research. The work by Bjerregaard and Mahadevan [6] presents a good survey of the current academic research on NoC and covers research in design methodologies, communication architectures as well as mapping issues. Kavaldjiev and Smit [7] mainly present some general communication architecture design decisions with only few concrete network examples. On the other hand, the work by Pop and Kumar [8] is exclusively dedicated to techniques for mapping and scheduling applications to NoC. In this introduction, we will focus mainly on the industrial MPSoC communication fabric solutions proposed by NXP, Silistix, and Arteris. They are today the most realistic alternatives to the on-chip shared bus architecture.

NXP (formerly Philips) was one of the first companies to propose a complete solution for a guaranteed throughput (GT) service in addition to a packet-based best effort (BE) service with the Æthereal NoC [9]. The GT service guarantees uncorrupted, lossless, and ordered data transfer, and both latency and throughput over a finite time interval. The current implementation is based on custom-made hardware FIFO queues, which allows considerable reduction of the area overhead [10–13]. The Æthereal network supports several bus-oriented transaction types: read, write, acknowledged write, test and set, and flush, as well as specific network-oriented connection types such as narrowcast, multicast, and simple. An Æthereal instance synthesized in 0.13 μm CMOS technology is presented by Goossens et al. and Rădulescu et al. [14,15]. It is based on six 32-bit ports and exploits custom designed queues (area overhead divided by more than a factor 16 with custom designed queues compared to RAM-based and register-based FIFO). The total router area is 0.175 mm^2. The bandwidth per port provided by the router reaches 16 Gbit per second. The network interface supports four standard protocols (master/slave, OCP, DTL, and AXI). The area of the network interface is 0.172 mm^2 in 0.13 μm technology. An automated design flow has also been developed to generate application-specific instance of the Æthereal network [16,17]. The design flow is based on an XML description of the network requirements (traffic characteristics, GT and BE requirements, and topology).

Silistix [18], a spin-off from the University of Manchester, commercializes a complete suite of EDA tools to generate, test, and synthesize asynchronous NoCs based on the CHAIN (CHip Area INterconnect) project [19]. The clockless circuits used in this network are very power-efficient, allowing the designers to target ultra-low power heterogeneous SoCs, such as the ones used in smart cards [20]. The CHAIN network supports OCP and AMBA communication interface protocols and offers different levels of bandwidth and latency guarantees based on priority arbiter [21,22].

The Arteris company proposes a complete commercial solution for SoC communication architecture. Their solution is based on the Danube NoC IP blocks library [23]. The Danube library offers three types of units: (1) Network Interface Units (NIU) connecting IP blocks to the network, (2) Packet Transport Units (PTU) constituting the network devices, and (3) physical links. PTU blocks include crossbars and FIFO queues. The Arteris NoC design flow is composed of two configuration environments: NoCexplorer and NoCcompiler. The NoCexplorer exploration tool is used to create a customized topology Danube NoC instance, defining how the PTUs are connected to each other. NoCexplorer captures the dataflow requirements of the IP blocks and helps the designer to choose the NoC topology based on an exploration of various topologies that match the required worst-case traffic pattern. It utilizes a data-flow simulation engine, a parameterizable data-flow source, and sink generators to model the system behavior. Based on the information of the NoC topology and on the configuration of basic building blocks provided by NoCexplorer, NoCcompiler is used for the actual RTL generation (SystemC or VHDL) of the corresponding NoCs. The Danube NoC IP library supports a custom proprietary network protocol called NTTP. Interface units for OCP, AMBA AHB, APB, and AXI protocols are also provided. The network can offer up to 100 GB per second throughput for a clock frequency of up to 750 MHz in 90 nm process technology. In the following section, we will describe in more details the Arteris NoC framework.

11.3 Arteris NoC: Basic Building Blocks and EDA Tools

11.3.1 NoC Transaction and Transport Protocol

The NoC Transaction and Transport Protocol (NTTP) proposed by Arteris adopts a three-layered approach with transaction, transport, and physical layers enabling different nodes in the system to communicate over the NoC. The transaction layer defines how the information is exchanged between nodes to implement a particular transaction. Transaction layer services are provided to the nodes at the periphery of the NoC by special units called Network Interface Units (NIUs). The transport layer defines the rules that apply as packets are routed through the NoC, by means of Packet Transport

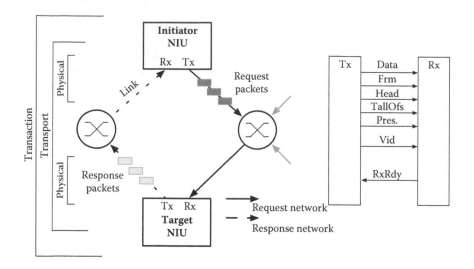

FIGURE 11.1
NTTP protocol layers mapped on NoC units and Media Independent NoC Interface—MINI.

Units (PTUs). The physical layer defines how packets are physically transmitted over an interface.

An NTTP transaction is typically made of request packets, traveling through the request network between the master and the slave NIUs, and response packets that are exchanged between a slave NIU and a master NIU through the response network. At this abstraction level, there is no assumption on how the NoC is actually implemented (i.e., the NoC topology). Transactions are handed off to the transport layer, which is responsible for delivering packets between endpoints of the NoC (using links, routers, muxes, rated adapters, FIFOs, etc.). Between NoC components, packets are physically transported as cells across various interfaces, a cell being a basic data unit being transported. This is illustrated in Figure 11.1, with one master and one slave node, and one router in the request and response path.

11.3.1.1 Transaction Layer

The transaction layer is compatible with bus-based transaction protocols used for on-chip communications. It is implemented in NIUs, which are at the boundary of the NoC, and translates between third-party and NTTP protocols. Most transactions require the following two-step transfers:

- A master sends request packets.
- Then, the slave returns response packets.

As shown in Figure 11.1, requests from an initiator are sent through the master NIU's transmit port, Tx, to the NoC request network, where they are routed to the corresponding slave NIU. Slave NIUs, upon reception of request packets

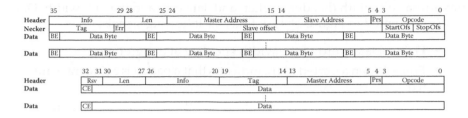

FIGURE 11.2
NTTP packet structure.

on their receive ports, Rx, translate requests so that they comply with the protocol used by the target third-party IP node. When the target node responds, returning responses are again converted by the slave NIU into appropriate response packets, then delivered through the slave NIU's Tx port to the response network. The network then routes the response packets to the requesting master NIU, which forwards them to the initiator. At the transaction level, NIUs enable multiple protocols to coexist within the same NoC. From the point of view of the NTTP modules, different third-party protocols are just packets moving back and forth across the network.

11.3.1.2 Transport Layer

The Arteris NTTP protocol is packet-based. Packets created by NIUs are transported to other parts of the NoC to accomplish the transactions that are required by foreign IP nodes. All packets are comprised of cells: a header cell, an optional necker cell, and possibly one or more data cells (for packet definition see Figure 11.2; further descriptions of the packet can be found in the next subsection). The header and necker cells contain information relative to routing, payload size, packet type, and the packet target address. Formats for request packets and response packets are slightly different, with the key difference being the presence of an additional cell, the necker, in the request packet to provide detailed addressing information to the target.

11.3.1.3 Physical Layer

The delivery of packets within the NoC is the responsibility of the physical layer. Packets, which have been split by the transport layer into cells, are delivered as words that are sent along links. Within a single clock cycle, the physical layer may carry words comprising a fraction of a cell, a single cell, or multiple cells. The link size, or width (i.e., number of wires), is set by the designer at design time and determines the number of cells of one word. NTTP defines five possible link-widths: quarter (QRT), half (HLF), single (SGL), double (DBL), and quad (QUAD). A single-width (SGL) link transmits one cell per clock cycle, a double-width link transmits two cells per clock cycle, and so on. Words travel within point-to-point links, which are independent from other protocol layers: a word is sent through a transmit port, Tx, over a link to a receive port, Rx. The actual number of wires in a link depends on the maximum cell-width (header, necker, and data cell) and the link-width. One link (represented in Figure 11.1) defines the following signals:

- **Data**—Data word of the width specified at design-time.

- **Frm**—When asserted high, indicates that a packet is being transmitted.

- **Head**—When asserted high, indicates the current word contains a packet header. When the link-width is smaller than single (SGL), the header transmission is split into several word transfers. However, the Head signal is asserted during the first transfer only.

- **TailOfs**—Packet tail: when asserted high, indicates that the current word contains the last packet cell. When the link-width is smaller than single (SGL), the last cell transmission is split into several word transfers. However, the Tail signal is asserted during the first transfer only.

- **Pres.**—Indicates the current priority of the packet used to define preferred traffic class (or Quality of Service). The width is fixed during the design time, allowing multiple pressure levels within the same NoC instance (bits 3–5 in Figure 11.2).

- **Vld**—Data valid: when asserted high, indicates that a word is being transmitted.

- **RxRdy**—Flow control: when asserted high, the receiver is ready to accept word. When de-asserted, the receiver is busy.

This signal set, which constitutes the Media Independent NoC Interface (MINI), is the foundation for NTTP communications.

Packet definition.   Packets are composed of cells that are organized into fields, with each field carrying specific information. Most of the fields in header cells are parameterizable and in some cases optional, which makes it possible to customize packets to meet the unique needs of an NoC instance. The following list summarizes the different fields, their size, and function:

Field	Size	Function
Opcode	4 bits/3 bits	Packet type: 4 bits for requests, 3 bits for responses
MstAddr	User Defined	Master address
SlvAddr	User Defined	Slave address
SlvOfs	User Defined	Slave offset
Len	User Defined	Payload length
Tag	User Defined	Tag
Prs	User defined (0 to 2)	Pressure
BE	0 or 4 bits	Byte enables
CE	1 bit	Cell error
Data	32 bits	Packet payload
Info	User Defined	Information about services supported by the NoC
Err	1 bit	Error bit

StartOfs	2 bits	Start offset
StopOfs	2 bits	Stop offset
WrpSize	4 bits	Wrap size
Rsv	Variable	Reserved
CtlId	4 bits/3 bits	Control identifier, for control packets only
CtlInfo	Variable	Control information, for control packets only
EvtId	User defined	Event identifier, for event packets only

For request packets, a data cell is typically 32 or 36 bits wide depending on the presence of byte enables (this is fixed at design time). For response packets, a data cell is always 33 bits wide. A possible instance of a packet structure is illustrated in Figure 11.2. Header, necker, and data cells do not necessarily have the same size. Different data cell widths and their relation to the cells are illustrated in Figure 11.3.

To provide services to IP cores, the transaction layer relies primarily on Load and Store transactions, which are converted into packets. The predominant packet types are Store and Load for requests, and Data and Acknowledge for responses. Control packets and Error Response packets are also provided for NoC management.

Quality of Service (QoS). The QoS is a very important feature in the interconnect infrastructures because it provides a regulation mechanism allowing specification of guarantees on some of the parameters related to the traffic. Usually the end users are looking for guarantees on bandwidth and/or end-to-end communication latency. Different mechanisms and strategies have been proposed in the literature. For instance, in Æthereal NoC [11,24] proposed by NXP, a TDMA approach allows the specification of two traffic categories [25]: BE and GT.

In the Arteris NoC, the QoS is achieved by exploiting the signal pressure embedded into the NTTP packet definition (Figures 11.1 and 11.2). The pressure

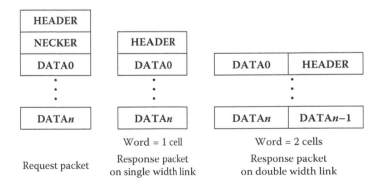

FIGURE 11.3
Packet, cells, and link width.

signal can be generated by the IP itself and is typically linked to a certain level of urgency with which the transaction will have to be completed. For example, we can imagine associating the generation of the pressure signal when a certain threshold has been reached in the FIFO of the corresponding IP. This pressure information will be embedded in the NTTP packet at the NIU level: packets that have pressure bits equal to zero will be considered without QoS; packets with a nonzero value of the pressure bit will indicate preferred traffic class.* Such a QoS mechanism offers immediate service to the most urgent inputs and variables, and fair service whenever there are multiple contending inputs of equal urgency (BE). Within switches, arbitration decisions favor preferred packets and allocate remaining bandwidth (after preferred packets are served) fairly to contending packets. When there are contending preferred packets at the same pressure level, arbitration decisions among them are also fair.

The Arteris NoC supports the following four different traffic classes:

- **Real time and low latency (RTLL)**—Traffic flows that require the lowest possible latency. Sometimes it is acceptable to have brief intervals of longer latency as long as the average latency is low. Care must be taken to avoid starving other traffic flows as a side effect of pursuing low latency.

- **Guaranteed throughput (GT)**—Traffic flows that must maintain their throughput over a relatively long time interval. The actual bandwidth needed can be highly variable even over long intervals. Dynamic pressure is employed for this traffic class.

- **Guaranteed bandwidth (GBW)**—Traffic flows that require a guaranteed amount of bandwidth over a relatively long time interval. Over short periods, the network may lag or lead in providing this bandwidth. Bandwidth meters may be inserted onto links in the NoC to regulate these flows, using either of the two methods. If the flow is assigned high pressure, the meter asserts backpressure (flow control) to prevent the flow from exceeding a maximum bandwidth. Alternatively, the meter can modulate the flows pressure (priority) dynamically as needed to maintain an average bandwidth.

- **Best effort (BE)**—Traffic flows that do not require guaranteed latency or throughput but have an expectation of fairness.

11.3.2   Network Interface Units

The Arteris Danube IP library includes NIUs for different third party protocols. Currently, three different protocols are supported: AHB (APB), OCP, and AXI. For each protocol, two different NIU units can be instantiated:

---

* Note that in the NTTP packet, the pressure field allows more then one bit, resulting in multiple levels of preferred traffic.

- **Initiator NIU**—third party protocol-to-NTTP, used to connect a master node to the NoC
- **Target NIUs**—NTTP-to-third party protocol, used to connect a slave node to the NoC

In the following, we will describe in more details both initiator and target NIU units for the AHB protocol, because this particular protocol has been used for all nodes in the MPSoC platform.

11.3.2.1 Initiator NIU Units

Initiator NIU units (the architecture of the AHB initiator is given in Figure 11.4) enable connection between an AMBA-AHB master IP and the NoC. It translates AHB transactions into an equivalent NTTP packet sequence, and transports requests and responses to and from a target NIU, that is, slave IP (slave can be any of the supported protocols). The AHB-to-NTTP unit instantiates a Translation Table for address decoding. This table receives 32-bit AHB addresses from the NIU and returns the packet header and necker information that is needed to access the NTTP address space: Slave address, Slave offset, Start offset, and the coherency size (see Figure 11.2). Whenever the AHB address does not fit the predefined decoding range, the table asserts an error signal that sets the error bit of the corresponding NTTP request packet, for further error handling by the NoC. The translation table is fully user-defined at design time: it must first be completed with its own hardware parameters, then passed to the NIU.

A FIFO memory is inserted in the datapath for AHB write accesses. The FIFO memory absorbs data at the AHB initiator rate, so that NTTP packets can

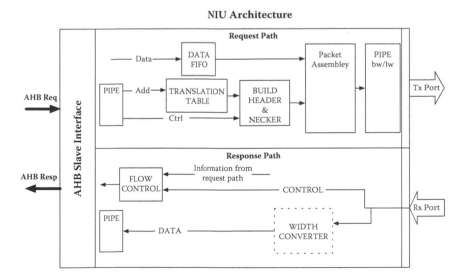

FIGURE 11.4
Network interface unit: Initiator architecture.

burst at NoC rate as soon as a minimum amount of data has been received. The width of the FIFO and the AHB data bus is identical, and the FIFO depth is defined by the hardware parameter. This parameter indicates the amount of data required to generate a Store packet: each time the FIFO is full, a Request packet is sent on the Tx port. Of course, if the AHB access ends before the FIFO is full, the NTTP request packet is sent. Because AHB can only tolerate a single outstanding transaction, the AHB bus is frozen until the NTTP transaction has been completed. That is

- During a read request, until the requested data arrives from the Rx port
- During a nonbufferable write request, in which case only the last access is frozen and the acknowledge occurs when the last NTTP response packet has been received
- When an internal FIFO is full

11.3.2.2 Target NIU Units

Target NIU units enable connection of a slave IP to the NoC by translating NTTP packet sequences into equivalent packet transactions, and transporting requests and responses to and from targets (the architecture of the AHB Target NIU is given in Figure 11.5). For the AHB target NIU, the AHB address space is mapped from the NTTP address space using the slave offset, the start/stop offset, and the slave address fields, when applicable (from the header of the request packet, Figure 11.2). The AHB address bus is always

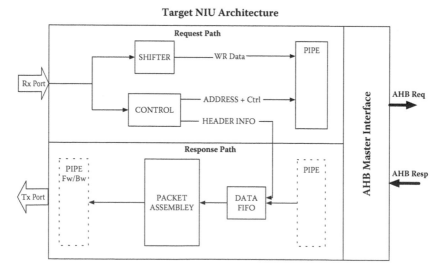

FIGURE 11.5
Network interface unit: Target architecture.

32 bits wide, but the actual address space size may be downsized by setting a hardware parameter. Unused AHB address bits are then driven to zero. The NTTP request packet is then translated into one or more corresponding AHB accesses, depending on the transaction type (word aligned or nonaligned access). For example, if the request is an atomic Store, or a Load that can fit an AHB burst of specified length, then such a burst is generated. Otherwise, an AHB burst with unspecified length is generated.

11.3.3 Packet Transportation Units

In the Arteris Danube library, packet transportation units represent different hardware modules used to route, transmit over a long distance, and adapt data flows with different characteristics in the heterogeneous NoC configuration. Typically the library describes the following elements:

- Switch (router)—enabling packet routing
- Muxes—allowing multiplexing of different flows over the same link
- Synchronous FIFOs—used for data buffering on critical links to avoid congestion
- Bisynchronous FIFOs—allowing synchronization between asynchronous domains
- Clock converters—connecting different domains timed by different but synchronous clocks
- Width and endian converters—adapting different link widths and endian conventions within the same NoC instance

In the following subsections, we will focus on describing the switch—the essential building block of the NoC interconnect system.

11.3.3.1 Switching

The switching is done by accepting NTTP packets carried by input ports and forwarding each packet transparently to a specific output port. The switch is characterized with a fully synchronous operation and can be implemented as a full crossbar (up to one data word transfer per port and per cycle), although there is an automatic removal of hardware corresponding to unused input/output port connections (port depletion). The switch uses wormhole routing, for reduced latency, and can provide full throughput arbitration; that is, up to one routing decision per input port and per cycle. An arbitrary number of switches can be connected in cascade, supporting any loopless network topology. The QoS is supported in the switch using the pressure information generated by the IP itself and embedded in NTTP packets.

A switch can be configured to meet specific application requirements by setting the MINI-ports (Rx or Tx ports, as defined by the MINI interface introduced earlier) attributes, routing tables, arbitration mode, and pipelining strategy. Some of the features can be software-controlled at runtime through the service network. There is one routing table per Rx port and one arbiter per Tx port. Packet switching consists of the following four stages:

1. **Choosing the route**—Using relevant information extracted from the packet, the routing table selects a target output port.

2. **Arbitrating**—Because more than a single input port can request a given output port at a given time, an arbiter selects one requesting input port per output port. The arbiter maintains input/output connection until the packet completes its transit in the switch.

3. **Switching**—Once routing and arbitration decisions have been made, the switch transports each word of the packet from its input port to its output port. The switch implementation employs a full crossbar, ensuring that the switch does not contribute to congestion.

4. **Arbiter release**—Once the last word of a packet has been pipelined into the crossbar, the arbiter releases the output, making it available for other packets that may be waiting at other input ports.

The simplified block diagram of the switch architecture is shown in Figure 11.6.

11.3.3.2 Routing

The switch extracts the destination address and possibly the scattering information from the incoming packet header and necker cells, and then selects an output port accordingly. For a request switch, the destination address is the slave address and the scattering information is the master address

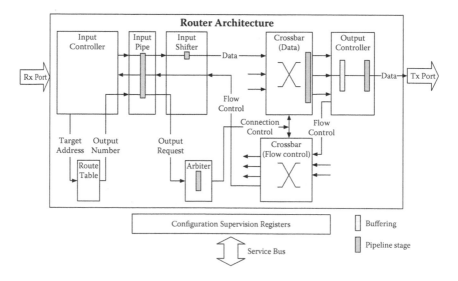

FIGURE 11.6
Packet transportation unit: Router architecture.

(as defined in packet structure, Figure 11.2). For a response switch, the destination address is the Master address and there is no scattering information. The switch ensures that all input packets are routed to an output port. If the destination address is wrong or if the routing table is not written properly, the packet is forwarded to a default output port. In this way, an NTTP slave will detect an error upon packet reception. The "default" output is the port of highest index that is implemented: port n, or port $n - 1$ if port n is depleted, or port $n - 2$ if ports n and $n - 1$ are depleted, and so on.

11.3.3.3 Arbitration

Each output port tracks the set of input ports requesting it. For each cycle in which a new packet may be transmitted, the arbiter elects one input port in that set. This election is conducted logically in two phases.

First, the pressure information used to define the preferred traffic class (QoS) of the requesting inputs is considered. The pressure information is explicitly carried by the MINI interface (signal Pres. in Figure 11.1), and indicates the urgency for the current packet to get out of the way. It is the maximum packet pressure backing up behind the current packet. The pressure information is given top priority by the switch arbiter: among the set of requesters, the input with the greatest pressure is selected. Additionally, the maximum pressure of the requesters is directly forwarded to the output port.

Second, the election is held among the remaining requesters (i.e., inputs with equal maximum pressure) according to the selected arbiter. Hardware parameters enable the user to select a per "output port" arbiter from the library, such as: random, round robin, least recently used (LRU), FIFO, or fixed priority (software programmable).

In general, the detection of packet tail causes the output currently allocated to that input to be released and become re-electable. Locked transactions are a notable exception. If packet A enters the switch and is blocked waiting for its output to become available, and if packet B enters the switch through a different input port, but aims for the same output port, then when the output port is released, at equal pressure, the selected arbitration mode must choose between A and B. The pressure information on an input port can increase while a packet is blocked waiting, typically because of a higher pressure packet colliding at the rear of the jam (packet pressure propagates along multiswitch paths). Thus, a given input can be swapped in or out of candidate status while it is waiting.

11.3.3.4 Packet Management

No reordering occurs in the switch. The incoming packets through input port A and the outgoing packets through output port B are guaranteed to be delivered on B in the order in which they arrive on A. Packets are processed sequentially on any given input port and no packet can be routed as long as its predecessor on the same input port has not been successfully routed. Because the switch implements a wormhole routing strategy, it can start the transmission of a packet before the packet has been completely received.

The switch routes incoming packets without altering their contents. Nevertheless, it is sensitive to Lock/Unlock packets: when a Lock packet is received, the connection between the input and the output as defined in the routing table is kept until an Unlock packet is encountered. The packets framed by Lock and Unlock packets, including the Unlock packet itself, are blindly routed to the output allocated on behalf of the Lock packet. The input controller extracts pertinent data from packet headers, forwards it to the routing table, fetches back the target output number, and then sends a request to the arbiter. After arbitration is granted, the input controller transmits the rest of the packet to the crossbar. The request to the arbiter is sustained as long as the last word of the packet has not been transferred. Upon transferring the last cell of the packet, the arbiter is allowed to select a new input.

Lock packets, on the other hand, are treated differently. Once a Lock packet has won arbitration, the arbitrated output locks on the selected input until the last word of the pending unlock packet is transmitted. Thus packets between lock and unlock packets are unconditionally routed to the output requested by the lock packet.

Depending on the kind of routing table chosen, more than one cycle may be required to make a decision. A delay pipeline is automatically inserted in the input controller to keep data and routing information in phase, thus guaranteeing one-word-per-cycle peak throughput. Routing tables select the output port that a given packet must take. The route decision is based on the tuple (destination address, scattering information) extracted from the packet header and necker. In a request environment, the Destination Address is the Slave Address and the Scattering Information is the Master Address. In a response environment, the Destination Address is the Master address and the Scattering Information is the Tag (Figure 11.2).

For maximum flexibility, the routing tables actually used in the switch are parameterizable for each input port of the switch. It is thus possible to use different routing tables for each switch input. Routing tables can optionally be programmed via the service network interface; in this case, their configuration registers appear in the switch register address map.

The input pipe is optional and may be inserted individually for each input port. It introduces a one-word-deep FIFO between the input controller and the crossbar and can help timing closure, although at the expense of one supplementary latency cycle.

The input shifter is optional and is implemented when arbiters are allowed to run in two cycles (the late arbitration mode is fixed at design time). The role of the shifter is to delay data by one cycle, according to the requests of the arbiter. This option is common to all inputs.

The arbiter ensures that the connection matrix (a row per input and a column per output) contains at most one connection per column, that is, a given output is not fed by two inputs at the same time. The dual guarantee—at most one connection per row—is handled by the input controller. Each output has an arbiter that includes prefiltering. For maximum flexibility, each port can specify its own arbiter from the list of available arbiters (random, round robin, LRU, FIFO, or fixed priority). A late arbitration mode is available to ease timing closure; when activated, one additional cycle is required to provide the arbitration result.

The crossbar implements datapath connection between inputs and outputs. It uses the connection matrix produced by the arbiter to determine which connections must be established. It is equivalent to a set of m muxes (one per output port), each having n inputs (one per input port). If necessary, the crossbar can be pipelined to enhance timing. The number of pipeline stages can be as high as $max(n, m)$.

The output controller constructs the output stream. It is also responsible for compensating crossbar latency. It contains a FIFO with as many words as there are data pipelined in the crossbar. FIFO flow control is internally managed with a credit mechanism. Although FIFO is typically empty, should the output port become blocked, it contains enough buffering to flush the crossbar. When necessary for timing reasons, a pipeline stage can be introduced at the output of the controller.

The switch has a specific interface allowing connection to the service network and a dedicated communication IP used for software configuration and supervision.

11.3.4 NoC Implementation Issues

11.3.4.1 Pipelining

Arteris Danube library elements contain a number of options that allow the designer to control the presence and the position of the pipelining registers in different NoC element instances. The global pipelining strategy implemented in the Arteris technology can be described as follows: time budgeting is voluntarily not imposed, to avoid a cycle delay in NoC units and making it possible to chain several units in the same combinatorial path. Flow control is carried by the RxRdy signal (Figure 11.1), moving from the receiver to the transmitter, and is said to be backward moving. All other signals move from the transmitter to the receiver, and are said to be forward moving. In NoC units, two major constraints have an impact on pipelining strategy:

- Backward signals (RxRdy) can be combinatorially dependent on any unit input.
- Forward signals can be combinatorially dependent on any other forward signal.

To avoid deadlock, a valid NTTP topology must be loopless. Consequently, although a legal unit assembly cannot contain a combinatorial loop, it can contain long forward or backward paths. It is the user's responsibility to break long paths, thus making sure that propagation delays remain reasonable. What is reasonable will depend on factors such as design topology, target frequency, process, or floor plan. The opportunity to break long paths is present on most MINI transmission ports, and is controlled through a parameter named fwdPipe: when set, this parameter introduces a true pipeline register on the forward signals, and effectively breaks the forward path. The parameter inserts the DFFs required to register a full data word as well as with control signals, and a cycle delay is inserted for packets traveling this path.

11.3.4.2 Clock Gating

Synthesis tools have the ability to optimize power dissipation of the final design by implementing local clock-gating techniques. All Danube IPs have been coded so that synthesizers can easily identify and replace DFF with synchronous load enable including clock-gating circuits. More than 90 percent of the DFF can be connected to a gater, thus minimizing the dynamic power dissipation of the NoC.

A complementary approach is to apply optimization techniques at higher levels of the design, thus disabling the clock to parts that are currently not in use. In the Arteris environment, global clock-gating can be applied at unit level (i.e., any NIUs or PTUs). Each instance of the Danube library monitors its internal state and goes in idle mode when

- All pipeline stages are empty.
- All state machines are in idle state.
- There is currently no transaction pending, in the case of an NIU.

Conversely, the unit wakes up, at no delay cost, upon reception of

- An NTTP packet
- A transaction on the associated IP socket, in the case of an NIU

The unit comprises all the logic that is necessary to control the global clock-gater, turning the clock off or on depending on the traffic. Note that the design can apply the local clock-gating technique, the global clock-gating technique, or both.

Local clock-gating implementation. NoCcompiler has an export option for local clock-gating. This feature has no impact on the RTL netlist, but generates a synthesis script with specific directives. Because clock gaters can be instantiated individually, typically by the synthesis tool per register stage of at least one configurable width, this process may instantiate many clock-gater cells per NoC unit. The remaining capacitive load on the clock pin(s) of the unit will be the sum of the clock-gater loads, plus the remaining ungated registers.

Global clock-gating implementation. As for local clock-gating, NoCcompiler has an export option for global clock-gating. This feature adds a clock-gater cell, and the associated control logic, to all NoC units. In addition to gating most of the unit DFFs, the global clock gating drastically reduces the capacitive load on the clock tree, because there are very few (usually a single one) such clock gaters instantiated per unit clock domain.

11.3.5 EDA Tools for NoC Design Tools

The Danube NoC IP library is integrated within the Arteris design automation suite, consisting of many different software tools that are accessible through the following two frameworks:

- **NoCexplorer**—used for NoC design space exploration (NoC topology definition)
- **NoCcompiler**—used for generation of the HDL RTL and cycle accurate SystemC NoC models

11.3.5.1 NoCexplorer

The NoCexplorer tool allows easy and fast NoC design space exploration through modeling and simulation of the NoC, at different abstraction levels (different NoC models can coexist within the same simulation instance). The NoC models and associated traffic scenarios are first described using scripting language based on a subset of syntax and semantics derived from the Python programming language. The NoC models can be very abstract, defined with only few parameters, or on the contrary they can be more detailed, thus being very close to the actual RTL model that will be defined within the NoCcompiler environment. One NoC model (or all of them) is then simulated for one (or all) traffic scenarios with a built-in simulation engine, producing performance results for further analysis. Typically, the designer can analyze bandwidths to and from all initiator and target nodes, the end-to-end latency statistics, the FIFO fillings, etc. These results are then interpreted to see if the NoC and associated architectural choices (NoC topology and configuration) meet the application requirements.

The NoCexplorer environment allows a very fast modeling and simulation cycle. NoC and traffic description depend heavily on the complexity of the system, but will require typically less than an hour, even for the specification of a complex system (provided the user is experienced). On the other hand, the actual simulation of the model will take less than a minute with a standard desktop computer, even for complicated systems containing dozens of nodes and including complex traffic patterns. This means that the designer can easily test different traffic scenarios for different NoC topology specifications until the satisfactory results are reached, before moving to the NoC specification for RTL generation. Note that the simulation cycle can be easily automated in more complex frameworks for wider benchmarking.

In the NoCexplorer framework, a typical NoC model will include the description of the following items:

- **Global system description**—also called Intent in Arteris jargon with the following definitions:

    Clock domains—different nodes in the system can operate at different frequencies

System sockets—the description of all initiator and target nodes in the system represented by their NIUs running at frequencies specified in the previous point

System connectivity—the specification describing which initiator node is allowed to communicate with which target node in the system (definition of the connectivity matrix)

Memory map—the selection of a target socket as a function of the initiator socket and the transaction address, with a possible intermediary transaction address translation

- **Traffic scenarios**—for a given system description, different traffic scenarios can be defined using built-in traffic generators.
- **NoC architecture definition**—this is in fact the NoC topology, specified through shared links (in fact these shared links are model routers), their parameters such as the link width, introduced latency, arbitration scheme, etc.

Note that to perform simulation, it is not necessary to define an architecture (i.e., NoC topology). Preliminary simulations can be performed on so-called "empty architecture" (i.e., the NoC at the most abstract level) assuming that the NoC is having an infinite bandwidth and no latency. Such simulation can be used to show the application to NoC architecture/topology adequation, modeled at the NIU level. Note that different aspects related to the different classes of traffic and QoS can also be expressed and modeled in the NoCexplorer environment.

Once the different traffic scenarios and architectures have been defined, the simulation may be run and results can be represented in either a tabular or graphical way. At this stage, the designer typically verifies that the required bandwidth has been achieved and the latency of different traffic flows is within expected latency range.

11.3.5.2   NoCcompiler

While the NoCexplorer tool enables fast exploration of the NoC design space using high-level NoC models, the NoCcompiler tool is used to describe the NoC at lower abstraction levels allowing automatic RTL generation after complete specification of the NoC. Typical NoC design flow using NoCcompiler can be divided into the following steps:

- **NoC configuration**—During the NoC exploration phase and using NoCexplorer, the topology of the NoC has been fixed for both request and response networks depending on the application requirements. This information is now used to describe the RTL NoC specification through the generation and configuration of different NoC instances such as NTTP protocol definition, initiator and target NIUs, routers, and eventually other elements provided by the Danube IP library. When generated, each new instance is verified for correctness and compatibility with other already generated instances, with which it may interact.

- **NoC assembly**—Different instances of the NoC elements (NTTP protocol configuration, NIUs for all initiator and target sockets, and router and interconnect fabric for request and response networks) are connected together using a graphical user interface. After this step, the NoCcompiler tool provides tools for automatic route verification and routing table generation as well as address decoding for third party-to-NTTP protocol conversion at NIUs. The time spent on both NoC configuration and assembly steps depends on the design complexity, but for an experienced user and a reasonably complex design, such as one described in Section 11.3.5, it will typically require one to two days.

- **NoC RTL generation**—When the complete NoC is specified (i.e., the configuration of all NoC elements plus generated decoding and routing tables), the corresponding RTL can be exported into standard hardware description languages such as VHDL or Verilog. Together with the HDL files, the framework can generate the necessary synthesis scripts for some of the most popular EDA tools (Synopsis and Magma). Starting from the same NoC specification, one can also generate a cycle-accurate NoC model in SystemC that can be used for faster simulation. At this stage, the tool can provide an area estimation for each of the NoC elements depending on the configuration and the complete NoC. This is quite a worst-case estimation (typically a 10–20 percent overestimate when compared with standard synthesis tools) and does not take into account the area due to the routing of wires, but can help the designer for preliminary NoC area estimations.

- **NoC RTL verification**—After the RTL generation, different automatic self-test diagnostic tests can be generated to validate a specific feature of the NoC using NoC Selftest Generator (NSG). First, each individual NoC element instance is tested to verify if it actually complies with its external specifications, and, second, the complete NoC specification is checked to ensure it performs according to the specifications. Typically these tests will include the following:

  Connectivity tests—used to validate address decoding at NIU level and routing tables (read/write operations to all targets memory).

  Minimum latency test—used to determine the end-to-end latency for WRITE and READ operations, based on short transactions (minimum transaction latency).

  Peak initiator to target throughputs—similar to the minimum latency test, except that the maximum obtainable throughput has been derived from bigger transactions.

Random transactions—used to validate the NoC arbitration and flow-control mechanisms. They allow all possible data flows to coexist, and ensure that no data is lost even when the NoC or the targets are heavily congested.

The simulation time of the NoC RTL verification step using NSG environment depends on the design complexity; but even for more complex designs, it remains in the range of one hour.

- **NoC documentation generation**—For an implemented and verified design it is possible to generate a documentation file, containing all the necessary information about the NoC instance configuration and properties.

The complete NoC design flow including NoCexplorer and NoCcompiler environments and their interaction is represented in Figure 11.7. Note that the traffic generated with built-in traffic generators used for the simulation of the NoC at the TL3 model using NoCexplorer (gray box in Figure 11.7 found in both NoCexplorer and NoCcompiler flows) can be exported and used for the simulation of the NoC TL0 model generated with the NoCcompiler (with both VHDL/Verilog and/or cycle-accurate SystemC model). The designer can then compare the performance predicted by the high-level model of the NoC with the actual performance of the RTL NoC model generated with the NoCcompiler. Furthermore, different models of the NoC can be exported to

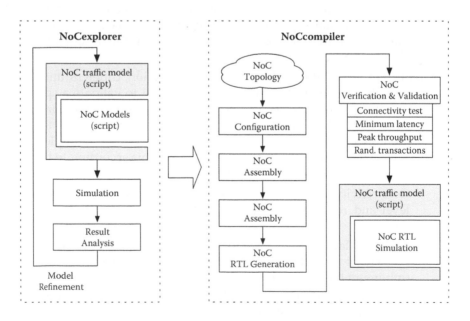

FIGURE 11.7
Arteris NoC design flow.

other EDA tools, such as CoWare, and used for transaction-level simulation of the complete SoC platforms. With such simulation frameworks, one can easily trade off between simulation speed and accuracy. This is a very useful feature, especially when the design involves larger and more complex MPSoC platforms running intensive applications, computationally.

11.4 MPSoC Platform

The global architecture of the MPSoC platform dedicated to modular, high-performance, low-power video encoding and decoding applications is shown in Figure 11.8(a). The platform is built using six CGA ADRES processors, separate data and instruction L2 memory clusters, one on-chip external DRAM memory interface (EMIF), and one node dedicated to the handling of incoming and outgoing video streams (FIFO). Configuration and control of the platform as well as audio processing are handled with one ARM926EJS core. The communication between different nodes in the system are handled with specific communication hardware units, called Communication Assists, interconnected using Arteris NoC.

The platform is intended to support multiple video coding standards, including MPEG-4, AVC/H.264, and SVC and should be able to handle different operating scenarios. This means that it is possible to process multiple video streams of different resolutions, combining both encoding and decoding operations using different compression algorithms at the same time. In the context of mobile multimedia devices design, this is often referred to as Quality of Experience. Therefore the proposed MPSoC appears as a flexible, general purpose computing platform for low-power multimedia applications.

For this particular MPSoC instance, the power budget of the platform for AVC encoding of the HDTV resolution video stream at 30 frames per second rate has been fixed at 700 mW. Such a constraint was imposed to maximize the autonomy of a mobile device and has been provided by our industrial partners. This absolute figure corresponds to the power that can be delivered with a fully charged standard cellular phone battery for approximately 10 hours.

Different dedicated implementations for video coding applications using advanced video coding standards have been proposed in the literature recently [26–29], some of them with power budgets lower than our goal [30]. However, the majority of these solutions are restricted to one particular compression algorithm and one compression direction (encoding or decoding), and they support only one resolution and frame rate, which are generally lower than the one we are targeting for this platform.

In the following, we will give a brief description of the different nodes of this platform and a more detailed description of the NoC. We will conclude with the system implementation results concentrating on area and power dissipation.

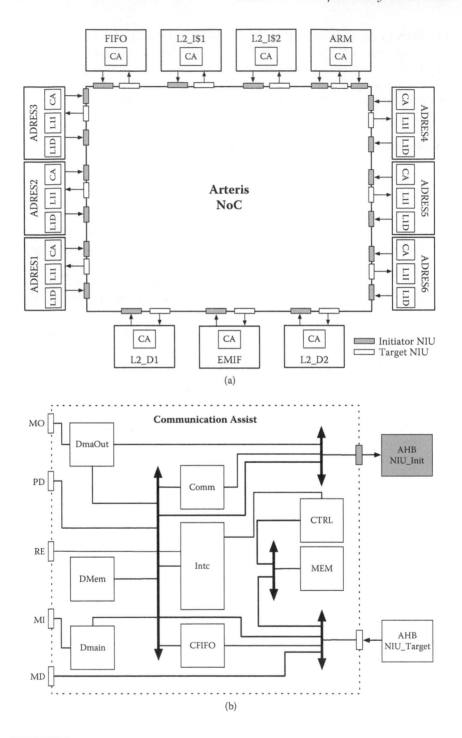

FIGURE 11.8
(a) Architecture of the MPSoC platform and (b) close-up of the communication assist architecture.

11.4.1 ADRES Processor

The ADRES is a CGA architecture template developed at IMEC [31–33]. The architecture of the ADRES processor is shown in Figure 11.9 and consists of an array of the following basic components: reconfigurable functional units

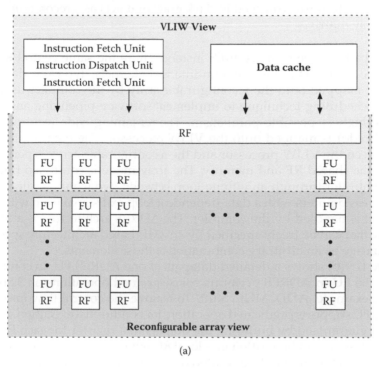

(a)

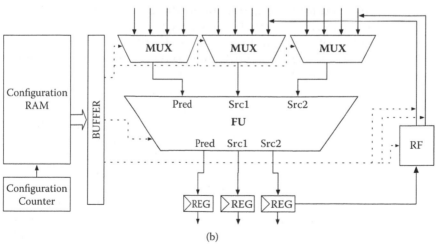

(b)

FIGURE 11.9
(a) Architecture of the ADRES CGA core and (b) close-up of the functional unit.

(RFUs), register files (RFs), and routing resources. The ADRES CGA processor can be seen as if it were composed of the following two parts:

- the top row of the array that acts as a tightly coupled Very Long Instruction Word (VLIW) processor (marked in light gray) and
- the bottom row (marked in dark gray) that acts as a reconfigurable array matrix.

The two parts of the same ADRES instance share the same central RF and load/store units. The computation-intensive kernels, typically data-flow loops, are mapped onto the reconfigurable array by the compiler using the modulo scheduling technique to implement software pipelining and to exploit the highest possible parallelism. The remaining code (control or sequential code) is mapped onto the VLIW processor. The data communication between the VLIW processor and the reconfigurable array is performed through the shared RF and memory. The array mode is controlled from the VLIW controller through an infinite loop between two (configuration memory) address pointers with a data-dependent loop exit signal from within the array that is handled by the compiler. The ADRES architecture is a flexible template that can be freely specified by an XML-based architecture specification language as an arbitrary combination of those elements.

Figure 11.9(b) shows a detailed datapath of one ADRES FU. In contrast to FPGAs, the FU in ADRES performs coarse-grained operations on 32 bits of data, for example, ADD, MUL, Shift. To remove the control flow inside the loop, the FU supports predicated operations for conditional execution. A good timing is guaranteed by buffering the outputs in a register for each FU. The results of the FU can be written in a local RF, which is usually small and has less ports than the shared RF, or routed directly to the inputs of other FUs. The multiplexors are used for routing data from different sources. The configuration RAM acts as a (VLIW) instruction memory to control these components. It stores a number of configuration contexts, locally, which are loaded on a cycle-by-cycle basis. Figure 11.9(b) shows only one possible datapath, as different heterogeneous FUs are quite possible.

The proposed ADRES architecture has been successfully used for mapping different video compression kernels that are part of MPEG-4 and AVC/H.264 video encoders and decoders. More information on these implementations can be found in studies by Veredas et al. , Mei et al. and Arbelo et al. [32,34,35].

For this particular MPSoC platform, all ADRES instances were generated using the same processor template, although the configuration context of each processor, that is, the reconfigurable array matrix, can be fixed individually at runtime. Each ADRES instance is composed of 4×4 reconfigurable functional units and has separate data and instruction L1 cache memories of 32 kB each. All ADRES cores in the system operate at the same frequency that can be either 150 MHz (the same as the NoC) or 300 MHz, depending on the computational load. This is fixed by the MPSoC controller node, the ARM core running a Quality of Experience Manager application.

11.4.2 Communication Assist

The communication assist (CA) is a dedicated hardware unit that provides means for efficient transfers of blocks of data (block transfers) over the NoC. A block transfer (BT) is a way of representing a request to move data from the source to the destination memory location. As such, it resembles a classical DMA unit. However, we make a distinction between the CA and the classical DMA engine, because the CA implements some very specific features that are not usually seen in traditional DMAs. Generally speaking, CA performs 2D data transfer of data arrays, where the array of data is defined by the start address, the width and the height of the memory location seen as a virtual 2D memory space. Such approach to memory access is much more adapted to the way images are actually stored in the memory and to how block-based compression algorithms such as MPEG-4 or AVC/H.264 access it. In fact, for the CA, even more common 1D data transfer is seen as a special case of a 2D transfer. Note that the CA also supports wrapped BTs: when the dimension of the data block is bigger, the array and the boundary are crossed (modulo addressing).

Architecturally speaking, the CA shown in Figure 11.8(b) has five ports: the Input Port is used by the processor for direct access to the Initiator NIU and to the internal CA resources; Output Port is a resume/interrupt signal for a processor; the Output Ports MO (Memory Out), MI (Memory In), and MD (Memory Direct connection) are used by the CA for accessing the local memory (L1 of the processor, L2/L3 in a memory node, and so on). On the network side, the CA is connected to an Initiator NIU for outgoing traffic and to a Target NIU for incoming traffic. Internally, the CA uses two DMA engines (DmaOut, connected to the Initiator NIU, and DmaIn, connected to the Target NIU), one control processor, one interrupt controller, and some memory. The standard AMBA AHB bus is used for most internal and external bus connections of the CA, and all interfaces are AHB-Lite.

The CA only performs posted write operations for the NoC; thus, only the request part of the network is used by the CA for these operations. Nevertheless, the response part of the NoC is needed for simple L/S operations, issued by the processor. The CA provides direct connection between the processor or memory and the NIUs. This connection is used for simple L/S operations, and the CA is transparent for these operations. When there are no L/S operations in progress, the CA uses the NIUs for BTs. The CA supports the following two different kinds of BTs:

- **BT-Write**—to move data blocks from a local memory, over the NoC and into a destination memory. It is the task of the CA to generate the proper memory addresses according to the geometrical parameters of the source data block. The CA will send the data over the network to some remote CA, as a stream of words using NoC transactions. This remote CA will process the stream and write the data into the memory by generating the proper memory addresses according to the geometrical parameters of the target data block. When the remote CA has finished writing the last byte/half-word/word in its local memory, it will acknowledge the BT to the originating CA.

- **BT-Read**—to move data blocks from a distant memory, over the network, and into a local memory. It is the task of the CA to request a stream of data from a remote CA according to the geometrical parameters of the source data block. The local CA will process the stream and write data into the local memory according to the geometrical parameters of the target data block. In this scheme there is no acknowledgment to send back. When all data is received from the remote CA, it is considered as an acknowledgment of the BT-Read. The DmaIn unit of the local CA and DmaOut unit of the remote CA are involved in the BT-Read, but not in the BT-Write.

In the context of the MPSoC platform, any node in the system (typically ADRES or ARM core) can set up a BT transfer for any other pairs of nodes (one master and one slave node) in the system. In such scenario, even a memory node can act as a master for a BT transfer; the memory can then perform a BT-Write or BT-Read operation to/from some other distant node. This is possible because memories, like processors, access the NoC through the CAs (provided that another node has programmed the CA). Also, each CA is designed to support a certain number of concurrent BTs. This means that any node can issue one or more BTs for any other pair of nodes in the system (CAs implement communication through virtual channels). The number of concurrent BTs is fixed at the design time, and in the case of this MPSoC platform we limit this number to four for processors and external memory node (EMIF) and to 32 for L2 memory nodes, which is a design choice made to balance performance versus area.* Finally, different CAs are part of the NoC clock domain and they operate at 150 MHz.

11.4.3 Memory Subsystem

The memory subsystem of the MPSoC platform is divided into three levels of memory hierarchy. The concept of BTs, implemented through CAs, allows seamless transfers of data between different levels of memory hierarchy. Because the CAs are controlled directly from the software running at each ADRES core using high-level pragmas defined in **C** language, it represents a powerful infrastructure for data prefetching. This can be used for further implementation optimizations, where optimal distribution and shape of the data among different memory hierarchy levels can be determined taking into

---

* Implementing the virtual channel concept in the CA comes with the implementation cost: area, and this can be costly, especially in the context of MPSoC system, where multiple instances of the CAs are expected. The choice of four concurrent BTs per processor is derived from the fact that for the majority of the computationally intensive kernels we foresee at most four concurrent BT, that is, at most four consecutive prefetching operation per processor and per loop. For memory nodes, we want to maximize the number of concurrent transfers taking into account the total number of concurrent BTs in the system (depending on number of nodes: six in this case).

account: memory hierarchy level size, access latency, power dissipation of each level, etc. (for more information refer to works by Dasygenis et al. and Issenin et al.).

In this MPSoC platform, the memory is built on the following three levels of memory hierarchy:

- L3 off-chip memory—one DRAM memory is accessed through the on-chip memory interface (EMIF).
- L2 memory—two single banked (1×512kB) instruction memories and two double-banked (2×256 kB) data memories.
- L1 memory—each ADRES core has 2×32 kB of memory for instructions and data.

Because the connection between the CA and the NoC is 32 bits data wide running at 150 MHz, the maximal throughput that can be achieved with one data memory node is 1.2 GB per second when both read and write operations are performed simultaneously to different memory banks (2×600 MB per second, 2.4 GB per second for the whole memory cluster). Because the instruction memory nodes are single banked, only 600 MB per second per node can be achieved (we do assume that the instruction memories will be used most of the time for reading, that is, the configuration of the system occurs every once in a while).

11.4.4 NoC

As shown in Figure 11.8, every node in the MPSoC is connected to the NoC through a CA using a certain number of NIUs, depending on the node type. ADRES CGA processors will require three NIUs: two NIUs are used for the data port (one initiator and one target NIU) and one initiator NIU is used for the instruction port. The ARM subsystem also counts three NIUs, two being connected to the corresponding CA, while one NIU is connected directly to the NoC for debugging purposes. Both data and instruction memories are connected to the NoC through a pair of NIUs: one initiator and one target NIU. The complete NoC instance, as shown in Figure 11.10, has a total of 20 initiator and 13 target NIUs. Note that all NIUs are using AHB protocol* and have the same configuration. Different initiator NIUs are single width (SGL), and can buffer up to four transactions. All transactions have a fixed length (4, 8, or 16 beat transactions, imposed by AMBA-AHB protocol) and introduce only one pipeline stage. Target NIUs are also single width (SGL) and introduce one pipeline stage in the datapath. Finally, the NoC packet configuration defines the master address (6 bits), slave address (5 bits), and

---

* The choice of the IP protocol for this MPSoC Platform can be argued. AHB does not support split and retry transaction, has a fixed burst length, and is therefore not very well adapted for the high-performance applications. We have chosen AHB only because all of our already developed IPs (namely ADRES processor and CA) have been using AHB interfaces.

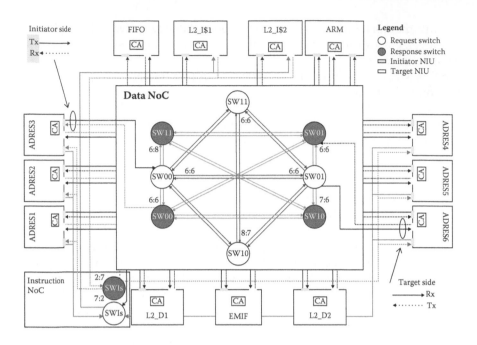

FIGURE 11.10
Topology of the NoC: Instruction and data networks with separated request and response network paths.

slave offset (27 bits) with a total protocol overhead of 72 bits for both header and necker cells of the request packet. The response packet overhead is 33 bits.

The adopted NoC topology shown in Figure 11.10 has been chosen to satisfy different design objectives. First, we want to minimize the latency upon instruction cache miss, because this will greatly influence the final performance of the whole system. Second, we want to maximize connectivity and bandwidth between different nodes because all video encoding applications are bandwidth demanding especially when considering high-resolution, high-frame rate video streams. Finally, we want to minimize the transfer latency, because of the performance and scalability requirements. For these reasons, the data and instruction networks are completely separated (in the following we will refer to these as Instruction and Data NoC) and each of these two networks is decomposed into separate request and response networks with the same topology. The data network topology consists of a fully connected mesh using 2 × 2 switches (routers) for both request (white switches) and response networks (gray switches). It allows connections between any pair of nodes in the systems with the minimum and maximum traveling distances of one and two hops, respectively.

The instruction network topology uses only one switch and enables ADRES instruction ports to access L2_I$1 and L2_I$2 memory nodes in only one hop, as shown in Figure 11.10. The only switch in this network is connected to the data NoC so that the instruction memories can be reached from any other node in the system. Typically, the application code is stored in the L3 memory and will be transferred to both L2_I\$s via EMIF. Note that the latency of such transfers will require three hops, but this is not critical because we assume that such transfers will occur only during the MPSoC configuration phase and will not interfere with normal encoding (or decoding) operations.

Different networks (data, instruction, request, and response) have 10 switches in all (Figure 11.10) with different numbers of input/output ports. All switches in the NoC have the same configuration and introduce one pipeline stage, whereas arbitration is based on the round-robin (RR) scheme representing a good compromise between implementation costs and arbitration fairness. The routing is fixed, that is, there is one possible route for each initiator/target pair, fixed at design time to minimize the gate count of the router.

All links in the NoC have the same size, they are single cell width (SGL), meaning that in the request path they contain 36 wires and for the response path 32 wires, plus 4 NTTP control wires. Because the NoC operating frequency has been set to 150 MHz, the maximal raw throughput (data plus NoC protocol overhead) is 600 MB per second per NoC link.

In this NoC instance, we also implemented a service bus, which is a dedicated communication infrastructure allowing runtime configuration and error recovery of the NoC. Because this MPSoC instance does not require any configuration parameters (all parameters are fixed at design time), the service bus is used only for application debugging. Any erroneous transaction within the NoC will be logged in the target NIUs logging registers. These registers can then be accessed at any time via service bus and from the control node. Appropriate actions can be taken for identification of the erroneous transaction. The service bus adopts token ring topology and is using only eight data and four control wires, minimizing the implementation cost of this feature. The access point from and to NTTP protocol is provided by the NTTP-to-Host IP (accessed from the ARM core in the case of the MPSoC platform). To simplify the routing, the token ring follows the logical path imposed by the floorplan: ARM, ADRES4, ADRES5, ADRES6, L2_D2, EMIF, L2_D1, ADRES1, ..., ARM (in order to simplify Figure 1.10, the service bus has not been represented).

11.4.5 Synthesis Results

The design flow of the complete MPSoC platform was divided into the following three steps:

1. **Logical synthesis** (using Synopsys Design Compiler)—Synthesis of the gate-level netlist from the VHDL description of all nodes in the platform. Because the ADRES core is quite large, the synthesis of this particular node is done separately.

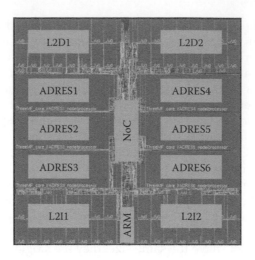

FIGURE 11.11
Layout of the MPSoC platform.

2. **Place and Route** (using Cadence SoC Encounter)—Placement and routing of the design. The input in this step is the Verilog gate-level netlist, which provides back annotation on the capacitance in the design.

3. **Power dissipation analysis** (using Synopsys Prime Power)— Evaluates the power consumption of the design based on the capacitance and activity of the nodes.

The results presented are relative to the TSMC 90 nm GHP technology library, the worst case using $V_{dd} = 1.08$ V and 125°C. The implemented circuit has been validated using VStation solution from Mentor Graphics. Figure 11.11 shows the layout of the MPSoC chip.

The complete circuit uses 17,295 kGates (45,08 mm^2), resulting in an 8×8 mm square die. Figure 11.12 provides a detailed area breakdown per platform node (surface and gate count) and the relative contribution of each node with respect to the total MPSoC area. Note that the actual density of the circuit is 70 percent, which is reasonable for the tool used. Maximum operating frequencies after synthesis are, respectively, 364 MHz, 182 MHz, 91 MHz for ADRES cores, and NoC and ARM subsystem. The NoC has been generated using the design flow described in Section 11.3.5 using NoCexplorer and NoCcompiler tools version 1.4.14. The typical size of basic Arteris NoC components for this particular instance as reported by the NoCcompiler estimator tool is given in Figure 11.13. This figure also gives the total NoC gate count, based on the number of different instances and without taking into account wires and placement/route overheads (difference of 450 kgates from actual placement and route results in Figure 11.12). The power dissipation breakdown of the complete platform and the relative contribution on a per instance base are given in Figure 11.14.

Node	Area $[mm^2]$	Size [kgates]	Relative [%]
ARM	0.89	317	1.8
ADRES	15	6630	38.2
EMIF	0.66	235	1.4
FIFO	0.54	191	1.1
L2D	13.48	4776	27.7
L2I	13.24	4696	27.2
NoC	1.27	450	2.6
Total	45.08	17295	100.0

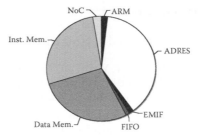

FIGURE 11.12
Area breakdown of the complete MPSoC platform.

Unit	Size [kgates]	Instances	Total [kgates]
NIU_I	4.6	20	92
NIU_T	6.2	13	80.6
Req. D	9	4	36
Req. I	3	2	6
Resp. D	9	4	36
Resp. I	3	2	6
Total			256.6

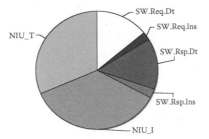

FIGURE 11.13
Area breakdown of the NoC.

Component	Power [mW]	Inst.	Total [mW]	Relative [%]
ADRES	91.1	6	546.6	84
L2D	20	2	40	6
L2I	15	2	30	5
ARM	10.5	1	10.5	2
NoC	25	1	25	4
Total			652.1	100

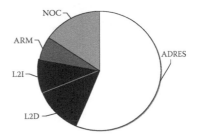

FIGURE 11.14
Power dissipation breakdown of the complete MPSoC and per component comparison.

11.5   Power Dissipation of the NoC for Video Coding Applications

In this section, we will derive the power dissipation of the NoC for the MPSoC platform described in the previous section. First, we will introduce the mapping of the MPEG-4 simple profile (SP) encoder and three different application mapping scenarios for the AVC/H.264 SP encoder. Then we will describe the power model of the NoC instance used in this MPSoC platform. Finally, in the last section, we will discuss the results and make a comparison with other state-of-the-art implementations presented in the literature.

11.5.1   Video Applications Mapping Scenarios

11.5.1.1   MPEG-4 SP Encoder

The functional block diagram of the low-power MPEG-4 SP encoder implementation presented by Denolf et al. [38] is shown in Figure 11.15. The figure indicates different computational, on-chip, and off-chip memory blocks represented, respectively, with white and light and dark gray boxes. For each link between two functional blocks, the figure indicates the number of bytes

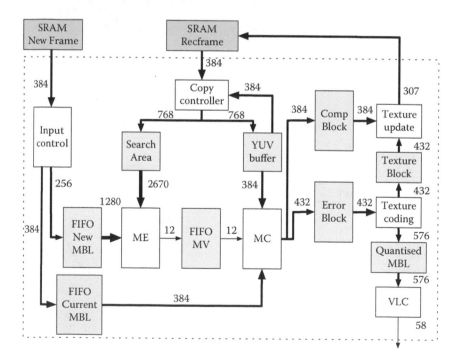

FIGURE 11.15

Functional block diagram of the MPEG-4 SP encoder with bandwidth requirements expressed in bytes/macroblock.

that have to be accessed for the computation of each new macroblock (MBL)*
expressed in bytes per macroblock units (B/MBL). The following three
columns show throughput requirements (expressed in MB/s) for CIF, 4CIF,
and HDTV image resolutions at 30 frames per second corresponding to 11880,
47520, and 108000 computed MBLs per second. For this particular implementation, the total power budget of the circuit built in 180 nm, 1.62 V technology
node for the processing of 4CIF images at 30 frames per second rate is 71 mW,
from which 37 mW is spent on communication, including on-chip memory
accesses.

The application mapping used in this implementation scenario can be easily
adapted (although it may be not optimal) to the MPSoC platform with the
following functional pipeline:

1. Input control of the video stream is mapped onto ADRES 1 core.

2. Motion Estimation (ME), Motion Compensation (MC), and Copy
   Controller (CC) are mapped on ADRES2, ADRES3, and ADRES4,
   respectively.

3. Texture update, Texture coding, and Variable Length Coding (VLC)
   are mapped on ADRES5 and ADRES6.

Table 11.1(a) summarizes the throughput requirements of the MPEG-4 SP
encoder when mapped on the MPSoC platform for different frame resolutions. The first two columns of the table indicate different initiator-target
pairs, the third column indicates the number of bytes required for the computation of each new MBL (expressed in bytes/MBL), and, finally, the last
three columns indicate the throughput requirements. For the NoC instruction, the throughput requirements have been estimated to be 150 MB per
second, by taking into account the MPEG-4 encoder code size, the ADRES
processor instruction size, and the L1 memory miss rate.

11.5.1.2 AVC/H.264 SP Encoder

The functional block diagram of the AVC/H.264 encoder is shown in Figure
11.16 with computational and memory blocks drawn as white and gray boxes.
As for the MPEG-4 encoder, each link is characterized with the number of
bytes that have to be accessed for the computation of each new MBL. Different
functional and memory blocks of the encoder can be mapped on the MPSoC
platform using the following implementation scenarios:

a. **Data split scenario.** The input video stream is divided into six equal
   substreams of data. Each substream is being processed with a dedicated ADRES subsystem.

---

* MBL is a data structure usually used in the block-based video encoding algorithms, such as
MPEG-4, AVC/H.264 or SVC. It is composed of 16×16 pixels requiring 384 bytes when an MBL
is encoded using 4:2:2 YC_bC_r scheme.

TABLE 11.1

MPEG-4 SP (a) and AVC/H.264 Data Split Scenario (b) Encoder Throughput Requirements When Mapped on an MPSoC Platform

Source	Target	B	CIF [MB/s]	4CIF [MB/s]	HDTV [MB/s]
FIFO	AD_1	384	4	18	40
ADS_1	$L2D_{1,2}$	640	7	30	66
$L2D_{1,2}$	$AD_{2,3}$	1664	19	77	172
AD2	L2D1	12	1	1	1
L2D1	$AD_{3,2}$	2682	31	125	276
L2D2	AD3	384	4	18	40
EMIF	AD4	384	4	18	40
AD_4	$L2D_{1,2}$	1536	18	72	158
L2D2	AD4	384	4	18	40
AD3	$L2D_{1,2}$	816	9	38	84
L2D1	AD5	384	4	18	40
L2D2	AD6	1008	12	47	94
AD6	$L2D_{1,2}$	1008	12	47	94
AD6	FIFO	58	1	3	6
L2D1	AD5	432	5	20	44
AD5	EMIF	307	3	14	32
$AD_{1,\dots,6}$	L2Is1,2		150	150	150
Total			**289**	**714**	**1377**

(a)

Source	Target	B	CIF [MB/s]	4CIF [MB/s]	HDTV [MB/s]
FIFO	EMIF	384	4.3	17.4	39.6
EMIF	ADi	64	0.7	2.9	6.6
L2D1	ADi	256	2.9	11.6	26.4
L2D2	L2D1	512	5.8	23.2	52.7
ADi	FIFO	10	0.1	0.5	1.2
ADi	EMIF	64	0.7	2.9	6.6
ADj	L2Is1		300	300	300
ADk	L2Is2		300	300	300
Total			**636.7**	**757.7**	**946.6**
Indexes: $i \in \{1, 6\}$, $j \in \{1, 3\}$, $k \in \{4, 6\}$					

(b)

b. **Functional split scenario.** Different functional blocks of the algorithm are mapped to the MPSoC platform as follows. Three ADRES subsystems (ADRES1, 2, and 3) are used for the computation of motion estimation (ME). Full, half, and quarter pixel MEs are each computed with a dedicated ADRES subsystem. ADRES4 is used for Intra prediction, DCT, IDCT, and motion compensation. The ADRES5 subsystem is dedicated for the computation of the deblocking filter and the last ADRES6 is reserved for entropy encoding.

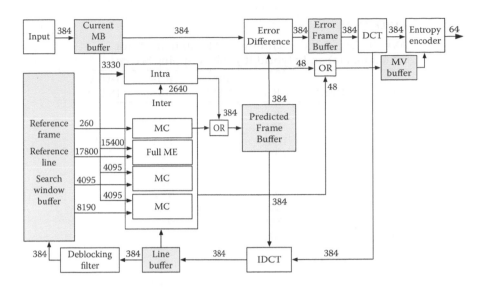

FIGURE 11.16
Functional block diagram of the AVC SP encoder with bandwidth requirements expressed in bytes/macroblock.

c. **Hybrid scenario.** In this implementation scenario, the most compu-tationally intensive task, the motion estimation, is computed with three ADRES subsystems, using data split. The remaining func-tional blocks of the encoder are mapped on the platform as in the functional split scenario.

The implementation of these scenarios will result in three different data flows. For different initiator-target pairs the corresponding throughput requirements (after application mapping) are presented in Tables 11.1(b) and 11.2(a), (b). Note that for clarity, some of the nodes in the system have been represented using indexes. These throughput requirements indicate the traffic within the NoC only. Local memory access, such as ADRES to L1 instruction or data memories, are not taken into account. This explains why the total NoC throughput requirements appear to be less than one suggested by the functional diagram represented in Figure 11.16.

The throughput requirements to the instruction memory have been deter-mined by taking into account the AVC/H.264 encoder code size, the ADRES processor instruction size, and the L1 memory miss rate. The instruction miss rate has been estimated at five, one, and two percent for data, functional, and hybrid mapping scenarios, respectively, resulting in total instruction through-put of 600, 150, and 250 MB per second.

A quick look at the throughput requirements allows some preliminary ob-servations. The data split scenario has the advantage of easing the imple-mentation and equal distribution of the computational load among different processors. The obvious drawback is heavy traffic in the instruction NoC,

TABLE 11.2

AVC/H.264 Encoder Throughput Requirements for
Functional (a) and Hybrid (b) Split Mapping Scenario

Source	Target	B	CIF [MB/s]	4CIF [MB/s]	HDTV [MB/s]
FIFO	EMIF	384	4.3	17.4	39.6
EMIF	ADi	256	2.9	11.6	26.4
AD1	AD2	1792	20.2	81.2	26.4
AD2	AD3	1792	20.2	81.2	52.7
AD3	AD4	512	5.8	23.2	1.2
L2D2	AD1	1792	20.2	81.2	6.6
AD4	ADj	384	4.3	17.4	39.6
AD5	L2D1	384	4.3	17.4	39.6
L2D1	L2D2	512	5.8	23.2	52.7
AD6	FIFO	60	0.7	2.7	6.2
ADk	L2Is1		75	75	75
ADl	L2Is2		75	75	75
Total			**241.4**	**518.2**	**986.7**

Indexes: $i \in \{1, 4\}$, $j \in \{5, 6\}$, $k \in \{1, 3\}$, $l \in \{4, 6\}$

(a)

Source	Target	B	CIF [MB/s]	4CIF [MB/s]	HDTV [MB/s]
FIFO	EMIF	384	4.3	17.4	39.6
EMIF	AD1	128	1.4	5.8	13.2
EMIF	AD4	384	4.3	17.4	39.6
L2D2	ADi	512	5.8	23.2	52.7
ADj	AD4	128	1.4	5.8	13.2
AD4	AD5	384	4.3	17.4	39.6
AD4	AD6	450	5.1	20.4	46.3
AD5	L2D1	384	4.3	17.4	39.6
AD6	FIFO	70	0.8	3.2	7.2
L2D1	L2D2	512	5.7	23.2	52.7
ADk	L2Is1		125	125	125
ADl	L2Is2		125	125	125
Total			**302.4**	**470.8**	**751.8**

Indexes: $i \in \{1, 3\}$, $j \in \{1, 3\}$, $k \in \{1, 3\}$, $l \in \{4, 6\}$.

(b)

caused by the encoder code size and the size of the L1 instruction memory.
The functional split solves this problem but at the expense of much heavier
traffic in the data NoC (which is more than doubled) and uneven computational load among different processors. Finally, the hybrid mapping scenario
offers a good compromise between pure data and pure functional split in
terms of total throughput requirements and even distribution of the computational load.

As for the MPEG-4 encoder, the application mapping scenarios do not pretend to be optimal. It is obvious that for lower frame resolutions, for example, it is not necessary to use all six ADRES cores. The real-time processing constraint could certainly be satisfied with fewer cores, with nonactive ones being shut down, thus lowering the power dissipation of the whole system.

11.5.2 Power Dissipation Models of Individual NoC Components

We will first introduce the power models of different NoC components; that is AHB initiator and target NIUs, switches and wires. These models have been established based on 90 nm, $V_{dd} = 1.08$ V technology node from the TSMC90G library and for threshold voltages $V_{thN} = 0.22$ V, $V_{thNP} = 0.24$ V. For the logic synthesis, we have used Magma BlastCreate tool (version 2005.3.86) with automatic clock-gating insertion option set on, and without scan registers. The traffic load has been generated using NoCexplorer application, using all available bandwidth (100% load) for each NoC component. Functional simulation produced the actual switching activity, dumped for formatting into Switching Activity Interchange Format (SAIF) files, that have been created using Synopsis VCS tool. Finally, the power analysis has been performed using the SAIF files and the Magma BlastPower tool (version 2005.3.133).

11.5.2.1 Network Interface Units

The experiments carried out on the initiator and target AHB NIUs showed that the power dissipation of the NIU can be modeled with the following expression:

$$P = P_{idle} + P_{dyn} \qquad (11.1)$$

where P_{idle} is the power dissipation of the NIU when it is in an idle state, that is, there is no traffic. The idle power component is mainly due to the static power dissipation and the clock activity, and depends on NIU configuration and NoC frequency. For a given configuration and frequency, the idle power dissipation component of the NIU is constant, so

$$P_{idle} = c_1 \qquad (11.2)$$

For AHB initiator NIU, we found that c_1 is 0.263 mW with 35 μW due to the leakage. For AHB target NIU, we found that $c_1 = 0.303$ mW with 52 μW for leakage. The slight difference in the power dissipation between these two can be explained by the fact that target NIU has an AHB master interface (connected to the slave IP), which is more complex than the slave interface found in initiator NIU (connected to the master IP).

In Equation 11.1, P_{dyn} designates the power dissipation component due to the actual activity of the NIU in the presence of traffic. This term reflects the power dissipated for IP to NTTP (and inverse) protocol conversion, data (de)palletization, and packet injection (or reception) into (or from) the NoC.

TABLE 11.3

Constant c_2 (Dynamic Power Dissipation
Component) of Initiator and Target AHB
NIU for Different Payload Size

	4 Bytes	16 Bytes	64 Bytes
Initiator c_2 [mW]	0.973	0.936	0.854
Target c_2 [mW]	0.945	0.909	0.830

Dynamic power component is also a function of the NIU configuration, NoC
frequency, payload size, and the IP activity. Experiments showed that for a
given frequency and configuration, P_{dyn} is a linear function of the mean usage
of the link A.

$$P_{dyn} = c_2 \cdot A \qquad (11.3)$$

expressed as a percentage of the NIU data aggregate bandwidth (600 MB/s
for 150 MHz NoC) that corresponds to the actual traffic to/from the IP.

To quantify the influence of different payload size on the dynamic power
dissipation component of the NIU, we measured the constant c_2 of the initiator and the target AHB NIUs with payloads of 4, 16, and 64 bytes (which
correspond to 1, 4, and 16 beat AHB bursts) and for NoC frequency of 150
MHz. The values of the constant c_2 are presented in Table 11.3. Note that the
dynamic power dissipation component decreases with the size of the payload,
due to less header processing for the same data activity.

Although the difference in the power dissipation of the NIU due to the
payload size is significant and should be taken into account in the more
accurate power model, we will assume in the following the fixed payload
of 16 bytes. This hypothesis is quite pessimistic in our case because the basic
chunks of data that will be transported over the NoC are MBLs. Because one
MBL requires 384 bytes, it can be embedded in six 16-beat AHB bursts, thus
minimizing the NTTP protocol overhead per transported MBL.

11.5.2.2 Switches

The power dissipation of a switch can be modeled in the same way as the NIU,
using Equations 11.1, 11.2, and 11.3. The activity A of a switch is expressed
as a portion of the aggregate bandwidth of the switch that is actually being
used. The aggregate bandwidth of a switch is computed with $min(n_i, n_o) \cdot l_{bw}$
where n_i, n_o are the number of input and output ports of a switch and l_{bw}
is the aggregate bandwidth of one link. The experiments have been carried
out to determine the values of the constants c_1 and c_2 for different arbitration
strategies and switch sizes (number of input/output ports). The influence of
the payload size on the power dissipation of a switch is small and will not be
taken into account in the following.

Because we are targeting low-power applications, we chose a round-
robin arbitration strategy for all NoC switches because it represents a good

TABLE 11.4

Constants c_1 and c_2 Used for Computation of the
Static and Dynamic Power Dissipation Components
of the Switch for Various Numbers of Input and
Output Ports

	$SW_{6\times6}$	$SW_{7\times8}$	$SW_{2\times7}$	$SW_{7\times2}$
c_1 [mW]	0.230	0.290	0.136	0.09
c_2 [mW]	1.324	1.668	0.781	0.516

compromise between the implementation cost and arbitration fairness. For
example, for a 6×6 switch with round-robin arbiter, the constants c_1 and c_2 are,
respectively, 0.230 and 1.324 mW and the synthesized switch has 5.7 kgates.
If we consider the same 6×6 switch with a FIFO arbitration strategy,* static
and dynamic power dissipation are, respectively, 15 and 66 percent higher,
while the gate count of the switch increases for about 72 percent (due to the
fact that the order of request has to be memorized).

The influence of the switch size with round-robin arbitration strategy on
the power dissipation is illustrated in Table 11.4. We show the values of c_1
and c_2 constants for typical switch sizes: 6×6, 7×8, 2×7, and 7×2.

11.5.2.3 Links: Wires

In Arteris NoC, each link (in the request or response network) can be seen as
a set of segments of fixed length, each segment being composed of a certain
number of wires with associated repeaters. Our experiments showed that the
power dissipation of one wire segment can be modeled with the following
expression:

$$P_s = w \cdot C \cdot A \cdot f_{NoC} \tag{11.4}$$

where w is the number of the wires in the segment (40 and 36 wires for
request and response networks, respectively), C reflecting the total equivalent
switching capacitance of the wire for a given technology node (including the
term corresponding to V_{dd}^2), A the activity of that link, and f_{NoC} the frequency
of the NoC.

The power dissipation of one NoC link of an arbitrary length is then modeled with

$$P_l = P_s \cdot l \tag{11.5}$$

where l is the length of the link expressed in [mm]. Total equivalent switching
capacitance C of 263 fF is obtained as the sum of the wire capacitance (140 fF for
1 mm wire), the input capacitance of the repeater attached to each wire (23 fF)
and the equivalent capacitance used to model the actual power dissipation of
the repeater (100 fF).

---

* In a FIFO arbitration scheme the order of requests will be taken into account, highest priority
being given to the least recently serviced requests.

Power dissipation of the wires in the request and the response network are computed separately, because all transactions in the NoC are supposed to be writes (CA to CA protocol). This implies that in the request network both data and control wires will toggle, while in the response network the toggling will occur only on control wires. While data wires in the request network toggle with the frequency depending on the activity of that link, the toggle rate of the control wires will take place every once in a while when compared to data wires. For the sake of simplicity, in the following we will assume that the control wires in the request network toggle with the same frequency as data wires. Such a hypothesis is quite pessimistic, but it can be used safely because there are only few control wires in a link and their influence on the overall power dissipation of the NoC is quite small. As explained above, for the power dissipation of the response network, we only count control wires because there will be no read operation. The activity of the control wires is fixed using the assumption that all packets will have 64 bytes of payload (16 beat AHB burst).

11.5.3   Power Dissipation of the Complete NoC

Because in Arteris NoC transactions between the same pair of initiator and target nodes will always use the same route, the total NoC power dissipation can be easily computed as the sum of the power dissipation of all the initiators and target NIUs (P_{NIU_I} and P_{NIU_T}) of the switches in request ($P_{SW_{Req}}$), of the response network ($P_{SW_{Res}}$), and of all links.

$$P = \sum P_{NIU_I} + \sum P_{NIU_T} + \sum P_{SW_{Req}} + \sum P_{SW_{Res}} + \sum P_L \qquad (11.6)$$

as a function of traffic. Given the application mapping scenarios of the MPEG-4 and AVC/H.264 SP encoder and the NoC topology, MPSoC platform specifications have been used to establish throughput requirements for every initiator–target pair for different implementation scenarios. To the pure data traffic, the NTTP overhead (72 and 32 bits for request and response packets) has been added to obtain the actual traffic in the NoC.

Note that in this model we will not take into account the NoC power dissipation due to the presence of the control node (ARM subsystem). This is true as long as we consider that this particular node will not interfere during one particular session of video encoding or decoding. In that case, the power dissipation is due to the presence of 30 NIUs (18 initiators and 12 targets).

For every switch in the network, we can define the total used bandwidth as the sum of all the bandwidths passing through that switch, according to the throughput requirements of the mapping scenario. The activity is then easily computed as a percentage of the aggregate bandwidth of that switch (based on switch topology, i.e, number of input and output ports). Note that because of the particular NoC configuration (only write operations), only request network switches will have the dynamic power dissipation component.

Because of the low activity of the switches in the response network (no data, control only), when compared with those in the request network, the power dissipation of these switches will be modeled with an idle power dissipation component only (the power dissipation of the request switches will be constant for different application mapping scenarios).

Based on the circuit layout (Figure 11.11) we can easily derive the total length of every link in the NoC for different mapping scenarios. The total length of 132, 102, 38, and 57 mm has been found for the MPEG-4 application and for three different scenarios for the AVC/H.264 encoder, respectively. Note that we assume the same length of the request and response networks for the same initiator–target pair. The power model of one wire segment presented earlier and the total length of the links can be combined to determine the power dissipation of the wires in the NoC. As we mentioned earlier, NoC links do not transport clock signal, so the power dissipation due to the insertion of the clock tree must be taken into account separately. Based on the layout, the total length of the clock tree has been estimated to be 24 mm. For such a length and for a frequency of 150 MHz, the power dissipation has been evaluated to 1 mW. This value is systematically added to the total power dissipation of the wires in the NoC.

The power model described above has been used to calculate the power dissipation of the Arteris NoC running at 150 MHz, for different mapping scenarios of the MPEG-4 and AVC/H.264 SP encoder and for typical frame resolutions (CIF, 4CIF and HDTV). Table 11.5 indicates leakage, static, and total idle power dissipation of different IPs in the NoC. Finally, if we take into account the NoC topology, we can easily derive the total idle power dissipation of this NoC instance (10.7 mW).

It is, however, worth mentioning that the new local (isolation of one NIU or router) and global (isolation of one cluster, the cluster being composed of multiple NIUs and switches) clock-gating methods implemented in the latest version of the Danube IP library (v.1.8.) enable significant reduction of the idle power dissipation component. Each unit (NIU, switch) is capable of monitoring its inputs and cutting the clock when there are no packets and when the processing of all packets currently in the pipeline is completed. When a new packet arrives at the input, the units can restart their operation in one clock cycle at most. Our preliminary observations show that the application

TABLE 11.5

Leakage, Static, and Total Idle Power Dissipation in mW for Different IPs of the NoC Instance

	Leakage	Static	Total	NoC
Initiator NIU	0.035	0.228	0.263	4.5
Target NIU	0.052	0.251	0.303	3.9
Switches			0.23	2.3
Total idle				10.7

TABLE 11.6

Power Dissipation of the NoC for MPEG-4 and
AVC/H.264 Simple Profiles Encoders for Different
Frame Resolutions (30 fps)

	MPEG-4	AVC/H.264 Data	AVC/H.264 Functional	AVC/H.264 Hybrid Split
CIF	13.6	18.6	15.64	14.61
4CIF	17.02	19.35	17.73	15.92
HDTV	22.34	21.37	21.27	18.16

of these local and global clock-gating methods can reduce the total idle power
dissipation of the NoC to only 2 mW.

Total power dissipation is presented in Table 11.6 and Figure 11.17. We also
show the relative contribution of different NoC IPs (NIUs, wires and switches)
to the total NoC power budget. The dynamic power component relative to
the instruction traffic is, respectively: 4.3, 1.4, and 2.2 mW, depending on the
mapping scenario.

The power dissipation of the NoC presented in this work can be compared
with the power dissipation of other interconnects for multimedia applications
already presented in the literature. Table 11.7 summarizes this comparison

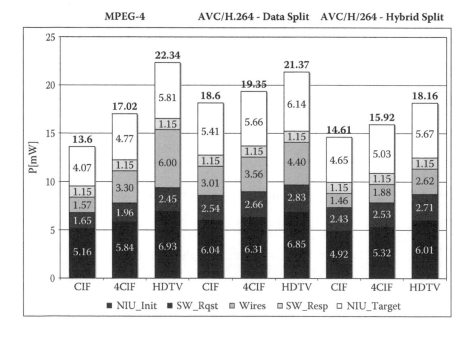

FIGURE 11.17
Power dissipation of the NoC for MPEG-4 and AVC/H.264 SP encoder: total power dissipation
and breakdown per NoC component.

TABLE 11.7

Comparison of the Communication Infrastructure Power Dissipation for Different Multimedia Applications

Design	Nodes	NoC Topology Routers,	BW [MB/S]	Process [nm, V]	Frequency [MHz]	Power [mW]	Scaled Power [mW]
1. MPEG-4 SP Encoder [38]	17	—	570	180,1.6	101	37	7
2. Various [39]	9	2, Star	3200	180,1.6	150	51	9
3. MPEG-4 SP Decoder [40,41]	12	Dedicated	570	130,1	NA	27	18
4. MPEG-4 SP Decoder [40,41]	12	Universal	714	130,1	NA	97	64
5. MPEG-4 SP Encoder	13	12, Mesh	714	90,1.08	150	—	17
6. AVC/H.264 Encoder	13	12, Mesh	470	90,1.08	150	—	15
7. AVC/H.264 Encoder	13	12, Mesh	760	90,1.08	150	—	19

with (1), (2), and (3) being the implementation presented here. The results are those for 4CIF resolution, chosen for closest bandwidth requirements. Note that for easier comparison we scaled down the power dissipation figures of the designs made in other technologies, to the 90-nm, 1.08-V technology node used in our implementation (last column), using the expression suggested by Denolf et al. [38], where V_{dd} is the power supply voltage and λ feature size

$$P_1 = P_2 \cdot \left[\left(\frac{V_{dd2}}{V_{dd1}} \right)^{1.7} \cdot \left(\frac{\lambda_2}{\lambda_1} \right)^{1.5} \right]^{-1} \tag{11.7}$$

The characteristics of the dedicated MPEG-4 SP encoder implementation are shown in line (4). Design 5 gives the power dissipation of a fairly simple SoC (only two routers), for all available bandwidth. Finally, the last two lines show the best and the worst cases of our work. Designs (6) and (7) show MPEG-4 decoder mapped on the MPSoC platform with ×pipes NoC with 12 routers in a mesh and optimized topology (without power dissipation in the NIUs).

Our results show that even for a nonoptimized NoC topology (full 2 × 2 for both request and response networks), chosen for maximum flexibility and applications that have high bandwidth requirements such as those of the MPEG-4 and AVC/H.264 encoders for HDTV and at 30 fps video rate (about 1 GB/s), the power dissipation of the NoC accounts for less than five percent of the total power budget of a reasonably sized NoC (13 nodes in all). Depending on the encoding algorithm used, the application mapping scenario, and the image resolution, the absolute power dissipation of the NoC varies from 14 to 22 mW, from which 10.7 mW are due to the idle power dissipation (no traffic), and could be further reduced with more aggressive clock-gating techniques. Note that an important part of the total power dissipation (from 60 to 70 percent) is due to the 30 NIUs (for 13 nodes only) and embedded smart DMAs circuits (Communication Assist - CAs engines). It is also interesting to underline that the increase in the throughput requirements leads to a relatively low increase in the dissipated power. If we consider the functional split, which is a worst case from the required bandwidth point of view, when moving from CIF to HDTV resolution, the data throughput will increase almost 400 percent (from 241 to 987 MB/s) but resulting in only 35 percent increase of the total power dissipation of the NoC.

The implementation cost of the NoC in terms of the silicon area is also more than acceptable, because it represents less than three percent of the total area budget (less than 450 kgates). When compared to other IPs in the system, on a one-to-one basis, the NoC represents eight percent of one ADRES VLIW/ CGA processor, twenty percent of one 256 kB memory and is forty percent bigger than the ARM9 core. This is acceptable even for the medium-sized MPSoC platforms targeting lower performances. As for the power dissipation, note that in this particular design and due to the presence of the CAs allowing block transfer type of communication, a considerable amount of the area is taken by the NIU units.

Finally, the complete design cycle (including the learning period for the NoC tools), NoC instance definition, specification with high- and low-level NoC models, RTL generation, and final synthesis took only two man months. This argument combined with the achieved performance in terms of available bandwidth, power, and area budget clearly points out the advantages of the NoC as communication infrastructure in the design of high-performance low-power MPSoC platforms.

References

[1] M. Millberg, E. Nilsson, R. Thid, S. Kumar, and A. Jantsch, "The Nostrum backbone—A communication protocol stack for Networks on Chip." In *Proc. of the VLSI Design Conference*, Mumbai, India, Jan. 2004. [Online]. Available: http://www.imit.kth.se/ axel/papers/2004/VLSI-Millberg.pdf.

[2] A. Jantsch and H. Tenhunen, eds., *Networks on Chip*. Hingham, MA: Kluwer Academic Publishers, 2003.

[3] N. E. Guindi and P. Elsener, "Network on Chip: PANACEA—A Nostrum integration," Swiss Federal Institute of Technology Zurich, Technical Report, Feb. 2005. [Online]. Available: http://www.imit.kth.se/~axel/papers/2005/ PANACEA-ETH.pdf.

[4] A. Jalabert, S. Murali, L. Benini, and G. D. Micheli, "xpipesCompiler: A tool for instantiating application specific Networks on Chip." In *Design, Automation and Test in Europe (DATE)*, Paris, France, February 2004.

[5] D. Bertozzi and L. Benini, "Xpipes: A Network-on-Chip architecture for gigascale Systems-on-Chip," *IEEE Circuits and Systems Magazine* 4 (2004), 18–31.

[6] T. Bjerregaard and S. Mahadevan, "A survey of research and practices of Network-on-Chip," *ACM Computing Surveys* 38(1) (2006) 1. [Online]. Available: http://www2.imm.dtu.dk/ tob/papers/ACMcsur2006.pdf.

[7] N. Kavaldjiev and G. J. Smit, "A survey of efficient on-chip communications for SoC." In *4th PROGRESS Symposium on Embedded Systems, Nieuwegein, Netherlands*. STW Technology Foundation (Oct. 2003): 129–140. [Online]. Available: http://eprints.eemcs.utwente.nl/833/.

[8] R. Pop and S. Kumar, "A survey of techniques for mapping and scheduling applications to Network on Chip systems," *ING Jönköping*, Technical Report ISSN 1404-0018 04:4, 2004. [Online]. Available: http://hem.hj.se/~poru/.

[9] E. Rijpkema, K. Goossens, and P. Wielage, "A router architecture for networks on silicon." In *Proc. of Progress 2001, 2nd Workshop on Embedded Systems*, Veldhoven, the Netherlands, October 2001.

[10] O. P. Gangwal, A. Rădulescu, K. Goossens, S. G. Pestana, and E. Rijpkema, "Building predictable Systems on Chip: An analysis of guaranteed communication in the Æthereal Network on Chip." In *Dynamic and Robust Streaming in and between Connected Consumer Electronics Devices*, Philips Research Book Series, P. van der Stok, ed., 1–36, Norwill, MA: Kluwer, 2005.

[11] K. Goossens, J. Dielissen, and A. Rădulescu, "The Æthereal Network on Chip: Concepts, architectures, and implementations," *IEEE Design and Test of Computers* 22(5) (September–October 2005): 21–31.

[12] C. Bartels, J. Huisken, K. Goossens, P. Groeneveld, and J. van Meerbergen, "Comparison of an Æthereal Network on Chip and a traditional interconnect for a multi-processor DVB-T System on Chip." In *Proc. IFIP Int'l Conference on Very Large Scale Integration (VLSI-SoC)*, Nice, France, October 2006.

[13] X. Ru, J. Dielissen, C. Svensson, and K. Goossens, "Synchronous latency-insensitive design in Æthereal NoC." In *Future Interconnects and Network on Chip*, Workshop at Design, Automation and Test in Europe Conference and Exhibition (DATE), Munich, Germany, March 2006.

[14] K. Goossens, S. G. Pestana, J. Dielissen, O. P. Gangwal, J. van Meerbergen, A. Rădulescu, E. Rijpkema, and P. Wielage, "Service-based design of Systems on Chip and Networks on Chip." In *Dynamic and Robust Streaming in and between Connected Consumer Electronics Devices*, Philips Research Book Series, P. van der Stok, ed., 37–60. New York: Springer, 2005.

[15] A. Rădulescu, J. Dielissen, S. G. Pestana, O. P. Gangwal, E. Rijpkema, P. Wielage, and K. Goossens, "An efficient on-chip network interface offering guaranteed services, shared-memory abstraction, and flexible network programming," *IEEE Transactions on CAD of Integrated Circuits and Systems* 24(1) (January 2005): 4–17.

[16] S. G. Pestana, E. Rijpkema, A. Rădulescu, K. Goossens, and O. P. Gangwal, "Cost-performance trade-offs in Networks on Chip: A simulation-based approach." In *DATE '04: Proc. of the Conference on Design, Automation and Test in Europe*. Washington, DC: IEEE Computer Society, 2004, 20764.

[17] K. Goossens, J. Dielissen, O. P. Gangwal, S. G. Pestana, A. Rădulescu, and E. Rijpkema, "A design flow for application-specific Networks on Chip with guaranteed performance to accelerate SOC design and verification." In *Proc. of Design, Automation and Test in Europe Conference and Exhibition* Munich, Germany, March 2005, 1182–1187.

[18] Silistix, "http://www.silistix.com," 2008.

[19] J. Bainbridge and S. Furber, "CHAIN: A delay insensitive CHip area INter-connect," *IEEE Micro Special Issue on Design and Test of System on Chip*, 142(4) (September 2002): 16–23.

[20] J. Bainbridge, L. A. Plana, and S. B. Furber, "The design and test of a Smartcard chip using a CHAIN self-timed Network-on-Chip." In *Proc. of the Design, Automation and Test in Europe Conference and Exhibition*, Paris, France, 3 (February 2004): 274.

[21] J. Bainbridge, T. Felicijan, and S. Furber, "An asynchronous low latency arbiter for Quality of Service (QoS) applications." In *Proc. of the 15th International Conference on Microelectronics (ICM'03), Cairo, Egypt*, Dec. 2003, 123–126.

[22] T. Felicijan and S. Furber, "An asynchronous on-chip network router with Quality-of-Service (QoS) support." In *Proc. of IEEE International SOC Conference*, Santa Clara, CA, September 2004, 274–277.

[23] Arteris, "A comparison of network-on-chip and busses," White paper, 2005.

[24] J. Dielissen, A. Rădulescu, K. Goossens, and E. Rijpkema, "Concepts and implementation of the Philips Network-on-Chip," *IP-Based SOC Design*, Grenoble, France, November 2003.

[25] E. Rijpkema, K. G. W. Goossens, A. Radulescu, J. Dielissen, J. L. van Meerbergen, P. Wielage, and E. Waterlander, "Trade-offs in the design of a router with both guaranteed and best-effort services for Networks on Chip." In *DATE '03: Proc. of the Conference on Design, Automation and Test in Europe*. Washington, DC: IEEE Computer Society, 2003, 10350.

[26] P. Schumacher, K. Denolf, A. Chilira-Rus, R. Turney, N. Fedele, K. Vissers, and J. Bormans, "A scalable, multi-stream MPEG-4 video decoder for conferencing and surveillance applications." In *ICIP 2005. IEEE International Conference on Image Processing*, Genova, Italy, 2005, 2 (September 2005): 11–14, II–886–9.

[27] Y. Watanabe, T. Yoshitake, K. Morioka, T. Hagiya, H. Kobayashi, H.-J. Jang, H. Nakayama, Y. Otobe, and A. Higashi, "Low power MPEG-4 ASP codec IP macro for high quality mobile video applications," *Consumer Electronics, 2005. ICCE. 2005 Digest of Technical Papers. International Conference*, Las Vegas, NV, (January 2005): 8–12, 337–338.

[28] T. Fujiyoshi, S. Shiratake, S. Nomura, T. Nishikawa, Y. Kitasho, H. Arakida, Y. Okuda, et al. "A 63-mW H.264/MPEG-4 audio/visual codec LSI with module-wise dynamic voltage/frequency scaling," *IEEE Journal of Solid-State Circuits*, 41(1) (January 2006): 54–62.

[29] C.-C. Cheng, C.-W. Ku, and T.-S. Chang, "A 1280/spl times/720 pixels 30 frames/s H.264/MPEG-4 AVC intra encoder." In *Proc. of Circuits and Systems, 2006. ISCAS 2006. 2006 IEEE International Symposium*, Kos, Greece, May 21–24, 2006, 4.

[30] C. Mochizuki, T. Shibayama, M. Hase, F. Izuhara, K. Akie, M. Nobori, R. Imaoka, H. Ueda, K. Ishikawa, and H. Watanabe, "A low power and high picture quality H.264/MPEG-4 video codec IP for HD mobile applications." In *Solid-State Circuits Conference, 2007. ASSCC '07. 2007 IEEE International Conference*, Jeju City, South Korea, Nov. 12–14, 2007, 176–179.

[31] B. Mei, "A coarse-grained reconfigurable architecture template and its compilation Techniques," Ph.D. dissertation, IMEC, January 2005.

[32] F.-J. Veredas, M. Scheppler, W. Moffat, and M. Bingfeng, "Custom implementation of the coarse-grained reconfigurable ADRES architecture for multimedia purposes." In *Field Programmable Logic and Applications, 2005. International Conference*, Tampere, Finland, August 24–26, 2005, 106–111.

[33] F. Bouwens, M. Berekovic, B. D. Sutter, and G. Gaydadjiev, "Architecture enhancements for the ADRES coarse-grained reconfigurable array," *HiPEAC*, 2008, 66–81.

[34] B. Mei, F.-J. Veredas, and B. Masschelein, "Mapping an H.264/AVC decoder onto the ADRES reconfigurable architecture." In *Field Programmable Logic and Applications, 2005. International Conference*, August 24–26, 2005, 622–625.

[35] C. Arbelo, A. Kanstein, S. López, J. F. López, M. Berekovic, R. Sarmiento, and J.-Y. Mignolet, "Mapping control-intensive video kernels onto a coarse-grain reconfigurable architecture: The H.264/AVC deblocking filter." In *DATE '07: Proc. of the Conference on Design, Automation and Test in Europe*. San Jose, CA: EDA Consortium, 2007, 177–182.

[36] M. Dasygenis, E. Brockmeyer, B. Durinck, F. Catthoor, D. Soudris, and A. Thanailakis, "A memory hierarchical layer assigning and prefetching technique to overcome the memory performance/energy bottleneck." In *DATE '05: Proc. of the Conference on Design, Automation and Test in Europe*. Washington, DC: IEEE Computer Society, 2005, 946–947.

[37] I. Issenin, E. Brockmeyer, M. Miranda, and N. Dutt, "DRDU: A data reuse analysis technique for efficient scratch-pad memory management," *ACM Transactions on Design Automation of Electronic Systems*, 12(2): 2007.

[38] K. Denolf, A. Chirila-Rus, P. Schumacher, et al., "A systematic approach to design low-power video codec cores," *EURASIP Journal on Embedded Systems* 2007, Article ID 64 569, 14 pages, 2007, doi:10.1155/2007/64569.

[39] K. Lee, S.-J. Lee, and H.-J. Yoo, "Low-power Network-on-Chip for high-performance SoC design," *IEEE Transactions on Very Large Scale Integration (VLSI) Systems*, 14 (2) (February 2006): 148–160.

[40] F. Angiolini, P. Meloni, S. Carta, L. Benini, and L. Raffo, "Contrasting a NoC and a traditional interconnect fabric with layout awareness." In *Proc. of Design, Automation and Test in Europe, 2006. DATE '06.*, 1, March 6–10, 2006, 1–6.

[41] D. Atienza, F. Angiolini, S. Murali, A. Pullini, L. Benini, and G. De Micheli, "Network-on-Chip design and synthesis outlook," *Integration, the VLSI Journal*, 41(2), (February 2008).

Index